Graduate Texts in Physics

Graduate Texts in Physics publishes core learning/teaching material for graduate- and advanced-level undergraduate courses on topics of current and emerging fields within physics, both pure and applied. These textbooks serve students at the MS- or PhD-level and their instructors as comprehensive sources of principles, definitions, derivations, experiments and applications (as relevant) for their mastery and teaching, respectively. International in scope and relevance, the textbooks correspond to course syllabi sufficiently to serve as required reading. Their didactic style, comprehensiveness and coverage of fundamental material also make them suitable as introductions or references for scientists entering, or requiring timely knowledge of, a research field.

Simone Di Mitri

Fundamentals of Particle Accelerator Physics

 Springer

Simone Di Mitri
Elettra—Sincrotrone Trieste SCpA
Basovizza, Trieste, Italy

ISSN 1868-4513   ISSN 1868-4521   (electronic)
Graduate Texts in Physics
ISBN 978-3-031-07664-0   ISBN 978-3-031-07662-6   (eBook)
https://doi.org/10.1007/978-3-031-07662-6

This Springer imprint is published by the registered company Springer Nature Switzerland AG
The registered company address is: Gewerbestrasse 11, 6330 Cham, Switzerland

To my son Giordano,
to Aunt Assunta
and Uncle Feliciano

Preface

This book offers an introduction to accelerator physics and technology, in connection to advanced accelerator complexes ranging from high energy colliders to coherent light sources. The book is targeted at both undergraduate and graduate students. Still, it can be of interest for young researchers and senior scientists not specialized in the field.

The book does not aim at being an omni-comprehensive manual, rather an agile textbook whose systematic study, in the order indicated by the Contents, will provide the Reader with the basics for the comprehension of a large variety of modern and most used charged particle accelerators, and with some preparation for accelerator design.

Particular attention is given to the physical interpretation of the mathematical expressions. These are limited to the essential but, as a peculiarity of this textbook, the majority of the derivations is explicitly given, in a rigorous but immediate manner whenever possible. A solid link to experimental cases, mostly related to high energy linear and circular accelerators for particle physics and X-rays production, is provided. Several guided exercises and case studies, named Discussions, interleave and complement the theoretical sections. The Discussions are intended to deepen the physical comprehension of previously derived expressions through quantitative examples. Everywhere in the book, italic text underlines the most relevant concepts, definitions of quantities, and theorems.

An essential bibliography is at the end of each chapter. It includes lectures given at CERN Accelerator Schools and other international Schools on particle accelerators, classical books' sections and articles. The bibliography was selected on the basis of the similarity of the contents and of their formulation to those presented in the Chapters, as well as for the clarity of exposition. A more comprehensive list of classical textbooks and manuals on particle accelerators is provided at the end of this book, for a more advanced and technical study. This additional bibliography is enriched by a (not exhaustive) list of breakthrough articles, some of them dating back to the beginning of this discipline.

The author draws on his almost 20 years-long experience in the design, commissioning and operation of accelerator facilities, on his 10 years-long teaching experience at the Engineering and the Physics Department of the University of Trieste, as well as at international Schools of accelerator physics and synchrotron radiation.

Basovizza, Trieste, Italy   Simone Di Mitri

Acknowledgements

This book was made possible by the expert guidance and warm encouragement I received over the last 20 years by my mentor, colleague and friend Massimo (Max) Cornacchia, who introduced me to the design of particle accelerators and to the international community of light sources.

This book is based on my knowledge of accelerator physics, and in particular on the fundamentals that I acquired through study, teaching and work in the field with recognized experts, here warmly acknowledged: Prof. Franco Cervelli, Prof. Paolo Rossi, Gaetano Vignola, Caterina Biscari, Mikhail Zobov, Gabriele Benedetti, Carlo Bocchetta, Lidia Tosi, Bruno Diviacco, Paolo Craievich, Claudio Serpico, Simone Spampinati, Marco Venturini, Peter Williams, Giuseppe (Pino) Dattoli, Alexander A. Zholents, William M. Fawley, Prof. Sytze Brandenburg, Prof. Sergio Tazzari, Prof. William Barletta, and Prof. Fulvio Parmigiani. I am in debt to Prof. S. Tazzari, who generously shared with me his lectures, and whose memory still inspires me in such activity. I am grateful to Max Cornacchia, Andy Wolski, Bruno Diviacco, Daniele Filippetto, Anna Bianco, Fulvio Parmigiani, Pino Dattoli, and Giovanni Perosa, for proofreading, enlightening discussions and suggestions. I thank all the students who helped in improving this book through their questions and comments. Still, any error or inaccuracy must be attributed exclusively to the author.

Elettra Sincrotrone Trieste and the University of Trieste are acknowledged for the opportunity given to me of joining the community of particle accelerators, and of sharing my experience with students and young researchers.

So as in any other activity of my life, during the writing of this book I received a solid support from my parents Giovanni and Teresa, my sisters Elisa and Diletta, my in-laws Franco and Tamara, and my wife Valentina: they are all profoundly acknowledged.

Contents

Special Relativity

In Classical Mechanics, position and velocity vectors add according to the Galilean transformations, which date back to the XVII century. Those assume an absolute time, i.e., the time coordinate is the same in all reference frames. With the advent of Newton's Mechanics in the XVIII century, a conceptualization of absolute time was given.

Newton's first law of Mechanics allows a rigorous definition of inertial reference frame. In such frame, any body which is not subject to any net external force tends either to maintain its state of rest, or to continue its uniform rectilinear motion. Newton's second and third law of Mechanics introduce, respectively, the inertial mass and the interaction of distinct bodies. They facilitated the comprehension of waves in Mechanics as an interaction between portions of matter (sea waves, sound, etc.).

After the discovery of Electromagnetism in the XIX century, the propagation of waves in a medium observed in Mechanics was identically transposed to electro-magnetic (e.m.) waves. Consistently with Classical Mechanics, the time was still an absolute coordinate shared by distinct reference frames. The medium was inferred to be an invisible and impalpable substance denominated ether, which thereby resulted in a privileged reference frame. The velocity of propagation of e.m. waves through ether was assumed to obey Galilean transformations.

Many experiments were conducted in the XIX and XX century to either confirm or contradict the existence of ether. One of the first and most famous experiment was by Michelson in 1881, then with Morley in 1887. In the experiment, an optical interferometer would allow the observer to note the variation of the arrival time of two distinct but synchronized light waves on a screen, by virtue of the different time of propagation of the waves along orthogonal paths in ether. The experiment was repeated several times (1881–1930) to reach higher and higher accuracy. The result is known and it brought to the rejection of the ether hypothesis.

© The Author(s), under exclusive license to Springer Nature Switzerland AG 2022
S. Di Mitri, *Fundamentals of Particle Accelerator Physics*, Graduate Texts in Physics,
https://doi.org/10.1007/978-3-031-07662-6_1

Michelson and Morley's experiment is here of some interest not only because it exemplifies the scientific method of Galilean memory, but also because its simple theoretical formulation makes apparent the need of lengths transformation according to the Lorentz-Fitzgerald's equations of Special Relativity. The transformations are not derived explicitly here, and the Reader is kindly sent to the References for a complete derivation starting from four postulates: (i) the transformation of spatial and time coordinates from one inertial reference frame to another is a linear map, (ii) the light speed c in vacuum is the maximum velocity and it is the same in all inertial reference frames, (iii) space is homogeneous, so that c is the same in all directions, and (iv) space is isotropic, that is, the physical properties of space are the same everywhere.

Special Relativity [1] plays a fundamental role in particle accelerators because accelerated charged particles easily reach relativistic velocities. Their motion can therefore only be understood within the framework of Special Relativity.

1.1   Relativistic Kinematics

1.1.1   Michelson and Morley's Experiment

A schematic top view of Michelson and Morley's interferometer is shown in Fig. 1.1. A light source S emits a wave which is partly transmitted and partly reflected by the central lens P. The light beam is split in two synchronized waves, propagating along orthogonal paths of length l_1 and l_2. The two waves are then reflected by the mirrors M_1 and M_2, respectively. They finally recombine at a screen, on which fringes of an interference pattern can be observed.

Let us assume for the moment that Earth and therefore the interferometer is moving through ether with velocity v along the direction l_1. The existence of ether would imply a retardation of light due to ether preassure (or wind) when the wave propagates from P towards M_1, so that the light speed in Earth's reference frame

Fig. 1.1 Schematic top view of the Michelson and Morley's experiment

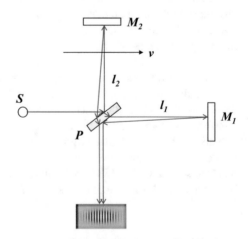

would be $c_{1,f} = c - v$. For the same reason, the light speed in the opposite direction would be $c_{1,b} = c + v$. Thus, the time wave-1 takes to reach the screen starting from the lens P is:

$$T_1 = \frac{l_1}{c - v} + \frac{l_1}{c + v} + T_P = \frac{2l_1}{c} \frac{1}{1 - \frac{v^2}{c^2}} + T_P \tag{1.1}$$

T_P is the time taken to travel from P to the screen, and it is the same for both waves.

In Earths's reference frame, ether gives wave-2 a velocity component v along the direction l_1. The same component is in fact given to the whole frame so that wave-2 will not miss either the mirror or the screen. The light velocity is the vectorial sum of c along l_2 and of v along l_1, or $c_2 = \sqrt{c^2 - v^2}$. The total time to travel from the lens P to the screen is:

$$T_2 = \frac{2l_2}{c\sqrt{1 - \frac{v^2}{c^2}}} + T_P \tag{1.2}$$

The phase (i.e., the arrival time) difference of the two waves at the screen in Earth's reference frame can now be calculated. If the wave central frequency is v, we end up with:

$$\Delta\phi_0 = v(T_1 - T_2) = \frac{2v}{c} \left(\frac{l_1}{1 - \frac{v^2}{c^2}} - \frac{l_2}{\sqrt{1 - \frac{v^2}{c^2}}} \right) \tag{1.3}$$

The assumption of Earth's motion through ether along the direction l_1 has now to be reconsidered, since it is a very specific situation. In order to make the result general, the eventual orthogonal components of Earth's velocity with respect to ether can be removed by repeating the experiment with the interferometer rotated by 90°, then by calculating the difference of the phase differences in the two configurations. It is simple to demonstrate that for the rotated system we only need to replace $l_1 \rightarrow -l_2$ and $l_2 \rightarrow -l_1$ in the previous expressions:

$$\Delta\phi_{90} = v(\tilde{T}_1 - \tilde{T}_2) = \frac{2v}{c} \left(\frac{l_1}{\sqrt{1 - \frac{v^2}{c^2}}} - \frac{l_2}{1 - \frac{v^2}{c^2}} \right) \tag{1.4}$$

We introduce the well-known notation $\beta = v/c$. If the theory of ether were true, the observer should observe a non-zero phase difference between the two experiments equal to:

$$\Delta\phi_0 - \Delta\phi_{90} = \frac{2v}{c} \left(\frac{l_1}{1-\beta^2} - \frac{l_2}{\sqrt{1-\beta^2}} - \frac{l_1}{\sqrt{1-\beta^2}} + \frac{l_2}{1-\beta^2} \right) =$$
$$= \frac{2v}{c}(l_1 + l_2) \left(\frac{1}{1-\beta^2} - \frac{1}{\sqrt{1-\beta^2}} \right) \approx \tag{1.5}$$
$$\approx \frac{2v}{c}(l_1 + l_2) \frac{1 - 1 + \beta^2/2}{1-\beta^2} \approx \frac{l_1 + l_2}{\lambda} \beta^2$$

The approximated expressions are taken for $\beta \ll 1$, owned to the fact that Earth's velocity is $v \approx 30 \cdot 10^3$ m/s, hence $\beta \approx 10^{-4}$.

In the experiment, Michelson and Morley had $l_1 = l_2 = 11$ m, the waves' central wavelength was $\lambda = 0.59$ μm. Hence, the variation of the interference fringes after $90°$ rotation of the system was expected to be $\Delta\phi_0 - \Delta\phi_{90} \approx 0.37$ rad. The sensitivity of the apparatus could detect a phase difference 100 times smaller. Nevertheless, no fringes variation was observed.

One could notice that the same (wrong) theoretical expectation would be obtained by calculating for example T_2 in the ether's reference frame. This is because Galilean composition of velocities assumes an absolute time, i.e., the time coordinate is the same in ether's and Earth's reference frame. Consequently, a time interval calculated in Earth's reference frame and another one in ether's reference frame can be summed. Special Relativity says that this is, of course, physically wrong, and the error becomes not negligible when relative velocities approach the light speed.

It is now apparent that a null phase difference is obtained if we simply assume that the light speed is constant along any direction, in any reference frame, or $c_1 = c_2 = c$, so that in the Earth's reference frame:

$$\Delta\phi_0 - \Delta\phi_{90} = \frac{2v}{c}(l_1 - l_2 - l_1 + l_2) = 0 \tag{1.6}$$

It is instructive to notice that the result must be the same in the reference frame of an observer external to Earth, and in which Earth and therefore the interferometer moves with a velocity v along l_1 (in fact, the number of fringes must be the same in any reference frame). In such frame we still have $T_2 = \dfrac{2l_2}{c\sqrt{1-\frac{v^2}{c^2}}}$. In order to make the variation of fringes null, one is brought to postulate a length contraction $l_1 \rightarrow l_1\sqrt{1-\beta^2}$ (same for l_2 in the rotated system), so that again:

$$\Delta\phi_0 - \Delta\phi_{90} = \frac{2v}{c}\frac{(l_1 - l_2 - l_1 + l_2)}{\sqrt{1-\beta^2}} = 0 \tag{1.7}$$

Equivalently, any time interval T_i measured in the Earth's reference frame is observed to be longer by a factor $\gamma = \frac{1}{\sqrt{1-\beta^2}} > 1$ (compare Eqs. 1.6 and 1.7). Fitzgerald interpreted the length contraction as a physical deformation of an object in movement through ether. We know today that, according to Special Relativity, time dilution and length contraction are physical processes intrinsic to the observation made in two different inertial reference frames.

Fig. 1.2 Inertial reference frames in relative motion

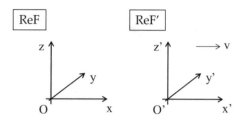

1.1.2 Lorentz-Fitzgerald's Transformations

The equations of Lorentz-Fitzgerald which describe the transformation of space-time coordinates from one inertial reference frame (ReF) to another are:

$$\begin{cases} x' = \gamma(x - v_x t) \\ y' = y, \\ z' = z \\ t' = \gamma(t - \frac{x v_x}{c^2}) \end{cases} \tag{1.8}$$

where we have assumed that ReF' is moving with respect to ReF with a velocity $\vec{v}$, that each reference frame has a right-handed triad of Cartesian coordinates (x, y, z), and that both triads are oriented with the x-axis along the direction of $\vec{v}$, see Fig. 1.2 (it is always possible to orientate the triads so that the latter condition is satisfied).

We remind that although Eqs. 1.8 apply to inertial reference frames, i.e., to reference frames in relative uniform rectilinear motion, accelerated motion is allowed internally to any ReF, with the prescription that velocities and accelerations have to obey to transformation rules consistent with Eqs. 1.8. Whenever one ReF is accelerated with respect to another, for example in uniformly accelerated linear motion or uniform circular motion, it is still possible to apply Eqs. 1.8 but at each individual timestamp. In most practical cases discussed in this book, ReF is intended to be the laboratory frame, while ReF' is intended to be the reference frame in which either a single particle or an ensemble of accelerated particles is (instantaneously) at rest.

1.1.3 Lengths and Time Intervals

In order to evaluate the *length contraction* of a moving body, the *proper length* $l_0 = x'_2 - x'_1$ is introduced. This is the length measured in ReF', where the body is at rest, see Fig. 1.3. Equations 1.8 are then applied to express the length in ReF, by

Fig. 1.3 Length and time interval in inertial reference frames

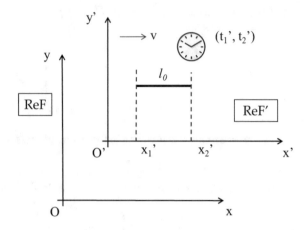

keeping in mind that the length of a moving body is measured by calculating the spatial coordinates of its extremes x_1, x_2 at the same time $t = t_1 = t_2$:

$$l_0 = \gamma(x_2 - v_x t_2) - \gamma(x_1 - v_x t_1) = \gamma(x_2 - x_1) - \gamma v_x(t_2 - t_1) = \gamma l$$

(1.9)

$$\Rightarrow l = \frac{l_0}{\gamma}$$

Similarly, the measurement of a *proper time interval* $\Delta t_0 = t_2' - t_1'$ in ReF' where the clock is at rest $(x_2' = x_1')$, translates into *time dilution* in ReF:

$$\Delta t = t_2 - t_1 = \gamma(t_2' - t_1') + \gamma \frac{v_x}{c^2}(x_2' - x_1') = \gamma \Delta t_0$$

(1.10)

$$\Rightarrow \Delta t = \gamma \Delta t_0$$

1.1.4   Velocities

Transformation of velocities can be calculated by applying a time-derivative to Eqs. 1.8 in ReF'. For the velocity component parallel to the relative motion of the reference frames:

$$u_x' = \frac{dx'}{dt'} = \frac{dx'}{dt}\left(\frac{dt'}{dt}\right)^{-1} =$$

$$= \frac{d}{dt}\left[\gamma(x - v_x t)\right] \cdot \left[\frac{d}{dt}\gamma\left(t - \frac{xv_x}{c^2}\right)\right]^{-1} =$$

$$= \gamma(u_x - v_x) \cdot \left[\gamma\left(1 - \frac{u_x v_x}{c^2}\right)\right]^{-1} =$$

$$= \frac{u_x - v_x}{\left(1 - \frac{u_x v_x}{c^2}\right)}$$

(1.11)

For the velocity components orthogonal to the relative motion (y, z):

$$u'_y = \frac{dy'}{dt'} = \frac{dy'}{dt}\left(\frac{dt'}{dt}\right)^{-1} =$$
$$= \frac{dy}{dt}\cdot\left[\gamma\left(1 - \frac{u_x v_x}{c^2}\right)\right]^{-1} = \qquad (1.12)$$
$$= \frac{u_y}{\gamma\left(1 - \frac{u_x v_x}{c^2}\right)}$$

Although the spatial coordinates orthogonal to the direction of the relative motion remain identical in the two reference frames, the transformation of time introduces a mix of velocity components in the orthogonal directions. Namely, the y- and z-components of velocity also depend from the relative motion of the reference frames.

1.2   Relativistic Dynamics

1.2.1   4-Vectors

Special Relativity postulates that physics laws are invariant, i.e., they are the same in all reference frames and so has to be their mathematical description. Equations invariant under transformation of coordinates are said to be written in *covariant form*. For example, Maxwell's equations of Electromagnetism are covariant under Lorentz-Fitzgerald's transformations [2].

A compact and practical form to write covariant transformations uses vectors in covariant and controvariant notation:

$$A'_\alpha = \frac{\partial x^\alpha}{\partial x'^\beta} A_\beta, \qquad\qquad A'^\alpha = \frac{\partial x'^\alpha}{\partial x^\beta} A^\beta \qquad (1.13)$$

where A' is intended in ReF', and summation is done over repeated indexes $\alpha, \beta = 0, 1, 2, 3$. These run over the time and spatial components of a 4-dimensional vector (henceforth, *4-vector*). The 4-vector space-time is $x^\mu = (ct, x, y, z)$.

The scalar product of two 4-vectors properly defined in Special Relativity is Lorentz-invariant and it is calculated according to the following metric:

$$A'^\mu B'_\mu = A_\mu B^\mu = (A_0 B^0 - \vec{A}\cdot\vec{B}) \qquad (1.14)$$

Given any controvariant vector $A^\mu = (A^0, \vec{A})$, the associated covariant vector is $A_\mu = (A^0, -\vec{A})$.

It is now immediate to show that the *proper time interval* $d\tau$ is Lorentz-invariant. At first, we calculate the space-time distance given by the scalar product of the 4-vector space-time. We demonstrate that, as expected, it is Lorentz-invariant:

$$s'^2 = x'^\mu x'_\mu = (ct')^2 - |\vec{x}'^2| =$$

$$= c^2\gamma^2 \left(t - \frac{xv_x}{c^2}\right)^2 - \gamma^2(x - v_x t)^2 - y^2 - z^2 =$$

$$= c^2\gamma^2 t^2 + \gamma^2 \frac{x^2 v_x^2}{c^2} - 2\gamma^2 x v_x t - \gamma^2 x^2 - \gamma^2 v_x^2 t^2 + 2\gamma^2 x v_x t - y^2 - z^2 =$$

$$= c^2 t^2 \gamma^2 (1 - \beta^2) - x^2 \gamma^2 (1 - \beta^2) - y^2 - z^2 =$$

$$= (ct)^2 - |\vec{x}|^2 = s^2$$

$$(1.15)$$

Equation 1.15 also holds for the differential form $ds^2 = dx^\mu dx_\mu = g_{\mu\nu} dx^\mu dx^\nu$ $= (cdt)^2 - |d\vec{x}|^2$, and $g_{\mu\nu} = (1, -1, -1, -1)$ is the metric tensor. In spite of its square notation, the 4-D space-time distance can be either positive or negative. When a particle is at rest in ReF', $d\vec{x}' = 0$ and therefore $dt' = d\tau$. With this, Eq. 1.15 gives $ds^2 = (cdt)^2 - (d\vec{x})^2 = (cdt')^2 - (d\vec{x}')^2 = c^2 d\tau^2$.

1.2.2   Momentum

The single particle 4-vector momentum is defined as $p^\mu = (\frac{E}{c}, \vec{p})$. The transformation rules for p^μ are $p'^\mu = \frac{\partial x'^\mu}{\partial x^\nu} p^\nu$. Its components result:

$$p'^0 = \frac{E'}{c} = \frac{\partial x'^0}{\partial x^0} p^0 + \frac{\partial x'^0}{\partial x^1} p^1 + \frac{\partial x'^0}{\partial x^2} p^2 + \frac{\partial x'^0}{\partial x^3} p^3 =$$

$$= \frac{\partial t'}{\partial t} \frac{E}{c} + c\frac{\partial t'}{\partial x} p_x + c\frac{\partial t'}{\partial y} p_y + c\frac{\partial t'}{\partial z} p_z =$$

$$= \gamma \frac{E}{c} - \gamma \frac{v_x p_x c}{c^2} + 0 + 0$$

$$\Rightarrow E' = \gamma (E - v_x p_x)$$

$$(1.16)$$

$$p'^1 = p'_x = \frac{\partial x'^1}{\partial x^0} p^0 + \frac{\partial x'^1}{\partial x^1} p^1 + \frac{\partial x'^1}{\partial x^2} p^2 + \frac{\partial x'^0}{\partial x^3} p^3 =$$

$$= \frac{\partial x'}{c\partial t} \frac{E}{c} + \frac{\partial x'}{\partial x} p_x + \frac{\partial x'}{\partial y} p_y + \frac{\partial x'}{\partial z} p_z =$$

$$= -\gamma \frac{v_x}{c^2} E + \gamma p_x + 0 + 0$$

$$\Rightarrow p'_x = \gamma (p_x - \frac{v_x E}{c^2})$$

$$(1.17)$$

$$p'^{2,3} = p'_{y,z} = \frac{\partial x'^{2,3}}{\partial x^0} p^0 + \frac{\partial x'^{2,3}}{\partial x^1} p^1 + \frac{\partial x'^{2,3}}{\partial x^2} p^2 + \frac{\partial x'^{2,3}}{\partial x^3} p^3 = p^{2,3}$$

$$\Rightarrow p'_{y,z} = p_{y,z}$$

$$(1.18)$$

By recalling the invariance of the proper time interval $d\tau$, we can define a 4-vector force as the time-derivative of the 4-vector momentum:

$$F^\mu = \frac{dp^\mu}{d\tau} = \frac{d}{d\tau}\left(\frac{E}{c}, \vec{p}\right) = \left(\frac{d(\vec{F} \cdot d\vec{s})}{cd\tau}, \frac{d\vec{p}}{d\tau}\right) = \left(\gamma \vec{F} \cdot \vec{\beta}, \gamma \vec{F}\right) \qquad (1.19)$$

The very last equality is by virtue of the time dilution $dt = \gamma d\tau$, which allows us to define the 4-vector force through quantities measured in the same reference frame, i.e., ReF.

Since $d\tau$ is the proper time interval in ReF', the particle is at rest in ReF', namely, the whole ReF' is anchored to the particle, which moves with velocity $\vec{u}$ with respect to ReF. Hence, γ in Eq. 1.19 for the ReF has to be evaluated in terms of the relative velocity $|\vec{u}|$ of the two reference frames. Moreover, $\gamma' = 1$ in ReF' because $\vec{u}' = 0$, and the 4-vector force evaluated in ReF' becomes $F'^\mu = \left(\gamma' \vec{F}' \cdot \vec{\beta}', \gamma' \vec{F}'\right) = \left(0, \vec{F}'\right)$.

1.2.3 Mass-Energy Equivalence

Since the scalar product $p^\mu p_\mu$ is Lorentz-invariant, the quantity it represents has to be independent from the particle's energy or velocity, charge or position. In other words, it has to be an intrinsic property of the particle, but still involved in its dynamics. A well-educated assumption for it is the particle's mass, which we multiply by c^2 to obtain units of energy:

$$c^2 p^\mu p_\mu = E^2 - |\vec{p}c|^2 = m_0^2 c^4 \qquad (1.20)$$

What is the relationship between the *inertial* or *rest mass* m_0 and the particle's kinetic energy T? In analogy to Classical Mechanics, we impose a linear dependence of E from T, through a constant C which we can interpret as the minimum energy level of a free particle. From Eq. 1.20 it follows:

$$E^2 = p^2 c^2 + m_0^2 c^4 \equiv (T + C)^2 = T^2 + 2TC + C^2 \qquad (1.21)$$

Since T is a function of p, and by equating member-to-member the second and fourth term of the previous equation, we obtain:

$$\begin{cases} pc = \sqrt{T^2 + 2TC} \\ C = m_0 c^2 \end{cases} \Rightarrow \begin{cases} pc = \sqrt{T^2 + 2Tm_0 c^2} \\ E = T + m_0 c^2 \end{cases} \qquad (1.22)$$

So, contrary to Newton's Mechanics, a free particle at rest has a non-zero minimum energy level equal to its rest mass energy, or *rest energy*. The *total* particle's energy is the linear sum of kinetic energy and rest energy.

In fact, we can use the kinetic energy to discriminate between non-relativistic, relativistic and ultra-relativistic regime of a particle's motion. Somehow arbitrarily, we impose the threshold for relativistic motion to be $T \approx m_0c^2$. Accordingly, a particle is said to be in non-relativistic (or classical) regime whenever:

$$T << m_0c^2 \quad \Rightarrow \quad T = \frac{|\vec{p}|^2}{2m_0} \propto p^2 \tag{1.23}$$

On the contrary, it is in the ultra-relativistic regime when:

$$T >> m_0c^2 \quad \Rightarrow \quad T \approx pc \propto p \tag{1.24}$$

If a direct linear proportionality between total energy and rest energy is introduced via a quantity (α in the following) that has to be dependent from the particle's velocity, the *relativistic mass* can be defined as $m = \alpha(v)m_0$:

$$\begin{cases} E^2 = p^2c^2 + m_0^2c^4 \\ \\ E = T + m_0c^2 \equiv \alpha(v)m_0c^2 \end{cases} \tag{1.25}$$

To find α, the second equality in Eq. 1.25 is substituted into the first one:

$$p^2c^2 = E^2 - m_0^2c^4 = \alpha^2 m_0^2c^4 - m_0^2c^4 \tag{1.26}$$

The definition of momentum with the prescription of a relativistic mass is $|\vec{p}| = \alpha m_0 \beta c$. Replacing it into Eq. 1.26 gives:

$$\alpha^2 \beta^2 m_0^2c^4 = \alpha^2 m_0^2c^4 - m_0^2c^4;$$

$$\alpha^2(\beta^2 - 1) = -1; \tag{1.27}$$

$$\alpha = \frac{1}{\sqrt{1-\beta^2}} = \gamma$$

Finally,

$$\begin{cases} E = \gamma m_0c^2, \\ \\ \vec{p} = \gamma m_0 \vec{\beta} c \\ \\ pc = \beta E \end{cases} \tag{1.28}$$

The expression $E = \gamma m_0c^2$ states the so-called Einstein's energy-mass equivalence. Figure 1.4 shows the quantities introduced so far as function of the particle's kinetic energy, for different particle species and therefore different rest energies.

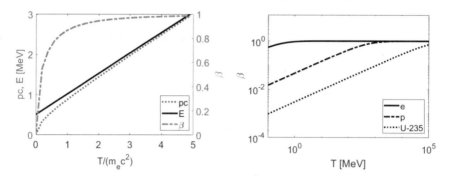

Fig. 1.4 Left: total momentum, total energy (left axis) and normalized velocity (right axis) as function of the kinetic energy in units of electron's rest energy. Right: normalized velocity as function of kinetic energy for an electron, a proton and an atom of Uranium-235

1.2.3.1 Discussion: Entering the Relativistic Regime

In the previous Section we introduced, somehow arbitrarily, an energy threshold according to which a particle enters the relativistic regime of motion, i.e., when its kinetic energy equals the particle's rest energy. What is the particle's velocity at such threshold energy? Does the velocity depend from the particle species?

Let us consider three nuclear and sub-nuclear species, an Hydrogen atom (mass number A = 1), an Helium atom (A = 4) and an electron. The velocity β can be calculated, for example, from the γ factor. This is in turn easily related to the particle's total energy via Einstein's formula for the energy-mass equivalence. For any species:

$$\gamma_{th} = \frac{E}{m_0 c^2} = \frac{T + m_0 c^2}{m_0 c^2} \equiv \frac{2 m_0 c^2}{m_0 c^2} = 2$$

$$\Rightarrow \beta = 1 - \frac{1}{\gamma^2} = 0.75$$

(1.29)

Hence, the definition of relativistic regime given above happens at the velocity $v = \frac{3}{4} c$ independently from the particle species.

1.2.3.2 Discussion: Pions' Lifetime

Relativistic time dilution finds routinely application in particle physics. As an example, a population of unstable particles reduces to $1/e$ of its initial value after a characteristic time interval called *lifetime*. This is defined in the reference frame in which the population is at rest. A particularly short lifetime can be brought to longer and therefore measurable time intervals by accelerating the particles close to the light speed. Acceleration can be either artificial, such as in particle accelerators, or natural, such as in cosmic rays.

As an example, let us consider a pion (a meson, i.e., a non-elementary particle constituted by a quark and an anti-quark); its rest energy is $m_\pi c^2 = 139.6$ MeV. It decays in other mesons, called kaons, with lifetime $d\tau = 26.029$ ns. What is the pions' lifetime in the laboratory frame, assuming that they are accelerated, and therefore gain a kinetic energy $T = 100$ MeV?

Since the lifetime of pions at rest is known ($d\tau$), and since it is by definition a proper time interval, the lifetime in the laboratory frame can be calculated as $dt = \gamma d\tau$. The knowledge of the pion's kinetic energy suggests to calculate γ through the total energy and, thereby, Einstein's relation:

$$E = T + m_\pi c^2 = 239.6 \, \text{MeV};$$

$$\Rightarrow \gamma = \frac{E}{m_\pi c^2} = 1.716 \tag{1.30}$$

which leads to $dt = \gamma d\tau = 44.674$ ns.

1.2.3.3 Discussion: A Particle's Point of View

A linear accelerator, simply *linac* in the literature, is a sequence of metallic structures in which charged particles are accelerated by a longitudinal electric field. The gained kinetic energy is linearly proportional to the accelerator length. The average energy gain per unit length is denominated *accelerating gradient*.

Let us consider the linac at the Stanford Linear Accelerator Center in California, USA. It is approximately 3 km long, and it is characterized by an average accelerating gradient of 20 MeV/m. Electrons are injected into the linac with an initial velocity $v_i = c/2$. Knowing that the electron's rest energy is $m_e c^2 = 0.511$ MeV, we wonder how long the accelerator is in the electrons' rest frame when they are injected. How long is it when electrons are at its middle point?

The relativistic effect of length contraction can be evaluated here by identifying the ReF with the laboratory frame, and the ReF' with the electrons' rest frame. Although electrons are accelerated, we can apply Lorentz-Fitzgerald's transformations at instantaneous time coordinates, i.e., at the injection point and at the linac middle point. Electrons see the linac moving towards them with the same velocity they are seen in the laboratory frame, but in the opposite direction. The knowledge of particles' velocity suggests to calculate the γ factor via kinematics, i.e., $\gamma = 1/\sqrt{(1 - \beta^2)}$, so that the linac length seen by the electrons will be contracted to $l_0 = l/\gamma$.

At the injection point:

$$\gamma_i = 1/\sqrt{1 - 0.25} = 1.1547 \Rightarrow l_{0,i} = 3 \, \text{km}/1.1547 = 2.6 \, \text{km} \tag{1.31}$$

At the linac midpoint, electrons have gained a kinetic energy of $\Delta E = 20$ MeV/m $\cdot$ 1.5 km $= 30$ GeV. Their total energy E_m at the linac midpoint is the sum of the gained energy and of their total energy at the injection point E_i. By virtue of Einstein's relation for the relativistic mass:

$$E_i = \gamma_i m_e c^2 = 0.59 \, \text{MeV}$$

$$E_m = E_i + \Delta E = 30000.59 \, \text{MeV} \tag{1.32}$$

$$\gamma_m = E_m/(m_e c^2) = 58709.6$$

which eventually leads to $l_{0,m} = 1.5 \, \text{km}/58709.6 = 2.6 \, \text{cm}$.

1.2.4 Invariant Mass

Let us consider an ensemble of N particles mutually interacting, but not subject to external forces. That is, the system is isolated. Each particle is characterized by a rest mass m_i, position and velocity vectors $\vec{x}_i$, $\vec{v}_i$. Position and velocity of the *center of mass* of the system (CM) are defined as:

$$\vec{x}_{cm} = \frac{m_1\vec{x}_1 + m_2\vec{x}_2 + \cdots + m_N\vec{x}_N}{m_1 + m_2 + \cdots + m_N} = \frac{1}{m_{tot}} \sum_{i=1}^{N} m_i\vec{x}_i$$

$$\vec{v}_{cm} = \cdots = \frac{1}{m_{tot}} \sum_{i=1}^{N} m_i\vec{v}_i = \frac{1}{m_{tot}} \sum_{i=1}^{N} \vec{p}_i = \frac{\vec{p}_{tot}}{m_{tot}}$$

(1.33)

We find $\vec{p}_{cm} = m_{tot}\vec{v}_{cm} = \vec{p}_{tot}$. Nevertheless, the energetic content of the system does not reduce to that one of the center of mass. Indeed, the center of mass total energy is:

$$E_{cm}^2 = |\vec{p}_{cm}c|^2 + (m_{cm}c^2)^2 = |\vec{p}_{tot}|^2c^2 + m_{tot}^2c^4$$

(1.34)

whereas the total energy of the system is:

$$E_{tot}^2 = \left(\sum_i E_i\right)^2 = \left(\sum_i \sqrt{(\vec{p}_ic)^2 + (m_ic^2)^2}\right)^2 \geq \left(\sum_i \vec{p}_ic\right)^2 + \left(\sum_i m_ic^2\right)^2$$

$$\Rightarrow E_{tot}^2 \geq |\vec{p}_{tot}|^2c^2 + m_{tot}^2c^4 \quad or \quad E_{tot} \geq E_{cm}$$

(1.35)

The inequality in Eq. 1.35 relies on the fact that, while the total energy of the system is contributed by the particles' momenta taken with their absolute values, the total energy of the CM is contributed by the vectorial sum of the momenta. In other words, it is possible to define a 4-vector total momentum p_{tot}^μ whose scalar product is the proper mass of the system, also called *invariant mass*. According to Eq. 1.35 it results:

$$p_{tot}^\mu = \sum_i p_i^\mu = \left(\sum_i \frac{E_i}{c}, \sum_i \vec{p}_i\right) = \left(\frac{E_{tot}}{c}, \vec{p}_{tot}\right)$$

$$c^2 p_{tot}^\mu p_{tot,\mu} = E_{tot}^2 - |\vec{p}_{tot}c|^2 \equiv (M_0c^2)^2 \geq \left(\sum_i m_ic^2\right)^2$$

(1.36)

What is the difference $(M_0 - m_{tot})c^2 \geq 0$ attributed to? It is the sum of the kinetic energies of the system components and, if present, of their potential or interaction energy. For example, in a ReF in which $\exists p_i$, $p_j \neq 0$ but the vectorial sum of the non-zero momenta is null ($\vec{p}_{tot} = 0$), the CM is at rest, namely, the CM's total energy is the total rest energy. However, the invariant mass is larger than that ($M_0 > m_{tot}$), in proportion to the absolute values of the individual non-zero momenta. As we will see, this is the case of head-on colliding beams in high energy accelerators, where new particles can be produced almost at rest from the interaction of counter-propagating accelerated beams, and with a rest energy equal to the sum of the energies of the accelerated beams.

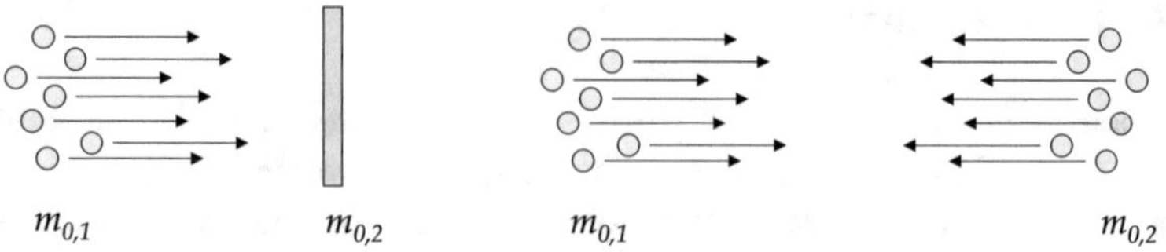

Fig. 1.5 Fixed target geometry (left) and head-on colliding beams (right)

On the contrary, $M_0 = m_{tot} \iff p_i = 0 \; \forall i$, and in this case $E_{tot} = E_{cm} = m_{tot}c^2$. This defines the invariant mass as the CM energy in the ReF where all system components are at rest. Such a special ReF, however, could not exist. Instead, one could notice that, since $\vec{p}_{cm} = \vec{p}_{tot}$, the velocity of the (virtual) particle associated to p^μ_{tot} is just v_{cm}. Then we have:

$$p_{tot}c = \beta_{cm} E_{tot} \quad \Rightarrow \quad M_0 c^2 = \sqrt{E^2_{tot}\left(1 - \beta^2_{cm}\right)} = \frac{E_{tot}}{\gamma_{cm}} \qquad (1.37)$$

Equation 1.37 introduces a general definition of the invariant mass as the total energy of the system evaluated in the CM reference frame (i.e., the ReF in which the CM is at rest, or $\gamma_{cm} = 1$). In high energy physics, the invariant mass is usually noted as $M_0 c^2 \equiv \sqrt{s}$.

1.2.5   Colliders

Colliders, either in linear or circular geometry, are particle accelerators devoted to the production of nuclear and sub-nuclear particle species. Colliders can be classified depending on the geometry of the collision at the interaction point, see Fig. 1.5. In the first class, an energetic particle beam hits a *fixed target*. In the second class, two *colliding beams* interact at one or multiple points along the accelerator. The choice of either one or the other geometry of collision is in most cases driven by the magnitude of the invariant mass of the system.

In order to discuss advantages and drawbacks of the two schemes, we aim at quantifying the invariant mass of the system (beam + target in the first case, beam + beam in the second) as function of the total energy of the accelerated beam(s). For simplicity, we consider accelerated beams whose particle's total energy is much larger than the rest energy.

In a fixed target geometry, the 4-vector momenta of the particle beams are $p^\mu_1 = (\frac{E_1}{c}, p_1)$ and $p^\mu_2 = (m_{0,2}c, 0)$. The invariant mass is (see Eq. 1.36):

$$\sqrt{s_t} = \sqrt{E^2_{tot} - p^2_{tot}c^2} = \sqrt{(E_1 + m_{0,2}c^2)^2 - p^2_1 c^2} =$$

$$= c^2 \sqrt{m^2_{0,1} + m^2_{0,2} + 2E_1 m_{0,2}/c^2} \approx \sqrt{2m_{0,2}c^2 E_1} \propto \sqrt{E_1} \qquad (1.38)$$

For two accelerated beams in head-on collision, $\vec{p}_1 = -\vec{p}_2$. We simplify the math by assuming identical species ($m_{0,1} = m_{0,2}$) and therefore same total energies ($E_1 = E_2$):

$$\sqrt{s_c} = \sqrt{E_{tot}^2 - p_{tot}^2 c^2} = \sqrt{(E_1 + E_2)^2 - (\vec{p}_1 + \vec{p}_2)^2 c^2} = (E_1 + E_2) = 2E_1 \tag{1.39}$$

In order to obtain the same invariant mass, the ratio of the energy of the accelerated beam in the "fixed target" scheme and in the "head-on collision" is:

$$\frac{\sqrt{s_t}}{\sqrt{s_c}} = \frac{\sqrt{2m_{0,2}c^2 E_{1,t}}}{2E_{1,c}} \equiv 1 \quad \Rightarrow \quad \frac{E_{1,t}}{E_{1,c}} = \frac{2E_{1,c}}{m_{0,2}c^2} \gg 1 \tag{1.40}$$

In conclusion, the invariant mass in the head-on collision has a more favourable scaling with the total energy of the accelerated beam than in the fixed target geometry. However, one should consider that the amount of power consumption for bringing the accelerated beams to the final energy level is up to doubled in head-on collision respect to the fixed target configuration. Fixed target geometries are usually convenient when the energy of the accelerated particles is comparable to the rest energy of the target particle: $E_{1,t}/E_{1,c} \approx 1 \Rightarrow E_{1,t} \approx m_{0,2}c^2$. This usually happens for values of the invariant mass $M_0 c^2 \leq 100s$ MeV, which is the typical energy range of nuclear physics experiments. Higher values of the invariant mass, up to TeV scale, are affordable only with beam-beam collisions.

1.2.6 Wave-Particle Duality

Planck's quantization of e.m. waves introduces light packets, made of electrically neutral particles called *photons* [3]. The energy of a single photon is $E = h\nu$, with ν the wave frequency and $h = 6.626 \cdot 10^{-34}$ Js the Planck's constant. From Eq.1.28, $pc = \beta E = E$ for a photon, hence:

$$\vec{p} = \frac{h\nu}{c}\hat{n} = \frac{h}{\lambda}\hat{n}. \tag{1.41}$$

Since photons travel at the light speed in vacuum, they are massless particles: $c^2 p^\mu p_\mu = (h\nu)^2 - (h\nu)^2 = 0$.

High energy (typically multi-GeV) particle accelerators configured as "light sources" find application in physics of matter by producing e.m. radiation at wavelengths comparable to, or smaller than, the spatial scale of the structure to be investigated. Speckle patterns observed as a result of radiation diffraction through the sample, on top of other processes such as light absorption, reflection or scattering, can be used to retrieve the structural properties of the sample. Equation 1.41 suggests that, for example, 1 nm spatial scale can be probed with photon energies around 1.24 keV or higher (x-rays).

Similarly, a wave function travelling at velocity $v < c$ can describe a massive particle of momentum $p = \gamma m_0 v$. In this case Eq. 1.41 defines the characteristic

Fig. 1.6 Photon energy (left axis, solid) and electron kinetic energy (right axis, dashed) versus De Broglie's wavelength

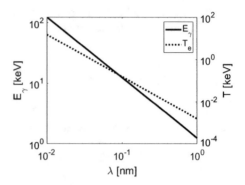

De Broglie's wavelength of the particle. As a matter of fact, low energy electrons are used in analogy to photons for diffraction experiments, where the higher the particle's momentum is, the smaller is the spatial scale that can be resolved by virtue of a corresponding shorter De Broglie's wavelength.

Figure 1.6 compares the electron beam's kinetic energy and the photon energy corresponding to a given wavelength. The wavelength associated to electrons is shorter than typical x-rays so that, e.g., $\lambda = 0.05$ nm (a fraction of an atom size) corresponds to $\sim$25 keV photons or $\sim$0.6 keV electron kinetic energy. However, by virtue of their massive and charged nature, electrons pose different constraints with respect to photons, to the characteristics of the sample to be studied, such as electron transparency and thickness smaller than 100 nm or so.

Low energetic electrons penetrate the sample more hardly compared to photons at the equivalent wavelength. The need of extremely well collimated electron beams usually limits the beam charge to the sub-pC scale and down to fC, with consequent lower signals than those produced by x-rays of comparable wavelength. This is partially leveraged by a much larger cross section for electrons compared to photons, etc. In summary, *electron diffraction* finds typical application in experiments tolerating relatively low signal per pulse, and associated to (sub-)picometer-scale wavelengths by exploiting tens of keV to few MeV's electrons.

1.2.7 Doppler Effect and Angular Collimation

The relativistic Doppler effect refers to the transformation of a wave's frequency from one ReF to another, when the relative velocity of the two frames approaches c. Its derivation, however, is tightly connected to the transformation of angles, which finds application for example in the calculation of astronomical distances.

In particle accelerators, the relativistic Doppler effect takes place when, for example, a charged particle (ReF') is subject to a centripetal force. In the laboratory frame (ReF), the charge emits radiation whose central frequency and direction of propagation depends from the particle's velocity. We want to quantify such a dependence.

Fig. 1.7 Wave emission in
inertial reference frames

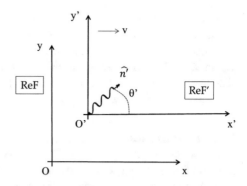

An e.m. wave is emitted by a particle instantaneously at rest in ReF'. We introduce the angles θ and θ' between the wave's direction $(\hat{n}')$ and the direction of relative motion of the two reference frames, see Fig. 1.7.

A wave phase is Lorentz-invariant because it can be written as the scalar product of the 4-vectors space-time and momentum:

$$\phi = p_\mu x^\mu = (Et - \vec{p}\vec{x}) = h\nu \left(t - \frac{\vec{p}\vec{x}}{E}\right) = h\nu \left(t - \frac{\vec{x}\hat{n}}{c}\right) \tag{1.42}$$

and $\hat{n} = (\cos\theta, \sin\theta, 0)$. It follows that $\phi = \phi'$ for the two phases in ReF and in ReF':

$$\nu t - y\frac{\sin\theta}{\lambda} - x\frac{\cos\theta}{\lambda} = \nu' t' - y'\frac{\sin\theta'}{\lambda'} - x'\frac{\cos\theta'}{\lambda'} \tag{1.43}$$

Lorentz-Fitzgerlad's transformations (see Eqs. 1.8) are then applied to the space-time coordinates:

$$\nu t - y\frac{\sin\theta}{\lambda} - x\frac{\cos\theta}{\lambda} = \nu'\gamma\left(t - \frac{xv_x}{c^2}\right) - y'\frac{\sin\theta'}{\lambda'} - \frac{\gamma(x - v_x t)}{\lambda'}\cos\theta' \tag{1.44}$$

and homogeneous terms grouped and made equal in the two reference frames:

$$\begin{cases} \nu t = \nu'\gamma t + \gamma\frac{v_x}{\lambda'}\cos\theta' t \\ \frac{\sin\theta}{\lambda}y = \frac{\sin\theta'}{\lambda'}y \\ -\frac{\cos\theta}{\lambda}x = -\nu'\gamma\frac{v_x}{c}x - \gamma\frac{\cos\theta'}{\lambda'}x \end{cases} \Rightarrow \begin{cases} \nu = \gamma\nu'\left(1 + \beta\cos\theta'\right) \\ \frac{\cos\theta}{\lambda} = \gamma\frac{(\cos\theta' + \beta)}{\lambda'} \end{cases} \tag{1.45}$$

The top-right expression of Eq. 1.45 is the so-called relativistic Doppler effect. Contrary to the classical Doppler effect, in case of orthogonal emission in ReF' ($\theta' = \pm\pi/2$), the frequency observed in ReF is always augmented by a factor γ.

The combination of the second and third left expressions in Eq. 1.45 brings to the so-called star light aberration or angular collimation effect:

$$\tan\theta = \frac{\sin\theta'}{\gamma(\cos\theta' + \beta)} \tag{1.46}$$

This implies that on-axis emission in ReF', either forward or backward, is seen as an on-axis emission in ReF too. However, a wave emitted in ReF' at $0 < |\theta'| < \frac{\pi}{2}$, is seen to be collimated in the laboratory frame in proportion to the source particle's velocity: for $\beta \to 1$ or equivalently $\gamma \gg 1$, $\theta \to \frac{1}{\gamma} \ll 1$. This is the case of e.m. radiation emitted by an ultra-relativistic charged particle in a dipolar magnetic field, or *synchrotron radiation*.

1.2.7.1 Discussion: Synchrotron Radiation

Demonstrate that the relativistic Doppler effect and angular collimation are two aspects of the same physical concept, i.e., the Lorentz's transformation of the longitudinal and the transverse momentum of an e.m. wave.

We recall the energy of a single photon, $E = h\nu$, and make use of the transformation of the 4-vector momentum in Eqs. 1.16 and 1.17. For example, let us calculate $\nu = \nu(\nu', \theta')$:

$$E = \gamma(E' + \vec{p}'\vec{v});$$
$$h\nu = \gamma(h\nu' + h\frac{\nu'}{c}v\cos\theta'); \tag{1.47}$$
$$\Rightarrow \nu(\theta') = \gamma\nu'(1 + \beta\cos\theta')$$

as already in Eq. 1.45. Alternatively, we can express $\nu = \nu(\nu', \theta)$:

$$E' = \gamma(E - \vec{p}\vec{v});$$
$$h\nu' = \gamma(h\nu - h\frac{\nu}{c}v\cos\theta); \tag{1.48}$$
$$\Rightarrow \nu(\theta) = \frac{\nu'}{\gamma(1 - \beta\cos\theta)}$$

From the conservation of the transverse momentum, $p'_{y,z} = p_{y,z}$:

$$\nu\sin\theta = \nu'\sin\theta' \tag{1.49}$$

This is combined with the transformation of the momentum component along the direction of relative motion:

$$p_x = \gamma(p'_x + \frac{\beta}{c}E');$$
$$\nu\cos\theta = \gamma(\nu'\cos\theta' + \beta\nu') = \gamma\nu'(\beta + \cos\theta') \tag{1.50}$$

$$\Rightarrow \tan\theta = \frac{\sin\theta'}{\gamma(\cos\theta' + \beta)}$$

as already in Eq. 1.46.

As an example, let us consider an ultra-relativistic electron in a uniform magnetic dipolar field. The electron total energy is, say, 0.5 GeV. Lorentz's force imposes a centripetal acceleration to the charge, which therefore emits radiation. In the electron's rest frame, most of radiation is emitted in the direction of acceleration (as

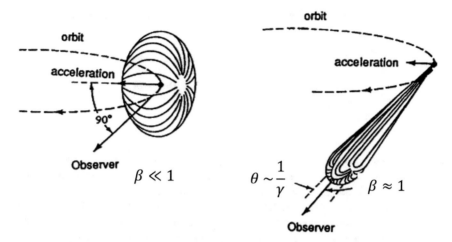

Fig. 1.8 Radiation emission from an ultra-relativistic charge in the presence of centripetal acceleration, in the particle's rest frame (left) and in the laboratory frame. (Image from Wikipedia public domain. Original picture in D.H. Tomboulian and P.L. Hartman, Phys. Rev. 102 (1956) 1423)

it happens for an electric dipole), therefore orthogonal to the instantaneous particle's velocity or $\theta' = \pi/2$, see Fig. 1.8. In the laboratory frame, since $\beta \to 1$, we obtain $\theta \approx 1/\gamma \approx 1$ mrad. Highly collimated radiation, tangent to the particle's orbit, driven by centripetal acceleration of an ultra-relativistic charge is called *synchrotron radiation*.

1.2.8 Forces

The transformation of forces is derived below for the simple case of a test particle at rest in ReF'. The condition $\vec{u}' = 0$ implies $\vec{u} = \vec{v}$ in ReF, with $\vec{v}$ the relative velocity of the two reference frames. For simplicity, we assume $\vec{v} = (v_x, 0, 0)$. The particle is subject to an external force $\vec{F}'$ in ReF'. The 4-vector force introduced in Eq. 1.19 is $F'^{\mu} = \left(\gamma' \vec{F}' \cdot \vec{\beta}', \gamma' \vec{F}'\right) = \left(0, \vec{F}'\right)$. We then apply the relativistic transformation in covariant form (see Eq. 1.13):

$$
F'^1 = F'_x = \frac{\partial x'^1}{\partial x^0} F^0 + \frac{\partial x'^1}{\partial x^1} F^1 + \frac{\partial x'^1}{\partial x^2} F^2 + \frac{\partial x'^1}{\partial x^3} F^3 =
$$

$$
= \frac{\partial x'}{c^2 \partial t} \gamma \vec{F} \cdot \vec{v} + \frac{\partial x'}{\partial x} \gamma F_x + 0 + 0 =
$$

$$
= -\frac{\gamma^2 v^2}{c^2} F_x + \gamma^2 F_x = F_x \gamma^2 (1 - \beta^2) = F_x
$$

$$
\Rightarrow F'_x = F_x
$$

(1.51)

For the components transverse to the relative motion of the two reference frames:

$$F'^2 = F'_y = \frac{\partial x'^2}{\partial x^0} F^0 + \frac{\partial x'^2}{\partial x^1} F^1 + \frac{\partial x'^2}{\partial x^2} F^2 + \frac{\partial x'^2}{\partial x^3} F^3 =$$

$$= \frac{\partial y'}{c^2 \partial t} \gamma \vec{F} \cdot \vec{v} + 0 + \frac{\partial y'}{\partial y} \gamma F_y + 0 = \gamma F_y \qquad (1.52)$$

$$\Rightarrow F'_y = \gamma F_y$$

It becomes apparent that the force acting on a particle, evaluated in the reference frame in which the particle is (at least instantaneously) at rest, is always greater than the force perceived by the same particle in any other reference frame.

The expressions for the general case $\vec{u}' \neq 0$ can be found in the References. In such case, the components of the force transversal to the direction of the relative motion are mixed, $F'_{y,z} = F'_{y,z}(F_{y,z})$. The component along the direction of the relative motion depends from all three components in ReF, $F'_x = F'_x(F_{x,y,z})$.

1.2.9 Fields

The transformation of electric and magnetic field is derived by considering the transformation of the Lorentz's force established between two charged particles q_1 and q_2 and, in particular, the force that q_2 exerts on q_1. Let us consider two cases.

First, both charges are at rest in ReF' ($u'_1 = u'_2 = 0$). Since $\vec{v} = (v_x, 0, 0)$, we have $u_1 = u_{1,x} = u_2 = u_{2,x} = v_x$. The force spatial components are first evaluated in ReF'. Then, they are transformed to ReF according to the prescriptions in Eqs. 1.51 and 1.52. Since q_2 is moving in ReF along the x-direction, it generates both electric and magnetic field in that frame, but the magnetic field has only y- and z-components:

$$\begin{aligned}
F'_x &= q_1 E'_x \equiv F_x = q_1 E_x \\
F'_y &= q_1 E'_y \equiv \gamma F_y = q_1 \gamma \left(E_y - u_{1,x} B_z \right) \\
F'_z &= q_1 E'_z \equiv \gamma F_z = q_1 \gamma \left(E_z + u_{1,x} B_y \right)
\end{aligned} \qquad (1.53)$$

It follows from Eq. 1.53 :

$$\begin{aligned}
E'_x &= E_x \\
E'_y &= \gamma (E_y - v_x B_z) \\
E'_z &= \gamma (E_z + v_x B_y)
\end{aligned} \qquad (1.54)$$

The inverse relationships are found for the second situation in which both charges are at rest in ReF ($u_1 = u_2 = 0$), and, q_1 is moving in ReF' ($u'_1 = u'_{1,x} = u'_2 = u'_{2,x} = -v_x$). In this case, the Lorentz's force exerted by q_2 on q_1 in ReF is purely electric, while in general it contains also magnetic components in ReF':

$$F_x = q_1 E_x \equiv F'_x = q_1 E'_x$$
$$F_y = q_1 E_y \equiv \gamma F'_y = q_1 \gamma \left(E'_y - u'_{1,x} B'_z \right) \quad (1.55)$$
$$F_z = q_1 E_z \equiv \gamma F'_z = q_1 \gamma \left(E'_z + u'_{1,x} B'_y \right)$$

In this case, ReF is the reference frame in which the test particle is at rest and therefore the force has to be maximum, namely, γ-fold larger than in ReF'. From these equations one finds:

$$E_x = E'_x$$
$$E_y = \gamma (E'_y + v_x B'_z) \quad (1.56)$$
$$E_z = \gamma (E'_z - v_x B'_y)$$

Equations 1.55 and 1.56 show that electric and magnetic fields in the direction of the relative motion of two reference frames do not mix their components. These mix, instead, in the directions orthogonal to the relative motion. In particular, a pure electric field in one frame is seen as a combination of electric and magnetic field in another moving frame.

Similar expressions and identical conclusions can be drawn for the magnetic field in the case of particles moving with velocities transverse to the relative motion of the reference frames. Only the field transformations are reported here for completeness:

$$B'_x = B_x$$
$$B'_y = \gamma (B_y + \tfrac{v_x E_z}{c^2}) \quad (1.57)$$
$$B'_z = \gamma (B_z - \tfrac{v_x E_y}{c^2})$$

1.2.10 Accelerations

The relativistic corrections to acceleration are introduced below. We first write the general expression of the spatial components of the 4-vector force, calculated as the time-derivative of the relativistic momentum:

$$\vec{F} = \tfrac{d\vec{p}}{dt} = m_0 c \tfrac{d}{dt}(\gamma \vec{\beta}) = \gamma m_0 \vec{a} + m_0 c \vec{\beta} \tfrac{d\gamma}{dt} =$$
$$= \gamma m_0 \vec{a} + m_0 c \vec{\beta} \gamma^3 (\vec{\beta} \cdot \dot{\vec{\beta}}) = \gamma m_0 \vec{a} + \gamma^3 m_0 \vec{v} \left(\tfrac{\vec{v} \cdot \vec{a}}{c^2} \right) \quad (1.58)$$

The last equality allows us to identify two contributions to the force, the first parallel to acceleration, the second parallel to the particle's velocity. The former case exemplifies particles accelerated by a longitudinal electric field, such as in a linac. The latter applies to particles accelerated by a centripetal force, such as in a magnetic dipolar field in a circular accelerator. From Eq. 1.58 one gets:

$$\begin{cases} F_\parallel = \gamma m_0 a_\parallel + \gamma^3 m_0 \beta^2 a_\parallel = \gamma m_0 a_\parallel (1 + \gamma^2 \beta^2) = \gamma^3 m_0 a_\parallel \\ \\ F_\perp = \gamma m_0 a_\perp \end{cases} \quad (1.59)$$

which says that, in the relativistic regime, the acceleration perceived by a particle is γ^3-times stronger in a linac and γ-times stronger in a circular accelerator, compared to the non-relativistic case. Since $\gamma' = 1$ in the particle's rest frame, Eq. 1.59 gives also the prescription for calculating the acceleration when passing from the particle's rest frame to the ReF where the particle moves at relativistic velocity.

As a by-product of Eq. 1.58, Einstein's expression of mass-energy equivalence can be retrieved. We start calculating the particle's instantaneous power, where the particle's energy variation is induced by an external force coupled to the particle's velocity:

$$\frac{dE}{dt} = \vec{F} \cdot \vec{v} = \gamma m_0 \vec{a} \cdot \vec{v} + \gamma^3 m_0 \beta^2 \vec{a} \cdot \vec{v} = \frac{c^2}{\gamma^2} m_0 \frac{d\gamma}{dt} + m_0 v^2 \frac{d\gamma}{dt} =$$

$$m_0 c^2 \frac{d\gamma}{dt} \left(\frac{1}{\gamma^2} + \beta^2 \right) = m_0 c^2 \frac{d\gamma}{dt} \tag{1.60}$$

By integrating in time the very first and last term we obtain $E = \gamma m_0 c^2$. Then, the definition of relativistic momentum allows us to write $pc = \beta E$ which, together with the previous relation, leads to $E^2 = (pc)^2 + (m_0 c^2)^2$.

1.2.10.1 Discussion: Electric and Magnetic Deflection

Charged particles can be bent by electric and magnetic external fields. Which of the two fields, assumed to be static, is better suited, assuming the same force is applied in the two cases?

The ratio of pure electric and magnetic force, for the same force magnitude, is:

$$\frac{|\vec{F}_e|}{|\vec{F}_m|} = \frac{q|\vec{E}|}{q|\vec{v} \times \vec{B}|} = \frac{E}{vB} \equiv 1 \quad \Rightarrow \quad \frac{|\vec{E}|}{|\vec{B}|} = \beta c \tag{1.61}$$

We conclude that while non-relativistic particles can be conveniently deflected by a pure electric field (typically not exceeding kV/m), ultra-relativistic particles would require $\approx$300 MV/m electric field versus 1 T magnetic field. Such high electric fields are not practical, while static magnetic fields in the range <2 T are routinely obtained with iron poles surrounded by coils at room temperatures (electromagnets) or permanent magnets.

1.2.10.2 Discussion: Coulomb Field of a Relativistic Charge

What is the spatial distribution of the Coulomb field of a particle moving with relativistic velocity? Assume for simplicity a source particle in *uniform rectilinear motion*. The source particle is at rest in ReF', which moves with velocity $\vec{v} = (v_x, 0, 0)$ with respect to the laboratory or ReF. The electric field is evaluated at the position of coordinates $\vec{r} = (x, y, z)$ in ReF.

The Coulomb field in ReF' is:

$$\begin{cases} \vec{E}' = \frac{q}{4\pi \epsilon_0} \frac{\vec{r}'}{r'^3} \\ r' = \sqrt{x'^2 + y'^2 + z'^2} \end{cases} \Rightarrow \begin{cases} E'_x = \frac{q}{4\pi \epsilon_0} \frac{x'}{r'^3} \\ E'_y = \frac{q}{4\pi \epsilon_0} \frac{y'}{r'^3} \end{cases} \tag{1.62}$$

According to Eqs. 1.55 and 1.56, the electric field evaluated in ReF and parallel to the relative motion of the reference frames is identical to that in ReF'. The orthogonal component, instead, corresponds to a force which has to be maximum in ReF (where the test particle is at rest), and therefore γ-fold larger than in ReF':

$$E_x = E'_x = \frac{q}{4\pi\epsilon_0}\frac{x'}{r'^3} = \frac{q}{4\pi\epsilon_0}\frac{\gamma(x-v_x t)}{[\gamma^2(x-v_x t)^2+y^2+z^2]^{3/2}} \sim \frac{1}{\gamma^2}$$

$$E_y = \gamma E'_y = \frac{q}{4\pi\epsilon_0}\frac{\gamma y'}{r'^3} = \frac{q}{4\pi\epsilon_0}\frac{\gamma y}{[\gamma^2(x-v_x t)^2+y^2+z^2]^{3/2}} \sim \frac{1}{\gamma^2}$$

(1.63)

The z-component of the field behaves identically to the y-component for symmetry. From the condition $B' = 0$ one also gets:

$$|\vec{B}| = \frac{|\vec{v}\times\vec{E}|}{c^2} \sim \frac{1}{\gamma^2} \tag{1.64}$$

Worth to notice, our findings still apply to an accelerated particle when the transformation of coordinates is applied at each timestamp.

Equations 1.63 and 1.64 show that the strength of the Coulomb field (or "near field" in the Liénard-Wiechert notation) is suppressed in the laboratory frame at high particle's energies. This effect is exploited, for example, in linacs devoted to acceleration of high charge density beams. To avoid particles' repulsion by Coulomb interaction at non-relativistic energies (so-called *space charge* force) and therefore dilution of the charge density, a particularly high accelerating gradient is adopted in the very first stages of acceleration to boost particles to ultra-relativistic energies.

The time coordinate in Eq. 1.63 acts as a "retarded time" in the sense of Liénard-Wiechert retarded potentials, because the source charge moves in ReF: since the radiated field takes some time to travel from the source to the observation point, the field at this point changes with time. In particular, Eq. 1.63 is evaluated at the time when the particle generated the field. However, for a more intuitive comprehension of the spatial distribution of the electric field lines, and therefore of the field intensity, it is convenient to choose $t = 0$, i.e., to evaluate the field at the time of observation. This is known in Special Relativity to produce the *apparent* situation of an "action at a distance", equivalent to that of a static picture in which the source particle is at rest on the perpendicular of the observation point to the (actual) velocity axis.

The representation of the field lines is given below in polar coordinates in ReF. The angle θ between the instantaneous particle's velocity and the direction of observation is introduced, so that $x = r\cos\theta$, and we set $z = 0$ for simplicity. The radial field component at the generic position $\vec{r}$ and $t = 0$ results:

$$|\vec{E}_r| = \sqrt{|\vec{E}_x|^2 + |\vec{E}_y|^2} = \frac{q}{4\pi\epsilon_0}\sqrt{\frac{\gamma^2(x^2+y^2)}{(\gamma^2 x^2+y^2)^3}} = \frac{q}{4\pi\epsilon_0}\frac{\gamma|\vec{r}|}{(\gamma^2 x^2+y^2)^{3/2}} =$$

$$= \frac{q}{4\pi\epsilon_0}\frac{\gamma r}{(\gamma^2 r^2\cos^2\theta+r^2\sin^2\theta)^{3/2}} = \frac{q}{4\pi\epsilon_0}\frac{\gamma r}{r^3[(1-\gamma^2)\sin^2\theta+\gamma^2]^{3/2}} =$$

$$= \frac{q}{4\pi\epsilon_0}\frac{\gamma r}{r^3[(-\beta^2\gamma^2)\sin^2\theta+\gamma^2]^{3/2}} = \frac{q}{4\pi\epsilon_0}\frac{1}{\gamma^2 r^2(1-\beta^2\sin^2\theta)^{3/2}}$$

(1.65)

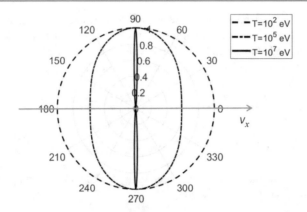

Fig. 1.9 Top-view of the electric field normalized intensity in Eq. 1.65, in polar coordinates. The source particle is an electron moving along the x-axis (arrow). The radial distance of the curves from the center of the plot is proportional to the field intensity, which is evaluated for a kinetic energy of 100 eV (dashed), 100 keV (dot-dashed) and 10 MeV (solid)

In the particle's rest frame, the Coulomb field has radial symmetry. However, when the particle moves at relativistic velocity, the field lines distribute asymmetrically in the plane of motion, in a "disc-like" configuration, as shown in Fig. 1.9. The lines density, i.e. the field absolute value, is maximum for $\theta = \pm \pi/2$. At such specific angle, by virtue of the choice $t = 0$, the electric field collapses to the non-relativistic expression, in accordance with our previous finding that the force perceived by a test particle is maximum in the frame where the test particle is at rest. The field is minimum at $\theta = 0, \pi$, suppressed by a factor $1/\gamma^2$ with respect to its maximum value.

References

1. R. Resnick, *Introduction to Special Relativity* (Wiley, New York, 1968)
2. J.D. Jackson, *Classical Electrodynamics*, 3rd edn. (Wiley, New York, 1999), pp. 514–578, 661
3. G. Gamow, *Thirty Years that Shook Physics: The Story of Quantum Theory* (Published by Dover Publications Inc., 2003). ISBN-13: 978-0486248950

Low Energy Accelerators

2

Particle accelerators [1,2] can be classified depending from several parameters such as geometry (linear vs. circular), accelerating technology (electrostatic vs. electro-dynamics, resonant vs. non-resonant), species of accelerated particles (leptons vs. barions, etc.), particle's energy (from keV to MeV, from GeV to TeV), and applications (high energy physics, physics of matter, medicine, industry, etc.).

Since the primary scope of an accelerator is to bring charged particles to high energy, it is natural to put some accent on that parameter. The average accelerating gradient, or energy gain per unit length (e.g., in MV/m), is a figure of merit which, in linacs, is inversely proportional to the accelerator size, hence to costs of construction.

Historically, the first accelerators exploited static electric field in linear geometry, thus named *electrostatic accelerators* [3,4]. Charged particles, typically electrons, protons or ions, were produced either by an external radioactive source, or extracted by an electrod brought to high static voltage ΔV. The accelerating electric field is conservative (irrotational), and the kinetic energy ΔT gained by each particle of charge q *internally* to the accelerator, is equal to the variation of the electric potential energy:

$$\oint \vec{E} d\vec{l} = 0 \Leftrightarrow \vec{\nabla} \times \vec{E} = 0 \;\Rightarrow \vec{E} = -\vec{\nabla} V \;\Rightarrow \Delta T = -q \Delta V = -\Delta U \quad (2.1)$$

The final energy of a free particle *injected* into an electrostatic accelerator results:

$$E_f = E_i - \Delta U = m_0 c^2 + T_i + \Delta T = m_0 c^2 + T_f \;\Rightarrow\; \Delta E = \Delta T = -q \Delta V \tag{2.2}$$

These accelerators are able to provide kinetic energies from tens of keV to tens of MeV. The main limitation to the maximum energy gain is from discharges happening when the electric field becomes higher than the dielectric rigidity of the medium in which the accelerator is immersed (commonly gas like N_2 and CO_2). Electrods were often shaped with spherical geometries to minimize the local field. A second limitation is related to the capability of the power supply to compensate for the total current emission, which includes both the main beam and spurious currents. In general, the maximum voltage of electrostatic accelerators is inversely proportional to the extracted beam current. Electrostatic accelerators produce either continuous or pulsed beam.

In 1930–1940s, the replacement of static voltages with rotational electric fields opened the door to the so-called *electrodynamic accelerators* [2]. These are supplied by radiofrequency (RF) power generators exploiting electric field varying with time. Such generators allowed higher accelerating fields at the tolerable expense of moderate average power.

Time-oscillating electric fields imply in most cases synchronization of the incoming particle beam with the accelerating field, which has to have the desired amplitude and phase at the arrival time of particles in the accelerating region. Electrodynamic accelerators are therefore further classified in *resonant* and *non-resonant*, depending if a synchronism condition has to be satisfied or not. Synchronization implies the production of pulsed beams, whose repetition frequency is therefore equal or a sub-multiple of the generator frequency. Each "packet" of particles is called *bunch* in the literature.

All these kinds of particle accelerators developed almost simultaneously, both in the linear and circular geometry. One of the advantages of circular accelerators, such as *betatrons*, *cyclotrons* and *synchrotrons*, is that the same accelerating gradient is applied to the circulating particles for, in principle, a very large number of turns. Hence, the energy gain per turn can be small. In reality, as we will see, the maximum beam energy is limited by the size of the accelerator (scaling with particle's energy), radiation emitted by the accelerated charges and, consequently, power consumption.

2.1 Electrostatic Accelerators

2.1.1 Cockcroft and Walton

Cockcroft and Walton invented the "multistage cascade generator" in 1932. In this electrostatic accelerator, the high voltage (<10 MV or so) is generated by means of a low-voltage oscillator followed by a series of diodes and capacitors, to multiply the initial potential energy, as shown in Fig. 2.1. The accelerator can be used either in continuous or pulsed mode, the latter one for reaching higher peak currents for the same average current level.

In the very first version, protons were produced from ionization of gaseous Hydrogen atoms through discharge. The gained kinetic energy of 0.4 MeV allowed protons (whose rest energy amounts to 938 MeV) to break the Coulomb barrier of Litium atoms in a thin foil, which then transmuted into Helium according to $p + Li_3^7 \rightarrow 2He_2^4$.

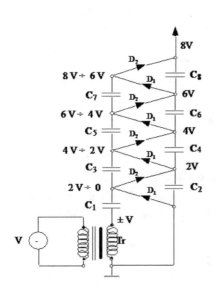

Fig. 2.1 Left: schematic of a Cockcroft-Walton voltage multiplier. Right: 870 kV Cockcroft-Walton proton injector at PSI, Switzerland, in operation since 1980s. The CW generator is on the left, the particle source is on the right. (Copyright Paul Scherrer Institute/Markus Fischer)

2.1.2 Van de Graaff

The Van de Graaff accelerator, first built in 1931, is an electro-mechanic system in which a rotating insulating belt is charged by effluvium (corona discharge) by a low-voltage generator ($\sim$10 kV). The high voltage ($<$10 MV) is generated by the mechanical deposition of charges from the belt onto an electrod. The electrod produces particle beam up to $\sim$10 MeV kinetic energy in continuous mode.

A more recent and energetic version of this accelerator is the so-called *Tandem* Van de Graaff. Let us consider negative ions of charge $-e$ produced by an external radioactive source. They are accelerated by an electrostatic voltage in an insulating gas, so gaining a kinetic energy equal to $e\Delta V$ per ion. At the point of maximum energy, the ions pass through a "stripper" (in the form of a thin foil or gas jet) which removes $(n + 1)$ electrons per ion. Each ion has now a positive charge $(-1 + n + 1)e = ne$, which can be further accelerated from the positive high voltage to ground. In short, the same voltage is used to accelerate twice, and in proportion to the number of electrons removed by the stripper, see Fig. 2.2. The total gained kinetic energy per ion is $\Delta T = e\Delta V + ne\Delta V = (n + 1)e\Delta V$. Since the stripper capacity is proportional to the atomic number of the stripper element, the gained kinetic energy is inversely proportional to the intensity of the accelerated beam surviving the stripper interaction.

The Tandem is particularly convenient to accelerate heavy ions because the high number of electrons to be possibly removed leads to a large multiplication factor of the accelerating voltage. For example, a static voltage $\Delta V = 15$ MV applied twice

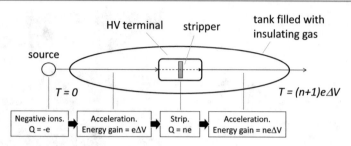

Fig. 2.2 Principle of operation of a Tandem Van de Graaff accelerator

to Au^- ions, in combination with a stripper capacity of $n + 1 = 14$, leads to a kinetic energy exceeding 200 MeV per ion. The Tandem is often used as injector of another class of accelerators commonly devoted to nuclear physics, the cyclotrons.

2.2 Electrodynamic Accelerators

2.2.1 Drift Tube Linacs

Electrodynamic accelerators exploit time-varying electric and magnetic field. Let us recall Maxwell's equations for the time-varying e.m. field in vacuum:

$$\vec{E} = -\vec{\nabla}V - \frac{\partial \vec{A}}{\partial t} \quad \Rightarrow \quad \vec{\nabla} \times \vec{E} = -\frac{\partial \vec{B}}{\partial t} \quad \Rightarrow \quad \oint_L \vec{E} d\vec{l} = -\frac{\partial \Phi(\vec{B})}{\partial t} \qquad (2.3)$$

with $\vec{A}$ the vector magnetic potential, $\vec{B}$ the magnetic flux density, and $\Phi(\vec{B})$ the magnetic flux through a region delimited by the closed path L. The l.h.s. equation suggests that there exists an electric potential generating $\vec{E}$ such that this field is made null by the presence of a magnetic potential outside the spatial region in which $\vec{\nabla}V \neq 0$. Namely, a time-varying longitudinal electric field can be produced under certain boundary conditions. At the same time, the r.h.s. expression says that by modulating in time the magnetic flux over a certain spatial region, a rotational electric field is generated such that a net energy gain (or loss) is obtained over the closed path concatenated with the magnetic flux. Namely, the electric field is not conservative (rotational).

The former of these two interpretations found application in the Ising and Wideröe's *Drift Tube Linac* (DTL), first developed in 1924–1927 (later on also engineered by Sloan and Lawrence). In a DTL, a series of metallic tubes mask the accelerated particles from any external force, like Faraday cages. Two successive tubes make a "gap", in which a longitudinal electric field oscillating with time at frequency f_{RF} is established by a generator at relatively low voltage, as sketched in Fig. 2.3-top left. Wideröe's DTL is classified as a "π-mode structure" because the electric field vector repeats identically every two cells, while it shows phase opposition at consecutive cells. This means that particles are bunched, but the bunches must be spaced with a free "bucket" in between.

Net acceleration is gained only if the electric field in the gap is synchronized with the particle's arrival time at any specific gap. The synchronism condition is $2f_{RF} = hf$, with f the frequency at which particles happen to be at successive gaps, and $h \in \mathbb{N}$. The factor 2 is owed to the fact that, for a π-mode structure, the field phase should vary by half period in the time particles take to reach the next gap.

For any particle velocity β_i through the i-th tube of length L_i, the condition of synchronization is:

$$2f_{RF} = hf = h\frac{\beta_i c}{L_i} \Rightarrow \begin{cases} L_i = \dfrac{h\lambda_{RF}\beta_i}{2} \\[2mm] G_i = \dfrac{\Delta V_i}{L_i} = \dfrac{2\Delta V_i}{h\lambda_{RF}\beta_i} \end{cases} \qquad (2.4)$$

As the particle gains energy and its velocity increases, the tube length has to increase too. The local accelerating gradient G_i is maximized by short wavelengths of the RF generator. For example, $f_{RF} \approx 30\,\text{MHz}$ implies a single tube not shorter than $\sim 10\,\text{m}$ when particles' velocity approaches c. A compact DTL forces to consider particles in non-relativistic motion (e.g., protons or ions) and/or short λ_{RF}. However, the wavelength of the oscillators cannot be made arbitrarily short, because when λ_{RF} approaches the size of the electrical components, these start emitting radiation, so diminishing the energetic efficiency of the generator.

An *Alvarez's structure* can be intended as a special case of DTL, in which the electrodes are part of a resonant macro-structure, as shown in Fig. 2.3-bottom left. The resonant frequency of the structure matches the frequency of the RF generator. In practice, the whole system is isolated in a metallic tank and, in most used configurations, the tubes are alternatively connected to an excited coaxial line and to the grounded outer wall. This accelerator is a 2π-mode structure because consecutive bunches can be accelerated in adjacent gaps.

Unlike the Wideröe's DTL, where the charges travel from one tube to the next one by passing through the RF generator, the Alvarez's structure is a resonant cavity, where the particles are immersed in a time-varying accelerating field established by the inner geometry of the structure.

2.2.2 Cyclotron

The principle of RF acceleration applied to a circular path brought to the invention of the *cyclotron*. In Lawrence's cyclotron (first built in 1932 at Berkeley, USA), the gap is constituted by two facing electrodes, called "dees" because of their geometry. The dees, which have analogous function of the tubes in the Wideröe's linac, are connected to an oscillator. The dees are immersed in a uniform vertical magnetic field. This generates the Lorentz's force to drive particles on a curved path (see Fig. 2.4-left plot). Particles originate at the center of the cyclotron, either from a radioactive source (electrons) or injected there by a lower energy Tandem accelerator.

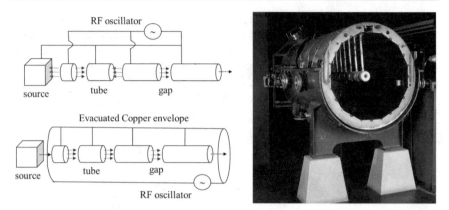

Fig. 2.3 Left: concept of Wideröe's DTL (top) and of Alvarez's structure (bottom). The electric field lines are shown at the same given time. Right: 32 MeV-Alvarez proton linac made operational in 1947 at the University of Southern California. (Photo reproduced under Creative Commons Zero license, National Museum of American History)

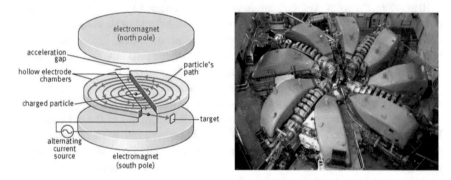

Fig. 2.4 Left: components of the Lawrence's cyclotron. (Image courtesy of A. A. El-Saftawy, Ph.D. Thesis, IOM, 2003). Right: sector-cyclotron at PSI, Switzerland. (Copyright Paul Scherrer Institute/Markus Fischer)

Particles' motion in a cyclotron can be understood by looking to the Lorentz's force acting on an instantaneous circular path across the dees, and recalling the relativistic correction to centripetal acceleration (see Eq. 1.59):

$$\gamma m_0 \frac{v_z^2}{R} = q v_z B_y \quad \Rightarrow \quad \begin{cases} \omega_L = \dfrac{q B_y}{\gamma m_0} \\[2ex] p_z = q B_y R \end{cases} \tag{2.5}$$

In practical units, $p_z [GeV/c] = 0.3 B_y[T] R[m]$.

The former equation introduces Lawrence's angular frequency, ω_L. As the particle is accelerated and its energy increased, ω_L decreases for a constant B_y. In order to maintain the synchronism with the RF generator, $\omega_R = h\omega_L$, the latter has to be modulated in frequency. This process cannot be indefinite, since each RF source has a well-defined bandwidth of functionality. Within such technological limitations, the cyclotron can provide up to few hundreds of MeV kinetic energy per nucleon, and it is called *synchro-cyclotron*.

The second equation shows that the larger the particle's longitudinal momentum is, the larger the radius of curvature becomes. Thus, particles follow spiralizing orbits starting from the center of the cyclotron, where they are injected, and they escape in correspondence of the ultimate bending radius. It is now clear that the maximum kinetic energy gained by the particle is basically constrained by the size of the cyclotron. Since the whole frame is immersed in a magnetic field of azimuthal symmetry, the maximum kinetic energy is proportional to the *volume* of the system.

A second version of Lawrence's cyclotron was developed, and is today largely used, which exploits a different mechanism of synchronization. Equation 2.5 shows that Lawrence's angular frequency can be kept constant, and so ω_{RF} can be, by properly modulating the magnetic field as function of the beam energy. In other words, as the particle's energy increases, the instantaneous value of B_y makes the particle shifting on a larger curved path. However, in correspondence of this new path, the magnetic field is made stronger such that $B_y/\gamma = const \ \forall r$.

But, how could $B_y(r)$ be made stronger at larger r? For example, by shaping the inner surface of the magnetic poles so to increase the field line density at larger distances from the cyclotron center, as sketched in Fig. 2.5. This solution, however, induces vertical *defocusing*, i.e., any vertically displaced particle is also subject to a horizontal component of the magnetic field, or field gradient. This pushes the particle further away from the horizontal plane (see the composition of velocity and field in Fig. 2.5 for a positive charge exiting the plane on the right and entering on the left). After several turns, particles can reach the dees' upper and lower boundaries, and eventually to get lost.

To overcome such unstable motion, the azimuthal symmetry of the magnetic poles is broken, and the poles are split into curved sectors, see Fig. 2.4-right picture. The curved shape of their entrance and exit face generates an overall vertical focusing force, i.e., particles which tend to escape in the vertical plane are brought back towards it, so performing vertical oscillations along their orbit. This kind of cyclotron

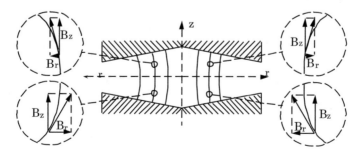

Fig. 2.5 Lateral section of the magnetic poles of a cyclotron. By virtue of the poles shaping, a charge vertically displaced with respect to the horizontal plane is subject to a horizontal component of the magnetic field. This produces a vertical defocusing Lorentz's force. (Image courtesy of S. Tazzari)

guarantees both horizontal and vertical stability. It is named *split-sector cyclotron* by virtue of the poles geometry, or *isochronous cyclotron* because of the constancy of the revolution period. The split poles geometry leaves free space in between the poles for RF generators, beam diagnostics, injection and extraction lines.

2.2.2.1 Discussion: Protons in a Synchro-Cyclotron

Cyclotrons are suited to accelerate protons or ions in a compact geometry, say, $R \sim 1 - 10$ m, for two reasons. First, heavy particles can reach relatively high total energies at low longitudinal momentum, which implies small bending radii. Second, the radius can be kept small by virtue of a multi-Tesla magnetic field generated by superconducting magnets. In both situations, particles move in a weakly relativistic regime.

As a case study, let us consider protons accelerated in a synchro-cyclotron of radius $R = 1$ m and immersed in a 4 T magnetic field. They are injected at the kinetic energy of 10 MeV. The energy gain per turn is determined by the peak accelerating voltage between the dees, e.g., 100 kV. What is the number of turns required to reach the extraction energy, and what the turn-by-turn variation of the orbit radius? What is the total relative variation of the RF angular frequency? What would these be in case of electrons, instead?

Equation 2.5 is used to calculate the maximum longitudinal momentum, i.e., the momentum in correspondence of the maximum radius of curvature: $p_{z,max} \cong 0.3 B_y R = 1.2$ GeV/c. It is easy to show that protons ($m_p c^2 = 0.938$ GeV) are still weakly relativistic at the extraction point, since their kinetic energy results:

$$T_{max} = E_{max} - m_p c^2 = \sqrt{(p_{max} c)^2 + (m_p c^2)^2} - m_p c^2 = 0.585 GeV < m_p c^2$$
(2.6)

Hence, we can write $T \cong \frac{p^2}{2m_p}$, and for the turn-by-turn variation of the orbit radius, respectively at the beginning and at the end of acceleration, we get:

$$\frac{\Delta p}{p} = \frac{\Delta R}{R} \cong \frac{1}{2}\frac{\Delta T}{T} = \frac{0.1 MeV}{2T} = [8.5 \times 10^{-5}, 5 \times 10^{-3}]$$
(2.7)

The relative energy gain per turn is small, and so is the relative variation of the curvature radius. Consequently, the number of turns to reach the maximum energy is large, $N_t = \frac{T_{max} - T_i}{\Delta T} = 5750$.

By virtue of the synchronism condition $\omega_{RF} \propto \omega$ (with ω the revolution frequency), the relative variation of the RF frequency over the whole accelerating loop is the same of the revolution frequency:

$$\frac{\Delta \omega_{RF}}{\omega_{RF,i}} = \frac{\Delta \omega}{\omega_i} = \gamma_i \Delta \left(\frac{1}{\gamma}\right) = -\frac{\Delta \gamma}{\gamma_f}$$
(2.8)

The initial and final total particle's energy is $E = T + m_p c^2 = 0.948$ GeV and 1.523 GeV, respectively, hence $\gamma_i = 1.01$ and $\gamma_f = 1.62$. The total relative (negative) variation of the frequency amounts to $\sim 40\%$.

In case we had electrons ($m_e c^2 = 0.511$ MeV), instead, they would behave as ultra-relativistic particles during the entire loop, since already at the injection $T_i \gg m_e c^2$, therefore $p \propto T$. The relative variation of longitudinal momentum and of curvature radius at the beginning and at the end of the accelerating loop is therefore approximately as twice as the one in Eq. 2.7 for non-relativistic particles.

The relative variation of the RF is calculated through Eq. 2.8, For electrons it results $\gamma_i = 21$ and $\gamma_f = 2349$, hence $\Delta\omega/\omega_i \approx 1$, which is not practical. We conclude that it is not possible to keep RF and electrons synchronous during the whole ramp.

2.2.3 Betatron

The "resonant" electrodynamic accelerators considered so far exploit the time-varying nature of the electric field introduced in the l.h.s. of Eq. 2.3. The *betatron*, invented by D. Kerst in 1940, is a circular electrodynamic accelerator which, instead, relies on the time-variation of the magnetic field flux. This generates an electric field tangent to the orbit, according to the r.h.s. of Eq. 2.3. However, only a specific spatial distribution of $B_y(t)$ guarantees uniform centripetal acceleration *and*, stability of the orbital motion. To show this, we put attention to the motion in the horizontal (bending) plane. The integral and differential equation for the azimuthal electric field is:

$$\begin{cases} \oint \vec{E}_\theta(r) d\vec{l} = -\frac{\partial \Phi(B_y)}{\partial t} = -\frac{\partial}{\partial t}\int_S (\vec{B}\cdot\hat{n})dS' = -S\frac{\partial}{\partial t}\left(\frac{1}{S}\int_S B_y dS'\right) = -\pi r^2\frac{\partial\langle B_y\rangle}{\partial t} \\ (\vec{\nabla}\times\vec{E})_z = \frac{1}{r}\left[\frac{\partial(rE_\theta)}{\partial r} - \frac{\partial E_r}{\partial\theta}\right] = \left[\frac{E_\theta}{r} + \frac{\partial E_\theta}{\partial r} - \frac{1}{r}\frac{\partial E_r}{\partial\theta}\right] = -\frac{\partial B_y}{\partial t} \end{cases}$$

$$(2.9)$$

The Top row in Eq. 2.9 describes the effect of a time-varying magnetic field *averaged* over a surface of area $S = \pi r^2$. The Bottom row in Eq. 2.9 describes the dependence of E_θ on the *local* value of $B_y(r)$, at any fixed r. The partial derivatives of the electric field in this equation are both zero. Indeed, since the curl is calculated for an infinitesimally small variation of r around its nominal value, E_θ is constant in proximity of r. Moreover, by virtue of the cylindrical symmetry of the system, E_r is independent from θ.

In both equations, if the variation of B_y with time is linear, the absolute value of the electric field E_θ is constant. The absence of RF acceleration classifies the betatron as a "non-resonant" accelerator.

By simplifying Eq. 2.9, we find for E_θ:

$$\begin{cases} 2\pi r E_\theta(r) = -\pi r^2\frac{\partial\langle B_y\rangle}{\partial t} \\ \frac{E_\theta}{r} = -\frac{\partial B_y}{\partial t} \end{cases} \Rightarrow \quad \frac{\partial B_y(r)}{\partial t} = \frac{1}{2}\frac{\partial\langle B_y(r)\rangle}{\partial t} \quad \forall r \qquad (2.10)$$

Fig. 2.6 Lateral section of a betatron and main components. (Original picture courtesy of S. Tazzari)

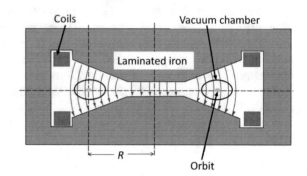

   In conclusion, the variation of the local magnetic field with time has to be half the time-variation of the field averaged over the orbital surface. This spatial field distribution can be guaranteed, for example, by shaping the magnetic poles as in Fig. 2.6. Such shaping is opposite to the one of the Lawrence's cyclotron in Fig. 2.5. Namely, the betatron naturally provides vertical focusing and therefore stability of the vertical motion (*betatron oscillations*).

   Like cyclotrons, betatrons are limited in maximum kinetic energy by the size or, better, the volume of the entire apparatus. But, unlike cyclotrons, the variation with time of the magnetic flux and the required azimuthal symmetry of this field over relatively large areas, prevent the adoption of superconducting magnets. This typically makes the energy gain per turn smaller in betatrons with respect to cyclotrons of comparable size. For this reason, cyclotrons and, in particular, sector-cyclotrons are often adopted for accelerating heavy particles, with application to nuclear physics experiments, medicine, and industry. Betatrons are most suited for electron acceleration.

2.2.4   Weak Focusing

The azimuthal symmetry of the magnetic field in cyclotrons and betatrons leads to the need of a spatial field gradient to keep the particles' motion stable. We now wonder what is the range of the field gradient needed to obtain stability, and what oscillation amplitude is induced by such gradient. We limit our analysis to linear focusing, i.e., the magnetic field is expanded in the particle's coordinates, but only the linear terms are kept. It is convenient to re-write the field gradient as a dimensionless quantity, by normalizing it to the value of the dipolar field component and to the curvature radius of the reference orbit:

$$B_y \approx B_0 + \left(\frac{\partial B_y}{\partial x}\right)_{y=0} x = B_0 \left(1 + \frac{R}{B_0}\frac{\partial B_y}{\partial x}\right)\frac{x}{R} \equiv B_0 \left(1 - n\frac{x}{R}\right)$$

$$(2.11)$$

$$\Rightarrow n = -\frac{R}{B_0}\frac{\partial B_y}{\partial x}$$

The coefficient n is called *field index*.

In the horizontal plane, the motion is stable as long as the Lorentz's force counteracts and overwhelms in absolute value the centripetal force, at each local curvature radius r. This condition is sufficient for the particle not to escape from the design orbit:

$$q v_z B_y = q v_z B_0 \left(1 - n \frac{x}{R}\right) > m \frac{v_z^2}{r};$$

$$\frac{p_z}{R} \left(1 - n \frac{x}{R}\right) > \frac{p_z}{r};$$

$$1 - n \frac{x}{R} > \frac{R}{r} \approx \frac{R}{R(1+x/R)} \approx 1 - \frac{x}{R}.$$

(2.12)

$$\Rightarrow n < 1$$

In the vertical plane, the stability is guaranteed by a weaker field at larger orbit radii:

$$\frac{\partial B_y}{\partial x} < 0 \Rightarrow n > 0$$ (2.13)

In conclusion, stability in both transverse planes is guaranteed by $0 < n < 1$. Because of the small value of n, the action of the magnetic force is said *weak focusing*.

The equation of motion in the horizontal plane, in polar coordinates in the laboratory frame, is:

$$m(\ddot{r} - \dot{\theta}^2 r) = -q v_z B_y$$ (2.14)

The three terms are expanded as follows:

$$
\begin{cases}
m\ddot{r} = m\ddot{x}. \\[2mm]
-m\dot{\theta}^2 r = -m\frac{v_z^2}{r} = -\frac{p_z^2}{mr} = -\frac{(q B_y R)^2}{m R(1+x/R)} = \\[2mm]
\qquad \approx -\frac{(q B_y)^2 R}{m}\left(1 - \frac{x}{R}\right) = -\frac{(q B_y)^2}{m} R + \frac{(q B_y)^2}{m} x. \\[2mm]
-q v_z B_y = -q \frac{p_z}{m} B_0 \left(1 - n\frac{x}{R}\right) = -\frac{(q B_0)^2 R}{m}\left(1 - n\frac{x}{R}\right) = \\[2mm]
\qquad = -\frac{(q B_y)^2}{m} R + \frac{(q B_y)^2}{m} n x.
\end{cases}
$$

(2.15)

By plugging all terms in Eq. 2.15 into Eq. 2.14, we obtain:

$$\ddot{x} + \left(\frac{q B_0}{m}\right)^2 (1-n)x = 0$$ (2.16)

A similar equation is found for the vertical plane, where we make use of $(\vec{\nabla} \times \vec{B})_z = 0$:

$$m\ddot{y} = -q v_z B_x = -q \frac{p_z}{m}\frac{\partial B_x}{\partial y} y = \frac{q^2 B_0 R}{m}\frac{\partial B_y}{\partial x} y = -\frac{(q B_0)^2}{m} n y;$$

$$\ddot{y} + \left(\frac{q B_0}{m}\right)^2 n y = 0$$

(2.17)

As expected, the motion in both planes is bounded and identical to a pure harmonic oscillator only if the condition $0 < n < 1$ is satisfied. The angular frequency $\omega_L = (q B_0/m)^2$ is the Lawrence's frequency. In both planes, it is modulated by a coefficient smaller than 1. This translates into a wavelength of oscillation in the horizontal plane:

$$\lambda_x = \frac{2\pi v_z}{\omega_L \sqrt{1-n}} = \frac{2\pi R}{\sqrt{1-n}} > 2\pi R \qquad (2.18)$$

In practice, a full oscillation is completed after more than one turn. This can also be seen as if the circular orbit translate in the bending plane under the effect of the magnetic field, in a periodic motion of period λ_x. The region of space accessible to the particles is comparable to the orbit radius R.

References

1. A. Sessler, E. Wilson, *Engines of Discovery* (World Scientific Publishing Co. Pte. Ltd., Singapore, 2007)
2. M. Conte, W.W. MacKay, *An Introduction to the Physics of Particle Accelerators*, 2nd edn. (Published by World Scientific, 2008), pp. 1–34. ISBN: 978-981-277-960-1
3. D.A. Bromley, The development of electrostatic accelerators. Nucl. Instrum. Methods Ser. **122**, 1 (1974)
4. E. Persico, E. Ferrari, S.E. Segre, *Principles of Particle Accelerators* (W.A. Benjamin Inc., 1968)

Radiofrequency Structures

3

The development of resonant accelerators during the first half of the XX century allowed particles to reach kinetic energies one order magnitude higher than in electrostatic devices. Nowadays, kinetic energies approaching and exceeding the GeV-level are obtained by means of Alvarez-type RF structures, or *RF cavities*.

The metallic boundary of the cavity allows the electric field to have a longitudinal component for acceleration. The internal geometry of the structure, often of cylindrical symmetry, is shaped according to the range of velocity exploited by the accelerated particles, and optimized to maximize the accelerating gradient per input e.m. power, while minimizing the probability of discharges ("breakdown rate"). RF structures installed in series and synchronized by an external common master clock constitute a linear accelerator or *RF linac*. Synchronization of RF field and particles' arrival time implies a pulsed time pattern of the particle beam, i.e., a series of bunches.

Linacs adopt RF structures made of many tens' of internal gaps, called *cells*. RF cavities made of single or few cells are used in high energy circular accelerators, where particles' energy is increased through the cavity over a large number of turns, thus a small amount of energy gain per turn is admitted.

In spite of the rotational electric field in RF structures, the process of energy gain can still be described in terms of an effective electric voltage ΔV, resulting from the path-integral of the time-varying electric field synchronized to the particle's arrival time, i.e., $\Delta T = -q\Delta V$ as already in Eq.2.2. The total energy of a charged particle initially in a field-free region, accelerated in a RF structure, and eventually extracted from it, is:

$$E_f = E_i - q\Delta V = m_0 c^2 + T_i + \Delta T = m_0 c^2 + T_f \qquad (3.1)$$

© The Author(s), under exclusive license to Springer Nature Switzerland AG 2022
S. Di Mitri, *Fundamentals of Particle Accelerator Physics*, Graduate Texts in Physics,
https://doi.org/10.1007/978-3-031-07662-6_3

3.1   Principles of Acceleration

3.1.1   Theorem of E.M. Acceleration

Acceleration of charged particles by an external rotational electric field requires boundary conditions which, as we will see below, can be satisfied by metallic surfaces. To demonstrate this [1], we first consider an e.m. wave in vacuum. The wave can also be thought as the *far field* component of an e.m. field derived from the Liénard-Wiechert retarded potentials. Maxwell's equations imply the wave equation:

$$\nabla^2 \vec{E} = -\frac{\omega^2}{c^2} \vec{E} \tag{3.2}$$

whose solutions can be expressed as sum of waves of the form:

$$\vec{E}(\vec{r}, t) = \vec{E}_0 e^{i(\vec{k}\vec{r}-\omega t)} \tag{3.3}$$

If $\vec{k}$ is real, Eq. 3.3 describes plane waves travelling in the direction of $\vec{k}$ at the velocity c. By applying Maxwell's equation $\vec{\nabla}\vec{E} = 0$, we obtain $\vec{k} \cdot \vec{E} = 0$, that is, the electric field is perpendicular to the direction of propagation (transversely polarized wave). In summary, e.m. waves remain in phase only with particles travelling at ultra-relativistic velocities. If the particles are not travelling parallel to $\vec{k}$, any interaction will be periodic and will add to nothing. If the particles move along $\vec{k}$, there is not net acceleration because $\vec{E}$ is orthogonal to the particle's velocity. This demonstrates the following theorem:

no combination of far fields in free vacuum can produce net linear acceleration.

A second demonstration, which recurs to energy conservation in Quantum Mechanics, is presented to point out that the theorem is not a consequence of a "classical" interpretation of Electromagnetism; rather, it concerns an intrinsic property of e.m. fields.

Far fields can be represented as a sum of photons. Acceleration of a charged massive particle by an e.m. field can be represented by the absorption of photons with no simultaneous emission. But, such an interaction is forbidden by energy and momentum conservation. This is evident if we calculate the Lorentz's invariant $p^\mu p_\mu$ of the particle before and after photon absorption, in the reference frame in which the particle is initially at rest. The particle's 4-momentum after photon absorption would be $p^\mu = (m_0 c^2 + E_\gamma, \vec{p}_\gamma)$, and thereby the particle's invariant results:

$$m_0^2 c^4 = \left(m_0 c^2 + E_\gamma\right)^2 - p_\gamma^2 c^2;$$
$$p_\gamma^2 c^2 = E_\gamma^2 + 2m_0 c^2 E_\gamma \tag{3.4}$$

Since for a photon $p_\gamma c = E_\gamma$, Eq. 3.4 holds only for $E_\gamma = 0$. Thus, there cannot be such acceleration by far field.

The above theorem concerns "net, linear" acceleration. What do these terms mean?

- "Net" means that acceleration should be experienced by a particle moving from and to a region of space free of e.m. field. In other words, local acceleration internal to a region occupied by e.m. field is still possible. But, when the motion is across a field-free region, the average accelerating effect is null.
- "Linear" means that the particle's energy change is linear with the amplitude of the accelerating field. There can be second order effects (e.g., Compton scattering) in which the acceleration is proportional to the square of the field amplitude. These cases, however, require extremely intense fields which are not part of conventional RF accelerators.

3.1.2 Pill-Box

Modern RF structures [2,3] are made of several cells, delimited by "disks" with central irises to allow particles to pass through. The disks allow the e.m. wave flowing through the cavity to constructively interfere, so generating a longitudinal component of the electric field, parallel to the particle's direction of motion (waveguide effect). Acceleration is therefore provided in respect of the theorem of e.m. acceleration.

The most simplified geometry of a cell, denominated *pill-box*, is shown in Fig. 3.1. Typical cell sizes are of the order of cm's in length and diameter; the iris radius usually ranges from few to 10 mm or so. The spatial distribution of electric and magnetic field lines inside a pill-box is derived below. In spite of more sophisticated cell geometries in real accelerators, the expression for the main component of the on-axis accelerating field remains quantitatively valid.

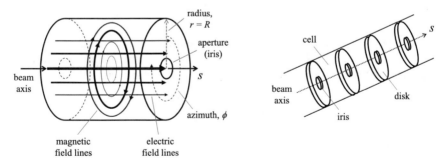

Fig. 3.1 Left: pill-box geometry and field lines. Right: periodic RF structure

Let us assume a cell of infinite electrical conductivity. We impose four boundary conditions:

1. The longitudinal electric field is required to be null at the cell's surface and maximum on-axis, where the beam is assumed to travel $\Rightarrow E_z(r = R) = 0$.
2. The time-varying E_z implies the presence of a similarly time-varying azimuthal magnetic field B_ϕ. However, we would like to have no magnetic field on-axis, because it would deflect particles via Lorentz's force $\Rightarrow B_\phi(r = 0) = 0$.
3. By cylindrical symmetry of the cell, the radial component of the electric field does not depend from z, $\frac{\partial E_r}{\partial z} = 0$.
4. By cylindrical symmetry of the cell, the radial component of the magnetic field does not depend from ϕ, $\frac{\partial B_r}{\partial \phi} = 0$.

Maxwell's differential equations for B_ϕ and E_z are:

$$\begin{cases} \left(\vec{\nabla} \times \vec{E}\right)_\phi = -\frac{\partial B_\phi}{\partial t} \\ \left(\vec{\nabla} \times \vec{B}\right)_z = -\frac{1}{c^2}\frac{\partial E_z}{\partial t} \end{cases} \Rightarrow \begin{cases} \frac{\partial E_r}{\partial z} - \frac{\partial E_z}{\partial r} = -\frac{\partial B_\phi}{\partial t} \\ \frac{1}{r}\left[\frac{\partial(rB_\phi)}{\partial r} - \frac{\partial B_r}{\partial \phi}\right] = -\frac{1}{c^2}\frac{\partial E_z}{\partial t} \end{cases} \quad (3.5)$$

We put to zero the derivatives as prescribed by points 3. and 4. above, further differentiate the top equation with respect to r and the bottom equation with respect to t, and finally substitute the second equation into the first one, to get a second order partial differential equation for E_z:

$$\begin{cases} \frac{\partial^2 E_z}{\partial r^2} = \frac{\partial^2 B_\phi}{\partial r \partial t} \\ \frac{1}{r}\frac{\partial B_\phi}{\partial t} + \frac{\partial^2 B_\phi}{\partial r \partial t} = \frac{1}{c^2}\frac{\partial^2 E_z}{\partial t^2} \end{cases} \Rightarrow \quad \frac{\partial^2 E_z}{\partial r^2} + \frac{1}{r}\frac{\partial E_z}{\partial r} = \frac{1}{c^2}\frac{\partial^2 E_z}{\partial t^2} \quad (3.6)$$

Equation 3.6 is a wave equation and it can be solved by separation of variables. We search a solution $E_z(r, t) = A(r)e^{i(\omega t + \phi_0)}$. By substituting this into the wave equation, we obtain an analogous equation for $A(r)$, whose solution is:

$$A(r) = a_0 J_0\left(\frac{\omega r}{c}\right) \quad (3.7)$$

with J_0 Bessel function of the first kind and 0-th order. The condition 1. above is satisfied by the first zero of J_0, which happens to be for the argument

$$\frac{\omega R}{c} = 2.405 \Rightarrow \omega \propto \frac{1}{R} \quad (3.8)$$

Hence, the higher the frequency of the accelerating field in an RF cavity is, the smaller the cell's outer radius has to be, with direct consequences on the mechanical accuracy required for machining the cell geometry. As an example, $f_{RF} = 1.5\,\text{GHz}$ (L-band RF) implies $R = 7.7\,\text{cm}$, $f_{RF} = 12\,\text{GHz}$ (X-band RF) leads to $R = 1\,\text{cm}$.

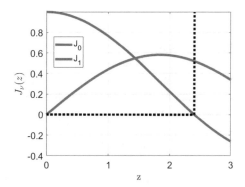

Fig. 3.2 Bessel functions of the first kind, of order 0 (blue) and 1 (orange), proportional to the amplitude of E_z and B_ϕ in a cell, respectively. Half cell is delimited by the dotted lines. The vertical line represents the lateral surface of the cell, where J_0 has its first zero; the cell axis is along z=0

The azimuthal magnetic field can be determined with analogous procedure (partial derivative ∂_t applied to the top equation of Eq. 3.5, ∂_r applied to the bottom equation, then substitution of one equation into the other). The solution of the wave equation is searched in the form $B_\phi(r, t) = C(r)e^{i(\omega t + \phi_0)}$. The separation of variables leads to an equation for the amplitude whose solution is $C(r) = c_0 J_1(\frac{\omega r}{c}) \equiv B_{\phi,0}$. J_1 is the Bessel function of first kind and 1-st order. In general, since the electric and magnetic fields belong to the same wave, it results $B_{\phi,0} \propto E_{z,0}$.

The Bessel functions $J_0(x)$, $J_1(x)$ are shown in Fig. 3.2. They represent the radial distribution of the peak values of E_z and B_ϕ in a cell. As expected, $E_z(r)$ is maximum on-axis and zero at the cell's surface. Since particle beams are commonly aligned to the cells' electrical axis ($r \approx 0$), we limit our attention to the on-axis accelerating field:

$$J_0(r \approx 0) \approx 1 \quad \Rightarrow \quad E_z \approx E_{z,0} \cos(\omega t + \phi_0) \tag{3.9}$$

On the contrary, $B_\phi(r)$ is zero on-axis, and maximum in proximity of the surface. The variation with time of the magnetic field induces currents on the metallic surface, which lead to power dissipation via thermal load.

E_z, B_ϕ are called *fundamental modes* because they correspond to the lowest frequency of oscillation among all those allowed by the cavity geometry. For the purpose of longitudinal acceleration, the sum of electric field amplitudes corresponding to different wave vectors ("higher order modes") is made negligible with respect to the fundamental mode, which therefore represents the largest part of e.m. energy stored in the cavity.

In general, Spatial configurations of the e.m field in a cell or a waveguide are classified as *Transverse Magnetic* (TM_{ijk}) or *Transverse Electric* (TE_{ijk}). TM (TE) stays for magnetic (electric) field transverse to the direction of propagation of the e.m. wave. The three numerical indexes classify the number of zeros of the magnetic (electric) field in the radial ($0 < r < R$), azimuthal ($0 < \phi < 2\pi$), and longitudinal coordinate ($0 < z < L$). The third index is often suppressed when $k = 0$. TM modes with a longitudinal component of the electric field are accelerating modes (see Fig. 3.1-left plot). TE modes are used, for example, to transversely deflect charged particle beams; such RF structures are often referred to as *transverse deflecting cavities*.

3.2   Periodic Structures

3.2.1   Travelling Wave

Since an accelerating structure with many cells can be approximated to a long periodic
system, the *Floquet's theorem* (whose demonstration is postponed) can be applied.
It results that the field E_z at two cross sections separated by one period (cell) differs
at most by a complex number. Namely, E_z is a periodic function of the longitudinal
coordinate s along the structure, and therefore it can be expanded in Fourier series:

$$E_z(r,t) = \sum_{n=-\infty}^{+\infty} a_n J_0(k_n,r)\cos(\omega t - k_n s + \phi_0), \qquad k_n = k_0 + \frac{2\pi n}{d} \qquad (3.10)$$

with d the distance between two consecutive irises, or cell length. Equation 3.10
describes the linear superposition of n "modes" of an e.m. *travelling wave* (TW),
all modes oscillating at the same frequency $\omega = 2\pi f_{RF}$ but with a different wave
number k_n. Consequently the phase velocity $v_{ph}^n = \omega/k_n$ is different for each mode.

Usually, the RF structure's inner geometry is built in a way that the fundamental
mode's amplitude a_0 is larger than the sum of amplitudes a_n of all other modes.
Moreover, $v_{ph}^0 \approx c$ for application to ultra-relativistic particles. This way, particles
are synchronous to the wavefront of the on-axis fundamental mode:

$$E_z^{TW}(r \approx 0, n = 0) = E_{z,0}^{TW} \cos(\omega t - ks + \phi_0) \qquad (3.11)$$

Synchronism between charged particles and fundamental accelerating mode
means that the field phase seen by the particle is the same at all the successive
cells, at the time the particle traverses them. For the synchronous particle we find:

$$\omega t_m + \phi_m(s) = \omega t_{m+1} + \phi_{m+1}(s) = \omega\left(t_m + \frac{d}{\beta_z c}\right) + \phi_{m+1}(s)$$

$$\Rightarrow \Delta\phi_m = \phi_m - \phi_{m+1} = \frac{2\pi f_{RF}}{\beta_z c}d = \frac{2\pi}{\beta_z}\frac{d}{\lambda_{RF}} \qquad (3.12)$$

$\Delta\phi_m$ is called *cell phase advance*. Since the field periodicity imposes $\frac{d}{\lambda_{RF}} \in \dot{\mathbb{Q}}$, $\Delta\phi_m$
assumes values like $\pi/3, 2\pi/3, 5\pi/4$, etc. when the structure is tuned for acceleration
of ultra-relativistic particles ($\beta_z \approx 1$).

In practice, some RF power generated by an external source (klystron or solid
state amplifier) is injected into the structure, where resonant modes build up. The
constructive interference of e.m. waves established by a suitable geometry of irises
privileges the fundamental accelerating mode. The e.m. energy flows through the
structure with a group velocity $v_g < c$, and it is extracted at the end of the structure
by a waveguide, to be eventually dissipated on a load.

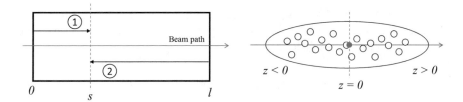

Fig. 3.3 Left: superposition of a forward (1) and backward (2) e.m. wave travelling in a periodic structure with closed end. The wavefront of the standing wave generated by the constructive interference of the two travelling waves in correspondence of the coordinate s is formed at the time the initial wave takes to travel a path long $l + (l - s) = 2l - s$. Right: bunch of particles with internal coordinate z; the bunch head is at positive z

3.2.2   Standing Wave

If the structure's end is closed by a reflective medium (e.g., a Cu plate), a stationary wave, also called *standing wave* (SW), can build up as the resultant of constructive interference of the forward and backward (reflected) travelling wave. With reference to Fig. 3.3-left plot and recalling Eq. 3.11, we express the accelerating field in a SW as the superposition of the forward-propagating wave and the reflected wave:

$$E_z^{SW}(r \approx 0, n = 0) = E_{z,0}^{TW} \cos(\omega t - ks + \phi_0) + E_{z,0}^{TW} \cos(\omega t - k(2l - s) + \phi_0)$$
(3.13)

and hereafter we put $\phi_0 \equiv 0$ for simplicity. Equation 3.13 can be written in Euler's notation, where we retain the $\mathbb{R}e$ part only of the oscillatory term. Hereafter, we continue assuming ultra-relativistic particles. Since the phase advance of the fundamental mode through the whole structure of length l is $\Delta\phi_l = 2\pi m$ (m integer), and assuming $p \in \mathbb{N}$ cells of length d, Eq. 3.12 allows us to write $l = pd = p\Delta\phi_l/k = 2\pi p'/k$, $p' \in \mathbb{N}$. Then we have:

$$e^{i(\omega t - ks)} + e^{i[\omega t - k(2l - s)]} = e^{i(\omega t - ks)} + e^{i(\omega t + ks)}e^{-i2kl} =$$
(3.14)
$$= e^{i(\omega t - ks)} + e^{i(\omega t + ks)}e^{-ip'4\pi} = 2\cos(\omega t)\cos(ks)$$

The SW peak field is re-defined as twice the peak field of each TW propagating through the structure, so that:

$$E_z^{SW}(r \approx 0, n = 0) = E_{z,0}^{SW} \cos(\omega t + \phi_0)\cos(ks)$$
(3.15)

3.2.3   Synchronous Phase

Since a bunch is made of a very large number of particles, it has practical sense to identify a reference or *synchronous particle* (either virtual or real, such as the bunch center of mass). The coordinates of all other particles are referred to it. If t_s is the arrival time of the synchronous particle at a certain s in the RF structure and z is the distance, internal to the bunch, of the generic particle from the synchronous particle, then we can write for the generic particle:

$$\omega t = \omega(t_s + \Delta t) = \omega t_s + \omega \frac{z}{v_z} = \omega t_s + kz \tag{3.16}$$

By definition, $z = 0$ is the internal coordinate of the synchronous particle, see Fig. 3.3-right plot. In such case, Eqs. 3.11 and 3.15 can be re-written in the following standard notation:

$$
\begin{aligned}
E_z^{TW} &= E_{z,0}^{TW} \cos(\omega t_s - ks + \phi_0 + kz) \equiv E_{z,0}^{TW} \cos(\phi_{RF}^{TW} + kz) \\
E_z^{SW} &= \tilde{E}_{z,0}^{SW} \cos(\omega t_s + \phi_0 + kz) \cos(ks) \cong E_{z,0}^{SW} \cos(\phi_{RF}^{SW} + kz)
\end{aligned}
\tag{3.17}
$$

In the SW, $\cos(ks) \cong const.$ was assumed. Namely, at any time t, E_z^{SW} is approximated to a constant amplitude along the cell.

The phase ϕ_{RF} is called *synchronous phase*. In jargon, *on-crest* acceleration refers to the choice of ϕ_{RF} which maximizes the energy gain (e.g., $\phi_{RF} = 0$ for $E_{z,0} \sim \cos \phi$). Any phase different from that, determines *off-crest* acceleration. Deceleration is, of course, possible as well.

In summary, the accelerating electric field inside periodic RF structures is usually reduced to the on-axis longitudinal fundamental mode TM_{010}. The expression of $E_z(t, s)$ for a TW and a SW differs for its dependence from the s-coordinate along the structure (compare Eqs. 3.11 and 3.15). In a SW, such a dependence is often neglected. In both cases E_z can be written as an amplitude times an oscillatory term, whose phase is the sum of the synchronous phase and the relative phase of the generic particle (see Eq. 3.17). It is intuitive to identify ϕ_{RF} as the arrival "time" of the bunch as a whole, while kz determines the spread in energy gain internal to the bunch, due to the slightly different arrival times of the individual particles at a specific s inside the structure.

3.2.4   Transit Time Factor

The energy gain of a particle in an SW structure is calculated below. The individual cell has longitudinal coordinates $[-g/2, g/2]$. By virtue of the synchronization between the particle's phase and the phase of the electric field, the total energy gain is the gain in a cell times the number of cells. The energy gain in a single cell is:

$$\Delta E^{SW}(g,t) = q \int_{-g/2}^{g/2} E_z^{SW} ds = q \tilde{E}_{z,0}^{SW} \int_{-g/2}^{g/2} ds \cos(\omega t + kz + \phi_0)\cos(ks) =$$

$$\approx q E_{z,0}^{SW} \int_{-g/2}^{g/2} ds \left[\cos(\tfrac{\omega s}{\beta_z c})\cos(kz + \phi_0) - \sin(\tfrac{\omega s}{\beta_z c})\sin(kz + \phi_0) \right] =$$

$$= q E_{z,0}^{SW} \tfrac{1}{\frac{\omega}{\beta_z c}} \sin(\tfrac{\omega s}{\beta_z c})|_{-g/2}^{g/2} \cos(kz + \phi_0) = q E_{z,0}^{SW} g \tfrac{\sin x}{x}\cos(kz + \phi_0) =$$

$$= q \Delta V_0^{SW}(g) T_{tr}\cos(kz + \phi_0)$$

$$\tag{3.18}$$

We defined $x = \frac{\omega g}{2\beta_z c}$, and made use of the approximation of slowly varying field amplitude internally to the cell, $\cos(ks) \cong const$. The integral of odd sin-like functions over an even path is zero.

The dimensionless *transit time factor* $T_{tr} = \sin x / x < 1 \; \forall x$ describes the reduction of the nominal energy gain $q\Delta V_0^{SW}$ due to the time interval $\Delta t = g/(\beta_z c)$ the particle takes to travel through the cell, and during which the electric field amplitude $E_z(t)$ reduces compared to the peak value. T_{tr} depends from the ratio g/λ_{RF}, and it is commonly in the range 0.85–0.95.

The calculation is repeated below for a TW structure (see the electric field in Eq. 3.11):

$$\Delta E^{TW}(g,t) = q \int_{-g/2}^{g/2} E_z^{TW} ds = q E_{z,0}^{TW} \int_{-g/2}^{g/2} ds \cos(\omega t - ks + kz + \phi_0) \cong$$

$$\approx q E_{z,0}^{TW} \int_{-g/2}^{g/2} ds \cos(kz + \phi_0) = q E_{z,0}^{TW} g \cos(kz + \phi_0) =$$

$$= q \Delta V_0^{TW}(g)\cos(kz + \phi_0)$$

$$\tag{3.19}$$

where we have taken the limit $\omega t - ks \to 0 \; \forall s, t$ when $\beta_z \to 1$. In conclusion, in a TW structure $T_{tr} = 1$ because the accelerating field is synchronous with the particle at any point along the structure ($v_{ph} \cong v_z \cong c$). In other words, the particle is always "surfing" the wavefront of the travelling wave.

3.3   RLC Circuit Model

In this chapter, the electric field amplitude or, equivalently, the effective peak voltage of a RF cavity is quantified in terms of macroscopic parameters related to the RF power injected into the cavity, and to the structure's geometry. To do this, the pill-box is modelled as a resonant RLC circuit [2,3].

3.3.1   Standing Wave

Let us consider a metallic pill-box, as shown in Fig. 3.1. An external RF source injects RF power into it, so generating an effective time-varying electric voltage of amplitude ΔV_0. The disks at the cell edges constitute a capacity, C. The cylindrical surface introduces an inductance, L. The cavity material is characterized by a resistive *shunt impedance*, R_s. The cell can therefore be analysed as a resonant RLC circuit supplied by an oscillator.

The circuit is said to be resonant because there exists a specific frequency at which the transmisssion of e.m. energy through the circuit is maximized. The frequency depends from the reactive impedances, $\omega = 1/\sqrt{LC}$. In other words, the geometry of the cell selects the frequency of the largest amplitude component of the e.m. field, i.e. the fundamental mode, consistently with Eq. 3.8.

The analysis of the RLC circuit is first done for the case of RF energy stored in the cell, i.e., the RF cavity has closed ends and the accelerating field behaves as in a standing wave. The energy stored in the circuit is $U_0 = \frac{1}{2}C\Delta V_0^2$. The shunt impedance is defined as the resistive part of the cell through which the power averaged over one RF cycle is dissipated according to Ohm's law:

$$R_s = \frac{\Delta V_0^2}{\langle P_d \rangle} \tag{3.20}$$

Let us introduce a figure of merit, the so-called *quality factor Q* of the cell, as the ratio of the power stored in the cavity in a RF cycle and the time-averaged dissipated power:

$$Q := \frac{U_0 \omega}{\langle P_d \rangle} = \frac{C\Delta V_0^2}{2} \frac{\omega R_s}{\Delta V_0^2} = \frac{R_s}{2\sqrt{L/C}} \tag{3.21}$$

When evaluated in the absence of charged beam traversing the cavity, Q is said "unloaded", and often noted as Q_0.

To extend the characterization of a single cell to a periodic structure, the aforementioned quantities are defined per unit length. Equation 3.21 becomes:

$$Q = \frac{\frac{dU_0}{ds}\omega}{\frac{d\langle P_d \rangle}{ds}} = \frac{u\omega}{\frac{d\langle P_d \rangle}{ds}} \tag{3.22}$$

and from Eq. 3.20:

$$\Delta V_0 = \sqrt{\langle P_d \rangle r_s l} \tag{3.23}$$

with $r_s = dR_s/ds$ and l the cavity total length.

Since the e.m. energy is first injected into the cavity and then dissipated through r_s, we expect a decrement with time from the initial value U_0, until a new RF pulse fills the cavity again. From the definition of instantaneous dissipated power, and from Eq. 3.21 evaluated at the generic time t, we get:

$$P_d = -\frac{dU(t)}{dt} = \frac{\omega U}{Q} \quad \Rightarrow \quad U(t) = U_0 e^{-\frac{\omega}{Q}t} \tag{3.24}$$

The ratio:

$$t_f = \frac{Q}{\omega} \tag{3.25}$$

is denominated *filling time*. Equation 3.24 shows that, specifically for a SW structure, Q is proportional to the number of RF cycles during which the stored energy is kept close to its initial, maximum value (the higher Q is, the longer is the time interval during which the RF field resonates in the cavity at large amplitude).

The peak value of the accelerating field is derived from Eq. 3.23:

$$\left(E_{z,0}^{SW}\right)^2 = \left(\frac{d\Delta V_0(s)}{ds}\right)^2 \approx \left(\sqrt{r_s \frac{d\langle P_d\rangle}{ds} \frac{ds}{ds}}\right)^2 = r_s \frac{d\langle P_d\rangle}{ds} \tag{3.26}$$

Let us now pay attention to the ratio r_s/Q, which is made explicit by means of Eqs. 3.26 and 3.22:

$$\frac{r_s}{Q} = \frac{\left(E_{z,0}^{SW}\right)^2 \frac{d\langle P_d\rangle}{ds}}{\frac{d\langle P_d\rangle}{ds} u\omega} = \frac{\left(E_{z,0}^{SW}\right)^2}{u\omega} \tag{3.27}$$

The ratio is proportional to the accelerating field amplitude squared, per averaged stored power. It quantifies the capability of the cavity of transforming a certain amount of e.m. energy into acceleration and, for this reason, it is intended to quantify the *efficiency of acceleration*.

Equation 3.21 shows that $r_s/Q \propto \sqrt{L/C}$. Since the cylindrical geometry of the cell suggests $L \propto R \propto 1/\omega$, it turns out that $C \propto 1/\omega$ and therefore r_s/Q is independent from ω, as well as from the material of the structure. Indeed, that ratio only depends from the *geometry* of the cavity. Consequently, $Q \sim R_s/G$, where G stays for a numerical factor only dependent from the cavity geometry.

3.3.2 Travelling Wave Constant Impedance

The RLC description of a SW structure can be identically applied to the case of a TW by replacing $\langle P_d\rangle \to -P$, where P is now the RF power travelling along the structure and finally absorbed by a load. The opposite sign means that P is not absorbed from the structure, but it actively contributes to acceleration by flowing through it. Equation 3.24 can be re-written as follows:

$$P = \frac{dU(t)}{dt} = \frac{dU}{ds}\frac{ds}{dt} = u \cdot v_g \tag{3.28}$$

where $v_g < c$ is the group velocity of the fundamental mode (i.e., the velocity at which the e.m. energy propagates through the structrue). In this case, the filling time is simply:

$$t = \frac{l}{v_g} \tag{3.29}$$

Commonly, $v_g \approx c \left(\frac{r_{in}}{r_{out}}\right)^4 \approx (0.01 - 0.1)c$, with r_{in} and r_{out} the inner and outer radius of the cell iris, respectively.

If the TW structure is perfectly periodic, i.e., all cells are identical, then $r_s = \frac{dR_s}{ds} = \frac{R_s}{l}$ and the structure is named *TW constant impedance* (TW-CI). As $P(s)$ is flowing through the structure, R_s reduces its initial value $P(s = 0) = P_0$ by unavoidable dissipation. To explicit the variation of P along the structure, we recall Eqs. 3.22 and 3.28:

$$\begin{cases} Q = -\frac{u\omega}{\frac{dP}{ds}} \\ P = u \cdot v_g \end{cases} \Rightarrow \frac{dP}{ds} = -\frac{\omega}{Q}\frac{P}{v_g} \Rightarrow P(s) = P_0 e^{-\frac{\omega}{Q v_g}s} \qquad (3.30)$$

Since the longitudinal gradient of the power flowing through the structure is negative, Q is still correctly defined as a positive quantity.

An analogous dependence for the electric field amplitude is found by relating it to the power, see Eq. 3.27:

$$\begin{cases} \frac{r_s}{Q} = \frac{E_{z,0}^2}{u\omega} \\ P = u \cdot v_g \end{cases} \Rightarrow E_{z,0}^2 = \frac{\omega r_s}{Q v_g} P \Rightarrow \left(E_{z,0}^{CI}\right)^2 = \frac{\omega r_s}{Q v_g} P_0 e^{-\frac{\omega}{Q v_g}s} \qquad (3.31)$$

In conclusion, the peak field in a TW-CI is *not* constant through the structure, but exponentially damped. For a normal-conducting TW-CI with typical parameters $f_{RF} = 3\,\text{GHz}$, $Q \approx 10^4$ and $v_g = 0.1c$, the field amplitude at the end of a 3 m-long structure is approximately 80% of its initial value.

The reduction of the field amplitude in a TW-CI is commonly expressed as function of a dimensionless constant denominated *attenuation factor*, which is basically determined by the structure's geometry:

$$\tau := \frac{\omega l}{2Q v_g} \Rightarrow \begin{cases} E_{z,0}^{CI}(s) = \sqrt{2\tau \frac{P_0 r_s}{l}} e^{-\tau \frac{s}{l}} \\ \left|\Delta V_0^{CI}(s)\right| = \left|-\int_0^l E_{z,0}^{CI}(s)ds\right| = \sqrt{2\tau P_0 r_s l} \left(\frac{1-e^{-\tau}}{\tau}\right) \end{cases} \qquad (3.32)$$

3.3.2.1 Discussion: Optimum Attenuation Factor

Equation 3.32 shows that the accelerating voltage can be maximized by a proper choice of τ or, in practice, by a suitable combination of the structure's parameters that define it. So, what is the optimum value of τ?

We need to find the maximum of ΔV_0^{CI} as function of τ, i.e., $\frac{d\Delta V_0}{d\tau} \equiv 0$. Doing so in Eq. 3.32, we find the following relation for τ:

$$\tau = \frac{1}{2}\left(e^\tau - 1\right) \Rightarrow \tau_{opt} = 1.26 \Rightarrow \left(\Delta V_0^{CI}\right)_{max} \approx 0.9\sqrt{P_0 r_s l} \qquad (3.33)$$

In conclusion, a proper choice of τ allows the effective accelerating voltage in a TW-CI structure to be only 10% smaller than the nominal peak voltage in a SW (compare with Eq. 3.23). Moreover, taking into account the typical value of T_{tr}, the two peak voltages are essentially at the same level for identical input power and shunt impedance. However, since the TW-CI shows lower peak electric field on average along the structure than in a SW, the former tends to be less subject to discharges. Especially when high accelerating gradients are requested.

3.3.3 Travelling Wave Constant Gradient

A periodic RF structure in which the peak electric field is constant along the structure is said *TW Constant Gradient* (TW-CG). As we will see, this requirement translates into a specific cells' geometry. Let us express the constancy of the accelerating field by recalling Eq. 3.26, now re-written for a TW structure:

$$\left(E_{z,0}^{CG}\right)^2 = -r_s \frac{dP}{ds} \equiv const. \Rightarrow P_l = P_0 + Cl \qquad (3.34)$$

The last equality defines $C = \frac{P(l)-P_0}{l}$. Expressed in terms of the generic coordinate s along the structure, it becomes:

$$P(s) = P_0 + Cs = P_0 + \left(\frac{P(l) - P_0}{l}\right)s \qquad (3.35)$$

$P(l)$ in Eq. 3.34 can be expressed in terms of an attenuation factor, in analogy to the power in a TW-CI (see Eq. 3.30 in the light of τ defined in Eq. 3.32): $P_l \equiv P_0 e^{-2\tau}$. We will show below that such factor is indeed the same factor introduced in Eq. 3.32. By substituting P_l in Eq. 3.35 we get:

$$P(s) = P_0 + \left(P_0 e^{-2\tau} - P_0\right)\frac{s}{l} = P_0\left[1 - \left(1 - e^{-2\tau}\right)\frac{s}{l}\right] \qquad (3.36)$$

The group velocity can be calculated from Eq. 3.28:

$$v_g(s) = \frac{P(s)}{u} = P(s)\frac{\omega}{Q\frac{dP}{ds}} = \frac{\omega}{Q}\frac{P_0\left[1-(1-e^{-2\tau})\frac{s}{l}\right]}{\frac{P_0}{l}(1-e^{-2\tau})} = \frac{\omega l}{Q}\left[\left(\frac{1}{1-e^{-2\tau}}\right) - \frac{s}{l}\right] \qquad (3.37)$$

In conclusion, the constraint of a constant field amplitude translates into a group velocity varying with s. Since v_g is determined by the cell geometry, Eq. 3.37 implies that the cell iris radius slightly changes along the structure or, in other words, the structure is quasi-periodic. In practice, as $P(s)$ decreases along the structure because of the dissipation of the e.m. energy via resistive impedance, v_g decreases at an identical rate. This means that the iris inner radius gradually shrinks along the structure. The iris variation along a structure of $\sim$100 cells and initial iris radius of $\sim$10 mm, can be smaller than 1 mm. One can also think of the gradual reduction of the iris radius as a way to keep the density of e.m. energy, hence the electric field amplitude, constant as the power decreases.

The electric field amplitude can be evaluated by recalling Eq. 3.34 and by replacing there the power per unit length (in turn evaluated from $P(s)$ in Eq. 3.35):

$$E_{z,0}^{CG} = \sqrt{r_s \frac{dP}{ds}} = \sqrt{\frac{\omega r_s}{Qv_g} P(s)} = \left\{ \frac{\omega r_s}{Q} \frac{P_0\left[1-\left(1-e^{-2\tau}\right)\frac{s}{l}\right]}{\frac{\omega l}{Q}\frac{\left[1-\left(1-e^{-2\tau}\right)\frac{s}{l}\right]}{\left(1-e^{-2\tau}\right)}} \right\}^{1/2} =$$

$$= \sqrt{\left(1-e^{-2\tau}\right)\frac{P_0 r_s}{l}}$$

(3.38)

which is constant in s, as expected. The effective accelerating voltage is:

$$\left|\Delta V_0^{CG}\right| = \left|-\int_0^l E_{z,0}^{CG}(s)ds\right| = \sqrt{\left(1-e^{-2\tau}\right)P_0 r_s l}$$

(3.39)

Since v_g is not constant with s, the filling time in a TW-CG has to be calculated more carefully:

$$t_f = \int_0^l \frac{ds}{v_g(s)} = \cdots = 2\tau\frac{Q}{\omega}$$

(3.40)

where the integral is evaluated for the expression of $v_g(s)$ in Eq. 3.37. By comparing Eq. 3.40 with Eq. 3.29, we end up with a definition of $\tau = \frac{\omega l}{2Qv_g}$, which is the attenuation factor introduced for a TW-CI in Eq. 3.32.

3.3.4 Comparison

The three types of periodic multi-cells RF structures introduced so far, i.e., SW, TW-CI and TW-CG, show different values of the peak accelerating field and voltage, as well as a different dependence of those quantities from the s-coordinate internal to the structure. If a large accelerating voltage is often required to maximize the particles' energy over a given length, a moderate peak field is generally desired to avoid discharges in the structure.

Table 3.1 compares the peak voltage (see Eqs. 3.23, 3.32 and 3.39) and the peak field (see Eqs. 3.26, 3.32 and 3.38) of SW, TW-CI and TW-CG structures, assuming identical shunt impedance, structure's length and input RF power, normalized to the quantity of a SW. Figure 3.4 plots those quantities versus τ. In reality, the SW peak voltage should be rescaled by the transit time factor (e.g., $T_{tr} \approx 0.9$). The peak field reported for the TW-CI is at the entrance of the structure ($s = 0$); its average average value along the structure results into the same scaling factor of the peak voltage.

It is worth reporting that the filling time is commonly much longer in a SW than in TWs, with the consequence that longer RF pulses, at typical repetition rates of

Table 3.1 Normalized peak accelerating voltage and peak field in SW, TW-CI and TW-CG structures. P is intended to be $\langle P_d \rangle$ in SW, P_0 in TW structures

	SW	$TW - CI$	$TW - CG$
$\frac{\Delta V_0}{\sqrt{P r_s l}}$	1	$\sqrt{\frac{2}{\tau}}\left(1 - e^{-\tau}\right)$	$\sqrt{1 - e^{-2\tau}}$
$\frac{E_{z,0}}{\sqrt{P r_s / l}}$	1	$\sqrt{2\tau}$	$\sqrt{1 - e^{-2\tau}}$

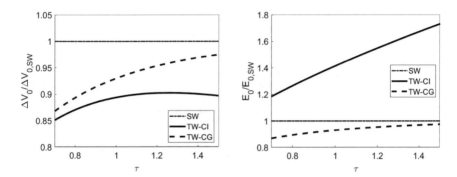

Fig. 3.4 Accelerating peak voltage (left) and field (right) for a SW, TW-CI and TW-CG structure, normalized to the value of a SW, as function of the attenuation factor. The transit time factor is not considered for the SW. See also Table 3.1

10–100 Hz, put the structure under mechanical and thermal stress, which can be source of shot-to-shot fluctuations of the RF performance. Such limited reliability is often compensated by making SW structures not longer than 1–2 m, and therefore limiting the total peak voltage. A TW-CI may be an attractive choice because of some easiness in the fabrication of perfectly identical cells. In the last decade, however, machining accuracy has reached the μm level, so reducing the fabrication cost of slightly different cells in a TW-CG.

3.3.5 Time Scales in RF Structures

The operation of RF structures covers ∼4 orders of magnitude in space, from the structure's length to typical bunch lengths, and ∼12 orders of magnitude in time, from the RF pulse repetition period to the bunch duration. They are recalled below and sketched in Fig. 3.5.

1. Power generators operate in several frequency ranges, such as L-band (1.3 or 1.5 GHz), S-band (2.998 or 2.856 GHz), C-band (around 6 GHz) and X-band (11.4 or 12 GHz). For the sake of discussion, let us consider the S-band range, whose *RF period* corresponds to approximately $T_{RF} = 1/f_{RF} \approx 333$ ps, or $\lambda_{RF} = c/f_{RF} = 0.1$ m.

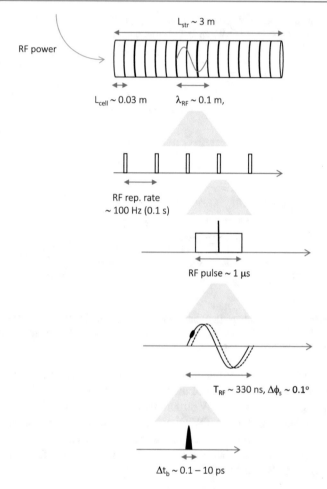

Fig. 3.5 Schematic (not to scale) of length and time scales associated to the operation of a NC TW S-band structure. From top to bottom: RF structure with $2\pi/3$-phase advance accelerating mode, RF pulse repetition rate, RF pulse duration synchronized to a particle bunch, bunch at the stable phase within an RF period, single bunch duration

2. The Typical operation of TW structures is at phase advance of $2\pi/3$ or so. Namely, one RF period is covered by few cells, which makes the length of each cell a fraction of the RF wavelength; for example, $L_{cell} \approx \lambda_{RF}/3 \approx 30\,\text{mm}$. The accelerating voltage is amplified by a large number of cells, say 100 or so, so that the typical *RF structure length* is around $L_{str} \approx 100 L_{cell} \approx 3$ m.
3. In order for the accelerated beam particles to see almost the same RF field amplitude and phase, the *bunch duration* should be much shorter than the RF period, e.g., $\Delta t_b \leq T_{RF}/10 \approx 30$ ps. As a matter of fact, electron bunches as short as tens' of ps can be further manipulated in linacs to reach the $\sim$ fs time scale.
4. The time interval corresponding to $1°$ S-band RF phase is $\Delta\phi_s = (2\pi f_{RF}\Delta t) \cdot (180/\pi) \equiv 1 \Rightarrow \Delta t(1°) \approx 1$ ps. The accuracy in *RF phase control* can similarly

be quantified in terms of a time interval (much) smaller than the bunch duration, so that the effective RF phase of all particles in a bunch is known with sufficient precision. For bunch duration at the ps-time scale, $\Delta\phi_s(\Delta t \leq 0.1\,ps) \leq 0.1°$. Nowadays state-of-the-art low-level RF controllers can guarantee RF phase stability down to 0.001–$0.01°$ S-band.

5. We have so far considered one bunch only injected into the accelerator. However, if the duration of the accelerating *RF pulse* is long enough, then multiple bunches can be injected in series to increase the effective beam repetition rate. The maximum duration of the RF pulse is commonly inversely proportional to the uniformity of the RF field along the whole pulse. The *minimum* duration is determined by the time interval required by the RF power to propagate through the structure and to allow the accelerating mode to resonate. This is just the *filling time*. For a TW structure, one may reasonably have $\Delta t_{RF} \geq t_f = L_{str}/v_g \approx 3m/0.1c = 0.1\ \mu s$. Indeed, RF pulse durations are commonly in the range 0.1–$1\ \mu s$.

One could then wonder how often a new single bunch (or train of bunches) can be injected into the RF structure or, in other words, what is the timing pattern of the RF pulse. This largely depends on the capability of the cooling system to absorb the RF average power dissipated on the cavity walls (low surface resistance), and that of keeping the accelerating field high over the duration of the pulse train (high quality factor). Technological choices group RF structures in two families.

Superconducting (SC) RF structures, commonly designed as SW in the L-band frequency range, are commonly made of either Nb-alloy or Cu-bulk covered by a film of Nb. The cavity is immersed in liquid He at 2–4 K, where Nb behaves as a superconductor. Such cavities tolerate CW RF power pulsed up to the MHz repetition rate. By virtue of the high quality factor $Q \sim 10^7 - 10^{11}$, the accelerating field amplitude remains approximately constant for at least $Q/\omega_{RF} \approx 10$ ms. The maximum accelerating gradient at the maximum repetition rate can be as high as 20–40 MV/m.

Normal-conducting (NC) RF structures, made of Cu, are characterized by $Q \sim 10^3 - 10^5$. They are water-cooled, and cannot normally operate above 1 kHz repetition rate or so. The accelerating gradient is in the range $\sim 10 - 100$ MV/m, inversely proportional to the RF repetition rate. These cavities are said to run in *pulsed mode* because the accelerating mode is present in the cavity only for relatively short time intervals, during which the beam is injected into the cavity.

It emerges that SC linacs are suitable for the acceleration of bunch trains (from 100 to 1000s bunches per train, with internal separation multiple of the RF period), and therefore useful for high beam repetition rates. This is at the expense of the maximum accelerating gradient, which can be $\sim$3 times lower than in NC structures. The latter ones are usually limited to 2-bunches operation in a single RF pulse.

References

1. R. Palmer, *Acceleration Theorems, Proceedings of the 6th Workshop on Advanced Accelerator Concepts*, BNL-61317 or CAP 1112-94C, Lake Geneva, WI, 1994 (1992)
2. M. Puglisi, *Conventional RF Cavity Design, Proceedings of CERN Accelerator School on RF Engineering for Particle Accelerators*, CERN 92-03, Geneva, Switzerland, vol. I (1992)
3. J. Le Duff, *Dynamics and Acceleration in Linear Structures, Proceedings of CERN Accelerator School*, CERN 94-01, Geneva, Switzerland, vol. I (1994), pp. 253–277

High Energy Accelerators

4

High energy accelerators driven by RF cavities can provide kinetic energies up to 5 orders of magnitude higher than DTLs, betatrons and cyclotrons. They can be grouped into three families: *synchrotrons* (circular geometry), *RF linacs* (single-pass, linear geometry), and *energy-recovery linacs* (race-track geometry). The use of radiofrequency implies that they are all resonant accelerators, synchronized to bunched beams. They are called *light sources* if devoted to emission of radiation, *colliders* if designed for collisions of accelerated beams, hence used for particle physics experiments.

To date, synchrotrons are the only accelerators capable of storing up to $\sim$ TeV-energy particle beams for a relatively long time (from a fraction of hour to days). This is made possible by three main advancements with respect to betatrons and cyclotrons.

1. Distinct accelerator components, installed along the reference closed orbit exploit distinct functions, thus decoupling the electric and the magnetic action: acceleration by RF cavities, beam guiding and focusing by magnets.
2. The synchronism established every turn between the RF electric field and the beam arrival time permits a small energy increase per turn, thereby a small orbit excursion (in combination with a properly ramped magnetic guiding when the beam energy is increased turn-by-turn). Consequently, the magnets can be installed in a series, aligned to the reference orbit.
3. The split of magnetic guiding along the orbit breaks the azimuthal symmetry exploited in cyclotrons and betatrons. This makes the maximum kinetic energy scaling only *linearly* with the orbit radius. Moreover, the smaller the orbit excursion and the beam transverse size is, the more compact the magnets can be, the stronger is the magnetic focusing they can provide.

© The Author(s), under exclusive license to Springer Nature Switzerland AG 2022
S. Di Mitri, *Fundamentals of Particle Accelerator Physics*, Graduate Texts in Physics,
https://doi.org/10.1007/978-3-031-07662-6_4

Unlike single or few pass accelerators, synchrotrons pose to the beam the additional constraint of periodic motion and long-term stability. The former modifies the boundary conditions of the equations of motion with respect to e.g., linacs. The latter implies persistent emission of radiation, which in turn affects the 6-D particle distribution. In the following, the longitudinal dynamics in a low energy linac is treated first, extended to ultra-relativistic motion then. The periodic longitudinal motion in a synchrotron is finally discussed. The second part of the Chapter treats single particle linear and nonlinear transverse dynamics. Nonlinear terms in the equations of motion are often neglected in linacs, while they are relevant for the determination of stable motion in synchrotrons.

4.1   General Features

The synchrotron was conceived by V. Veksler (1944) and E. MacMillan (1945), following the betatron and, as this one, with the aim of overcoming energy limitations of the Lawrence's cyclotron. Synchrotrons are circular accelerators in which the beam's energy is increased turn-by-turn ("energy ramp"), until the target energy is reached. A synchrotron is specified to be a *storage ring* when the accelerator is supplied by a "full-energy" injection system, and the beam energy is constant in the main ring.

In its simplest configuration, the injection stage to a storage ring is made of a short linac followed by a smaller synchrotron called *booster ring*, as illustrated in Fig. 4.1. In most recent storage ring light sources, the electron booster ring is installed in the same tunnel of the main ring. In few cases, beam injection is made directly into the storage ring from a high energy linac. Storage ring light sources produce radiation from IR to hard x-rays. The typical beam energy range is 2–8 GeV, their circumferences span ~0.2–2 km.

A circular collider complex can involve different particle species (leptons, hadrons, ions), manipulated at different energy levels and in different collisional configurations. The stages of beam production, acceleration and injection into the main storage ring can therefore be many and diverse. The Large Hadron Collider (LHC) is the largest circular collider in the world, with a circumference of approximately 27 km and beams' kinetic energies approaching the TeV scale. The Future Circular Collider (FCC) project is a storage ring targeting ~10 TeV invariant mass energy, over a circumference of ~100 km.

By virtue of the small orbit excursion and small transverse beam sizes, the beam stored in a synchrotron propagates into a *vacuum chamber* of diameter few to tens' millimeters wide. The chamber is interrupted by, or better it incorporates, RF cavities, diagnostic elements, valves and vacuum pumps to keep the inner pressure very low, so to minimize scattering of the stored particles with residual ions.

The chamber is surrounded by permanent magnets and electromagnets. The latter are commonly made of Cu coils surrounding Fe yokes, to maximize the magnetic field flux in proximity of the beam's orbit. The Lorentz's force exploited in *dipole magnets* bends the beam in the horizontal plane, so defining a reference closed orbit.

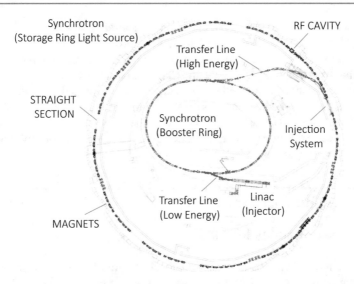

Fig. 4.1 Accelerator complex of the Elettra2.0 storage ring light source upgrade project, in Italy. The main ring circumference is 260 m long (on scale). Electrons are generated in a thermo-ionic gun, accelerated in two RF structures to 100 MeV, then ramped in energy by a booster ring over multiple turns. They are finally injected into the main ring at the energy of 2.4 GeV. (Original picture courtesy of S. Lizzit)

Multipole magnets, and in particular quadrupoles, control the beam's transverse sizes by forcing off-axis particles to travel in proximity of the reference orbit ("focusing").

Straight sections connecting consecutive dipole magnets make the synchrotron closed orbit a polygonal. The straight sections can host components very specific to the accelerator's scope. For example, an arrangement of particularly strong quadrupole magnets, called *final focusing*, is adopted in colliders in proximity of the interaction point to squeeze the beam sizes and therefore increase the charge density locally. Special dipolar arrays, named *Insertion Devices* (IDs), are installed in the straight sections of light sources to produce radiation with specific spectral features.

The split of functionalities that contributed to the success of synchrotrons, applies similarly to high energy RF linacs, where few meters-long RF structures alternate to quadrupole magnets. The maximum kinetic energy reached so far in a RF linac is ∼20 GeV over a length of ∼3 km, at the RF repetition rate of 120 Hz. Repetition rates as high as 1 MHz can only be reached at SC RF linacs, typically at the expense of peak accelerating gradient.

Energy-recovery linacs (ERLs) are made of parallel and relatively long straight sections, hosting one or two linacs, connected by magnetic arcs. Unlike single-pass linacs, the race-track geometry allows the beam to be recirculated, and therefore accelerated, once or in a few passes. Before the beam is dumped, it can be decelerated, so that a large fraction of the beam's e.m. power can be recovered and stored in the linac, and there used for acceleration of a newly injected beam.

In the last decade, the test of very high gradient (∼GeV/m) particle accelerators has been carried out worldwide. The electric field providing longitudinal acceleration ("wake field") is generated by the ionization of a neutral plasma in a small capillary traversed by a laser or a charged particle beam. A probe bunch is then injected into the plasma to surf the wake field ("plasma-wake field accelerators"). An analogous process has been proposed for a dielectric channel in place of the plasma capillary. In this case, acceleration is provided by the interaction of the probe bunch with the e.m. field associated to the image currents excited by the leading bunch on the dielectric surface ("dielectric-wake field accelerator"). All these non-conventional accelerators are not treated in this book, and the Reader is referred to the additional bibliography for an introduction to the subject.

4.2   Longitudinal Dynamics

4.2.1   Phase Stability in a Linac

If a particle moves in a linac in weakly relativistic regime, its velocity increases with the kinetic energy (see Fig. 1.4). In this case, the beam's injection phase is important because it determines the relative motion of particles in a bunch, by keeping the particles ensemble bounded or unbounded, depending from their relative spread in velocity. The proper choice of injection phases which allow stable motion constitutes the so-called *phase stability* [1].

The concept is illustrated in Fig. 4.2. The curve shows the energy gain in a RF structure, see e.g. Eq. 3.17. The phase convention is such that the bunch head, i.e. earlier arriving particles, is at smaller phases (a_1, a_2). Energy gain is only for points in the positive (upper) half-plane.

In a RF period, and limiting ourselves to the half-plane of energy gain, the synchronous phase can be chosen at either positive or negative slope of the "RF curve". If the positive slope is chosen, i.e., the synchronous phase of P_2 is $3\pi/2 < \phi < 2\pi$, leading particles in the bunch (a_2) enter the RF cavity with a phase which makes them to gain less energy than the synchronous particle (the relative energy variation results $\delta < 0$). Therefore, their velocity relative to the synchronous particle is

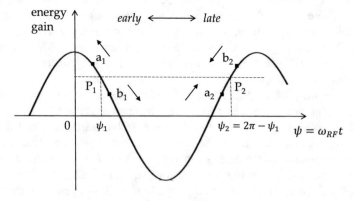

Fig. 4.2 Phase stability in a linac at weakly relativistic velocities

smaller (particles are slower) and, as the beam propagates through the accelerator, they become closer to the bunch centre. Similarly, trailing particles (b_2) gain more energy ($\delta > 0$), therefore they get a larger velocity than the synchronous particle (particles are faster), and they eventually approach the bunch centre. This repeats as leading and trailing particles exchange their positions inside the bunch, i.e., the particle's motion is bounded. The opposite happens if the synchronous particle P_1 lies on the negative slope of the energy gain curve: in this case leading (a_1) and trailing particles (b_1) progressively move far from the synchronous particle. After some time, particles get lost in the accelerator.

In summary, if the particles' initial phase and energy are close to those of the synchronous particle, and if this sits on a stable RF phase, *oscillations* in the *longitudinal phase space* (ϕ, δ) can happen, which ensure the bunch motion to be stable.

Phase stability is derived below from the single particle's equations of motion. Reduced variables in the longitudinal phase space (ϕ, δ) are adopted, which describe the motion of the generic particle relative to the coordinates of the synchronous particle:

$$\begin{cases} \Delta t = t - t_s \\\\ \phi = \psi - \psi_s = \omega \Delta t = kz \\\\ w = \Delta E - \Delta E_s \end{cases} \tag{4.1}$$

ΔE is the particle's energy gain. In the following, we will assume $v_x, v_y << v_z \cong |\vec{v}|$, and the independent spatial coordinate along the linac is indicated with s. We have:

$$\begin{cases} \dfrac{d\phi}{ds} = \omega \dfrac{d}{ds}(t - t_s) = \omega \left(\dfrac{1}{v_z} - \dfrac{1}{v_{z,s}} \right) \approx -\omega \dfrac{v_z - v_{z,s}}{v_{z,s}^2} \\\\ w = \frac{1}{2}m_0(v_z^2 - v_{z,s}^2) = \frac{1}{2}m_0(v_z + v_{z,s})(v_z - v_{z,s}) \approx m_0 v_{z,s}(v_z - v_{z,s}) \end{cases} \tag{4.2}$$

The approximate equalities are for $v \approx v_s$. The second equation is substituted into the first one, and the first derivative of the reduced energy is calculated:

$$\begin{cases} \dfrac{d\phi}{ds} = -\dfrac{\omega}{m_0 v_{z,s}^3} w \\\\ \dfrac{dw}{ds} = \dfrac{d}{ds}(\Delta E - \Delta E_s) = q E_{z,0} \dfrac{d}{ds} \left[\int_0^s ds \cos \psi - \int_0^s ds \cos \psi_s \right] = \\\\ = q E_{z,0} [\cos(\phi + \psi_s) - \cos \psi_s] \approx -q E_{z,0} \Delta (\cos \psi_s) = q E_{z,0} \sin \psi_s \cdot \phi \end{cases} \tag{4.3}$$

The approximate equality in the second equation is for $\psi \approx \psi_s$. Another derivation with respect to s leads to:

$$\begin{cases} \dfrac{d^2\phi}{ds^2} = -\dfrac{\omega}{m_0 v_{z,s}^3} \dfrac{dw}{ds} = -\dfrac{q E_{z,0} \omega \sin \psi_s}{m_0 v_{z,s}^3} \phi \equiv -\Omega_l^2 \phi \\\\ \dfrac{d^2 w}{ds^2} = q E_{z,0} \sin \psi_s \dfrac{d\phi}{ds} = -\dfrac{q E_{z,0} \omega \sin \psi_s}{m_0 v_{z,s}^3} w \equiv -\Omega_l^2 w \end{cases} \tag{4.4}$$

where the first derivatives were replaced according to their expressions in Eq. 4.3. We introduced the *longitudinal angular frequency* Ω_l:

$$\Omega_l := \sqrt{\frac{q E_{z,0} \omega \sin \psi_s}{m_0 v_{z,s}^3}} \tag{4.5}$$

In conclusion, the longitudinal motion of weakly relativistic particles in a linac is described by equations of a quasi-harmonic oscillator (the oscillation frequency still depends from the particle's velocity), which ensures limited orbit excursion in the phase space. The Motion is stable if the following conditions are met:

1. all particles have phase and velocity close to those of the synchronous particle, i.e., the spread in phase and relative energy is $<< 1$;
2. the synchronous phase is chosen such that $\Omega_l^2 > 0$, or $q E_{z,0} \sin \psi_s > 0$ for an accelerating field which goes like $\sim \cos \psi$. For example, the convention $q E_{z,0} > 0$ leads to energy gain for $\cos \psi_s > 0$, and the synchronous phase has to be chosen in the range $0 < \psi_s < \pi/2$.

4.2.2 Adiabatic Damping

The dependence of Ω_l from $v_{z,s}$ implies that, as the bunch's average velocity increases during acceleration, the oscillation frequency reduces. If the variation of particle's velocity is so slow that $\Delta v_{z,s} << v_{z,s}$ in a period of oscillation, the motion in the longitudinal phase space can be approximated to that of a pure harmonic oscillator. Then, the particle's orbit in the phase space can be represented by an ellipse, whose extreme points have coordinates $[0, w_{max}]$ and $[\phi_{max}, 0]$. The ellipse area is $\pi \phi_{max} w_{max}$, and it is approximately constant over one period.

Recalling Eq. 4.3 and assuming $v_{z,s} \approx const.$, we find:

$$\begin{cases} \dfrac{dw}{ds} = q E_{z,0} \sin \psi_s \cdot \phi \\[2mm] \dfrac{d\phi}{ds} = -\dfrac{\omega}{m_0 v_{z,s}^3} w \end{cases} \Rightarrow \begin{cases} ds = \dfrac{dw}{q E_{z,0} \sin \psi_s \cdot \phi} \\[2mm] \phi d\phi = -\dfrac{\omega}{m_0 v_{z,s}^3 q E_{z,0} \sin \psi_s} w dw \equiv \alpha w dw \end{cases} \tag{4.6}$$

By integrating the lower equation on the r.h.s. in the ranges $[0, \phi_{max}]$ and $[0, w_{max}]$, we get:

$$\begin{cases} \phi_{max}^2 - \alpha w_{max}^2 = C_1 \equiv 0 \\[2mm] \phi_{max} w_{max} = C_2 \neq 0 \end{cases} \Rightarrow \begin{cases} \phi_{max} \propto \alpha^{1/4} \propto \dfrac{1}{v_{z,s}^{3/4}} \\[2mm] \dfrac{w_{max}}{E_s} \propto \dfrac{\alpha^{-1/4}}{v_{z,s}^2} \propto \dfrac{1}{v_{z,s}^{5/4}} \end{cases} \tag{4.7}$$

Equation 4.7 shows that, in the non-relativistic regime, both the linac *phase acceptance* ϕ_{max} and the *relative energy acceptance* w_{max} decrease with the beam's average velocity.

When the particle enters the ultra-relativistic regime $v_z \approx v_{z,s} \to c$, Eq. 4.2 predicts $\frac{d\phi}{ds} \to 0$, $\frac{d^2w}{ds^2} \propto \frac{d\phi}{ds} \to 0$. Namely, the longitudinal oscillations tend to disappear. If the bunch is short enough with respect to the RF wavelength, and it is accelerated on-crest not to sample much of the RF field "curvature", all particles will have approximately the same RF phase, i.e., they will all gain the same amount of energy. Consequently, the initial *absolute* energy spread is preserved, while the *relative* energy spread decreases with the beam energy, $\delta = \frac{\Delta \bar{E}(s)}{E_s} \propto \frac{1}{\gamma(s)}$.

Since both the linac energy acceptance in the non-relativistic case and the beam's relative energy spread in the ultra-relativistic limit were derived in the approximation of slow velocity variation, their reduction with the beam's mean energy is called *adiabatic damping*.

4.2.2.1 Discussion: Proton Injector

What is the number of longitudinal oscillations of a proton beam injected into a 30 m-long S-band linac ($f_{RF} = 3$ GHz)? The initial kinetic energy is $T_i = 10$ MeV. The beam is accelerated by a peak accelerating gradient $E_{z,0} = 25$ MV/m, and the synchronous phase is $\psi_s = \pi/4$.

Owing to the fact that $\Omega_l^2 \sim 1/v_{z,s}^3$, the highest frequency of longitudinal oscillations is at the injection point (lowest beam energy), the smallest frequency is at the end of acceleration. Since Eq. 4.4 was derived for the independent variable s, the calculation of the frequency in units of inverse time requires Ω_l^2 to be multiplied by $v_{z,s}^2$:

$$\Omega_l(t) = \frac{2\pi}{T_s} = \sqrt{\frac{q E_{z,0} \omega \sin \psi_s}{p_{z,s}}} \tag{4.8}$$

The frequency at the beginning and at the end of the accelerator can be calculated once the longitudinal momentum at the beginning and at the end of acceleration is found, respectively. This can be evaluated in turn via the kinetic energy:

$$\begin{cases} T_i = 10 \,\text{MeV} \\ \\ T_f = T_i + e E_{z,0} L \cos(\frac{\pi}{4}) = 540.3 \,\text{MeV} \end{cases} \tag{4.9}$$

$$\Rightarrow \begin{cases} p_{s,i} c = \sqrt{E_i^2 - (m_p c^2)^2} = \sqrt{T_i^2 + 2 T_i m_p c^2} = 137.3 \,\text{MeV} \\ \\ p_{s,f} c = \sqrt{T_f^2 + 2 T_f m_p c^2} = 1142.6 \,\text{MeV} \end{cases} \tag{4.10}$$

By replacing these momenta in Eq. 4.8, we find $\Omega_l^i(t)/(2\pi) = 136$ MHz and $\Omega_l^f(t)/(2\pi) = 47$ MHz.

Although these instantaneous values look extremely high, one has also to consider that the particles, though in weakly relativistic regime, take extremely short time (compared to the human scale) to pass through the linac. Indeed, the instantaneous velocity, either at the injection or at the extraction point, is $\beta c = pc^2/E$, which amounts to 0.145 c and 0.773 c, respectively. Hence, the time a proton would take to pass 30 m at the smallest and largest velocity would be, respectively, 0.69 μs and 0.13 μs. At the end, the number of longitudinal oscillations the protons have effectively completed during acceleration is expected to be within the interval [47 MHz $\cdot$ 0.13 μs–136 MHz $\cdot$ 0.69 μs] $\approx$ [6 − 94].

An exact calculation of the number of oscillations should take into consideration the variation of longitudinal velocity and momentum through the linac, assuming a linear variation of the particle's total energy with s, i.e., $\gamma(s) = \gamma_0 + \frac{qE_{z,0}}{m_pc^2}s \equiv \gamma_0 + \alpha s$, and $\gamma_0 = 1.01$. In this case, we calculate the number of oscillations as the ratio of the travelling time and the instantaneous oscillation period:

$$N_{osc} = \frac{\Delta t}{T_l} = \int_0^L \frac{ds}{v_z(s)} \frac{\Omega_l(t)}{2\pi} = \frac{1}{2\pi} \int_0^L \frac{c\,ds}{\beta_z(s)} \sqrt{\frac{\omega q E_{z,0} \sin\psi_s}{p_z(s)}} =$$

$$= \frac{\sqrt{\omega q E_{z,0} \sin\psi_s}}{2\pi c} \int_0^L \frac{ds}{\beta_z(s)} \sqrt{\frac{c}{\beta_z(s)E(s)}} = \frac{1}{2\pi} \sqrt{\frac{\omega q E_{z,0} \sin\psi_s}{m_p c^3}} \int_0^L \frac{ds}{\beta_z^{3/2}(s)\gamma(s)} =$$

$$= \frac{1}{2\pi} \sqrt{\frac{\omega q E_{z,0} \sin\psi_s}{m_p c^3}} \int_0^L ds \frac{\gamma^2}{(\gamma^2-1)^{3/2}} = \frac{1}{2\pi} \sqrt{\frac{\omega q E_{z,0} \sin\psi_s}{m_p c^3}} \int_0^L ds \frac{(\gamma_0+\alpha s)^2}{[(\gamma_0+\alpha s)^2-1]^{3/2}} \approx 43$$

$$(4.11)$$

According to Eq. 4.11, $N_{osc} \sim \frac{1}{\gamma\sqrt{m_0c^2}} \sim \frac{\sqrt{m_0c^2}}{E}$. Thus, the number of oscillations of lighter particles is smaller than for heavier particles, for the *same* total energy. Namely, lighter particles become ultra-relativistic sooner, and their relative longitudinal position is frozen earlier.

4.2.2.2 Discussion: Electronic Capture in a RF Gun

An RF cavity characterized by $v_{ph} \approx c$, denominated *RF Gun*, can be used as very first accelerating stage of non-relativistic electrons, as long as the accelerating field is high enough to allow the electrons to enter the ultra-relativistic regime in less than one RF period. In other words, the relative shift of the e.m. wave with respect to the synchronous particle ("slippage") in one period has to be small enough to allow an energy gain much larger than the particle's initial energy. In most advanced RF Guns, an infrared (IR) laser impinges on the metallic or semiconductor surface (the "cathode") of the inner back face of the cavity. Electrons are emitted by photoelectric effect, commonly at kinetic energies lower than 10 eV. What is the minimum peak electric field in the RF Gun to "capture" the electrons?

The electron-wave slippage length per unit of RF phase is:

$$\frac{dl}{d\phi} = \frac{(v_{ph} - v_0)dt}{d\phi} = \frac{c(1 - \beta_0)dt}{d\phi} = \frac{c}{\omega}(1 - \beta_0) = \frac{\lambda_{RF}}{2\pi}(1 - \beta_0) \quad (4.12)$$

In order for the energy gain during such slippage to be much larger than the initial particle's energy (i.e., in proximity of the cathode surface), the peak field has to be:

$$\Delta E \approx q E_{z,0} \frac{dl}{d\phi} = q E_{z,0} \lambda_{RF} \frac{1-\beta_0}{2\pi} \gg \gamma_0 m_0 c^2$$

$$\Rightarrow E_{z,0} \gg \frac{2\pi m_0 c^2}{q \lambda_{RF}} \left(\frac{\gamma_0}{1-\beta_0} \right)$$

(4.13)

For electrons accelerated in an S-band Gun ($f_{RF} = 3$ GHz) and therein emitted with an initial kinetic energy $T_0 \approx 10$ eV, one finds:

$$\gamma_0 = \frac{m_e c^2 + T_0}{m_e c^2} = 1.00002$$

$$\Rightarrow \beta_0 = \sqrt{1 - \frac{1}{\gamma_0^2}} = 0.006$$

(4.14)

$$\Rightarrow E_{z,0} \gg 32\text{MV/m}$$

State-of-the-art RF Guns in the S-band to X-band frequency range, run at peak fields as high as 60–250 MV/m.

4.2.3 Momentum Compaction

Dipole fields are used to bend beam particles in linacs as well as in synchrotrons [2]. According to Eq.2.5, for any given bending field, particles of different longitudinal momenta will follow different curvature radii, and therefore their path lengths will be different. The variation of relative longitudinal momentum δ, curvature radius R and magnetic field B_y, are related each other by

$$\frac{dp_z}{p_z} = \frac{dR}{R} + \frac{dB_y}{B_y} = \frac{dR}{R} \left(1 + \frac{R}{B_y} \frac{dB_y}{dR} \right) \equiv \frac{1}{\alpha_c} \frac{dR}{R}$$

$$\Rightarrow \alpha_c := \frac{dR/R}{dp_z/p_z} = \frac{dL/L}{\delta}$$

(4.15)

L is the path length of the on-energy, synchronous particle. α_c is called *momentum compaction* and it quantifies the variation of path length of an off-energy particle ($\delta \neq 0$) relative to the path length of the synchronous particle.

The transverse coordinates of the generic particle w.r.t. the synchronous particle are described hereafter in the *Frenet-Serret frame of coordinates*. The motion of the synchronous particle defines the reference orbit, whose longitudinal curvilinear coordinate is s. At any s, the longitudinal velocity of the particle is tangent to the instantaneous orbit of radius $R(s)$, namely, $\vec{v}_z = v_z \hat{s}$. The transverse plane is orthogonal to $\hat{s}$, such that the generic particle is distant x and y from the synchronous particle, in the bending plane and in the plane orthogonal to it, respectively. The reference system is illustrated in Fig. 4.3-left plot.

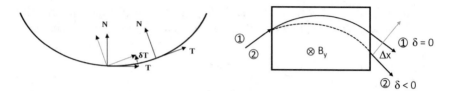

Fig. 4.3 Left: Frenet-Serret coordinate system. Right: top-view of orbit dispersion of an off-energy particle in a dipole magnet

A particle initially aligned to the synchronous particle but with a lower longitudinal momentum, is bent on an orbit of smaller curvature radius. The distance along the x-axis between the two particles, at any point s of the reference orbit, is $x(s)$, see Fig. 4.3-right plot. We define *linear momentum-dispersion function*, simply dispersion hereafter, the quantity:

$$D_x(s) := \frac{x(s)}{\delta} \tag{4.16}$$

The difference in curvature radius of the two orbits in Fig. 4.3, where particle-1 is intended to be the synchronous particle, results:

$$dR = \frac{C_2 - C_1}{\theta_b} = \frac{1}{\theta_b}\left(\int ds_2 - \int ds_1\right) = \frac{1}{\theta_b}\int d\theta\,[(R_1 + x) - R_1] =$$

$$= \frac{1}{\theta_b}\int x\,d\theta = \langle x\rangle_\theta \tag{4.17}$$

C_1, C_2 are the path lengths of the two particles from the common origin to any arbitrary s, and θ_b is the total bending angle of the synchronous particle. By recalling Eq. 4.15, we find:

$$\alpha_c = \frac{dR/R}{\delta} = \frac{1}{R}\frac{\langle x\rangle_\theta}{\delta} = \frac{\langle D_x\rangle_\theta}{R} = \frac{1}{R\theta_b}\int D_x\,d\theta = \frac{1}{C}\int \frac{D_x(s)}{R(s)}\,ds \tag{4.18}$$

$C = R\theta_b$ is the total nominal path length, and $R(s)$ the *local* curvature radius evaluated along the nominal path (derived from the change of variable $d\theta = ds/R(s)$). Clearly, if the orbit is closed, the integral is closed as well, C is the ring circumference, and α_c is the average value of the ratio D_x/R along the closed orbit. A magnetic lattice in which all dipole magnets have the same curvature radius is said *isomagnetic*.

It should be noted that α_c can be either positive, null or negative, and that it receives non-zero contribution only from the dispersion defined along curved paths, e.g., inside dipole magnets. This implies that $D_x \neq 0$ along a drift (for which $R(s) \to \infty$) does not contribute to α_c at first order in δ. However, if D_x is non-zero in a drift, quadrupole magnets can modify its value at a successive dipole magnet, so that, at the end, α_c can be tuned via a suitable manipulation of D_x both inside and outside the dipole magnets.

Owing to the relation $D_x \sim \frac{dx}{dp_z} \sim \frac{dR}{dp_z} \sim \frac{1}{B_y}$ (the last relation is from Eq. 2.5), the dispersion function results to be an intrinsic property of the magnetic lattice.

Since also the curvature radius is a property of the lattice ($R\theta_b = l_b$), it turns out that $\alpha_c \sim \frac{D_x}{R}$ inherits the magnetic lattice.

The term "linear" attributed above to D_x and α_c refers to the order in δ taken to describe the particle's motion: $x = D_x \delta + o(\delta^2)$. In general, higher orders can be considered, so that higher order energy dispersion and momentum compaction can be defined, e.g., $\alpha_c = \alpha_1 + \alpha_2 \delta + \alpha_3 \delta^2 + \cdots$.

4.2.4 Transition Energy

Off-energy ultra-relativistic particles in a synchrotron travel on slightly different orbits, which result in slightly different revolution frequencies. This amounts to $\omega_s = \beta_{z,s} c / R_s$ for the synchronous particle in an isomagnetic lattice. The relative variation of revolution frequency per unit deviation of the longitudinal momentum is called *slip factor*:

$$\eta := \frac{d\omega/\omega_s}{dp_z/p_{z,s}} \tag{4.19}$$

From the definition of ω_s we have:

$$\frac{d\omega}{\omega_s} = \frac{d\beta_z}{\beta_{z,s}} - \frac{dR}{R_s} = \frac{d\beta_z}{\beta_{z,s}} - \alpha_c \frac{dp_z}{p_{z,s}} = \frac{dp_z}{p_{z,s}} \left[\left(\frac{dp_z}{d\beta_z} \right)^{-1} \frac{p_{z,s}}{\beta_{z,s}} - \alpha_c \right] =$$

$$= \frac{dp_z}{p_{z,s}} \left[\frac{1}{\gamma^3 m_0 c} \frac{p_{z,s}}{\beta_{z,s}} - \alpha_c \right] = \delta \left(\frac{1}{\gamma^2} - \alpha_c \right) \tag{4.20}$$

$$\Rightarrow \eta = \frac{1}{\gamma^2} - \alpha_c$$

and we calculated $\frac{dp_z}{d\beta_z} = \frac{d(\gamma\beta_z)}{d\beta_z} m_0 c = \gamma^3 m_0 c$. Typically, $0 < \alpha_c \ll 1$ in synchrotrons.

Equation 4.20 suggests that, for any given α_c determined by the magnetic lattice, there exists a value of the beam energy $\gamma_{tr} = 1/\sqrt{|\alpha_c|}$, named *transition energy*, making $\eta(\gamma_{tr}) = 0$. The transition energy determines a swap of the sign of revolution frequency variation per unit of relative energy deviation.

If the beam's energy is approximately constant along one or several turns and such that $\eta > 0$ ($\gamma < \gamma_{tr}$ or "below transition"), particles at $\delta > 0$ will take a shorter time to make a turn than the synchronous particle, by virtue of their higher angular frequency. If $\eta < 0$ instead ($\gamma > \gamma_{tr}$ or "above transition"), particles at $\delta > 0$ will take a longer time. This latter result is also named "negative mass" behaviour because more energetic particles become slower.

Phase stability in a synchrotron is ensured by a different synchronous phase depending whether the beam is below or above transition. As shown in Fig. 4.4, ψ_s of a beam above transition has to be chosen in a way that leading particles (a_1) gain more energy ($\delta > 0$) than the synchronous particle. Doing so, they take a longer time to make a turn, thus they arrive later at the RF cavity on successive turns, and therefore they move towards the synchronous particle. A similar bounded motion

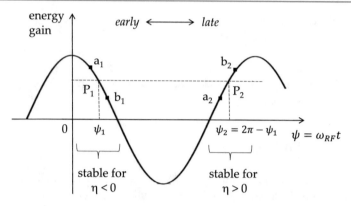

Fig. 4.4 Left: phase stability in a synchrotron below ($\eta < 0$) and above transition energy ($\eta > 0$)

also happens for the trailing particles (b_1), and *synchrotron oscillations* are established. For the same reason, ψ_s of a beam below transition has to be chosen such that leading particles (a_2) gain less energy than the synchronous particle, etc.

The working point $\eta = 0$ is critical for the longitudinal stability. When the beam energy crosses the transition energy during an energy ramp, the synchronous phase chosen below transition becomes unstable (see Fig. 4.4). A fast switch of the synchronous phase must therefore be implemented to keep the beam motion stable above transition.

A lattice with $\eta = 0$ can nevertheless be conceived. It is said to be *isochronous* because particles take the same time to make a turn independently from their energy deviation, see Eq. 4.19. The synchronous phase, however, has to be chosen $\psi_s = 2\pi n$, $n \in \mathbb{N}$ (on-crest acceleration). Otherwise, since the particles' arrival time at the RF cavity is frozen, they will continue gaining more (less) energy than the synchronous particle. The beam energy spread would then grow indefinitely, until far off-energy particles cannot be safely kept on a closed orbit, and get lost. At the on-crest phase, instead, and if the bunch is short enough not to sample too much RF curvature, the energy gain per turn can be made approximately equal for all particles, the bunch length and the relative energy spread remain approximately constant.

In synchrotrons, α_c is typically in the range $10^{-5} - 10^{-3}$. For example, for electron's total energy of 2 GeV, $\frac{1}{\gamma^2} \approx 10^{-7}$ and $\eta \approx -\alpha_c < 0$. For protons at the same energy, however, $\frac{1}{\gamma^2} = 0.2$ and $\eta \cong \frac{1}{\gamma^2} > 0$. While electrons are always injected into synchrotrons above transition by virtue of their small rest energy, protons are often injected below transition. These have to rapidly cross the transition energy until they reach the target energy. For high intensity proton beams, the effect of inter-particle Coulomb interactions ("space charge" forces) changes sign at transition, causing a sudden change and oscillation of the bunch length, hence a dilution of the charge density, an increase of the beam energy spread, and possible particles loss. Moreover, if the instability is not suddenly damped, space charge-induced energy spread can produce a spread in γ_{tr} internal to the bunch, so that particles would not cross the transition simultaneously.

Common remedies, denominated *fast γ_{tr} jump* schemes, foresee the artificial increase of the transition crossing speed by means of fast pulsed quadrupole magnets. These are arranged in doublets in the proton synchrotrons PS and SPS at CERN. The modification to the dispersion function is intended to increase α_c, thus to lower γ_{tr}, while ideally keeping the transverse dynamics unperturbed outside the doublets. While the beam energy ramp in a proton synchrotron can be of the order of $d\gamma/dt \approx 10 - 100\ s^{-1}$, the required variation of the magnets' strength can be implemented in milliseconds, so that $d\gamma_{tr}/dt \approx 10^3 - 10^4\ s^{-1}$.

4.2.4.1 Discussion: Momentum Compaction of a Drift Section

By virtue of Eq. 4.19, η behaves as a generalized momentum compaction. Namely, it is able to quantify the relative longitudinal shift of off-energy particles even for $\alpha_c = 0$, such as in a non-dispersive drift section. In this case, what is the relative particles' slippage, as function of their relative momentum deviation?

The deviation in revolution frequency in the definition of η can be intended here as the deviation in arrival time at a given s-coordinate along the drift or, equivalently, the deviation in longitudinal position Δz after a given time interval T_s. In the approximation of small deviation of the longitudinal momentum, $\delta \ll 1$:

$$\frac{d\omega}{\omega_s} = -\frac{\Delta T}{T_s} = -\frac{\beta_{z,s}}{\beta_{z,s}}\frac{\Delta T}{T_s} \approx \frac{L-L_s}{L_s} = -\frac{\Delta z}{L_s} = \eta\delta = \frac{\delta}{\gamma_s^2}$$

$$\Rightarrow \frac{\Delta z}{\delta} = -\frac{L_s}{\gamma_s^2}$$

(4.21)

$\Delta z < 0$ means that, assuming two particles occupying initially the same position, a more energetic particle moves ahead of the on-energy particle after a path length long L_s. The longitudinal slippage of particles along a drift section due to their relative spread of velocities is suppressed by Special Relativity by a factor $1/\gamma^2$.

The slip factor is usually neglected in linacs at high beam energies. For typical values $\delta \le 0.1\%$ in electron linacs, particles' longitudinal shift along tens' of meters is at nm scale, $\Delta z \sim \frac{L\delta}{\gamma^2} \le 0.01\mu$m. The effect can be relevant for bunches produced already at μm-scale length and at energies as low as ~ 10 MeV, such as in ultrafast electron diffraction sources, in which few microns bunch lengthening can be accumulated for $\delta \sim 1\%$ over a 10 cm-long drift section.

4.2.4.2 Discussion: Magnetic Bunch Length Compression

Free-electron lasers are nowadays the most advanced light sources in x-rays. Electron bunches of high charge density are accelerated in a RF linac to multi-GeV final energy, then made to wiggle in magnetic devices, named "undulators", to emit highly energetic, collimated radiation pulses. The radiation intensity is exponentially amplified along the undulator in proportion to the electron bunch peak current. For this reason, the initial bunch duration, typically in the range of 5–20 ps fwhm, is compressed at intermediate linac energies to obtain peak currents at (multi-)kA level.

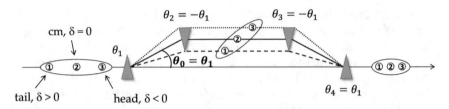

Fig. 4.5 Three-particle model of a bunch travelling through a 4-dipoles magnetic chicane. The leading particle (3) is at lower energy than the synchronous particle (2), e.g. identified with the bunch center of mass (cm). The trailing particle (1) is at higher energy. By virtue of the different path lengths, the bunch head and tail eventually catch up with the center of mass (cm), and the bunch duration is shortened. (Original picture courtesy of M. Venturini)

Bunch length compression is commonly accomplished in a 4-dipoles chicane (or a number of them), as shown in Fig. 4.5. We want to show that if a z-correlated momentum spread $\delta(z)$ is imparted to the ultra-relativistic beam, the bunch can be time-compressed by a large factor in virtue of the chicane momentum compaction α_c.

By simplifying the bunch longitudinal dynamics to a two-particle model, the initial bunch length is just $\Delta z_i = l_{b,i} = s_2 - s_1$, with s_1 taken along the reference trajectory. By virtue of the difference in longitudinal momentum and of the definition of α_c in Eq. 4.15, the two particles will run over different path lengths. Their relative distance at the chicane exit will be:

$$\Delta z_f = l_{b,f} = (s_2 + L_2) - (s_1 + L_1) = (s_2 - s_1) + (L_2 - L_1) = l_{b,i} + \Delta L$$

$$\Rightarrow l_{b,f} = l_{b,i} + \Delta L = l_{b,i} + \alpha_c L_1 \delta$$

$$(4.22)$$

The relative momentum spread is generated by the linac upstream of the chicane. The energy gain is:

$$\Delta E(z) = e\Delta V_0 \cos \phi_{RF} = e\Delta V_0 \cos(\omega_{RF} t) = e\Delta V_0 \cos(k_{RF} z) \qquad (4.23)$$

From this, the z-correlated relative momentum spread is calculated in the ultra-relativistic approximation:

$$\delta = \frac{\Delta p_z}{p_{z,0}} \approx \frac{\Delta E}{E_0} = \frac{1}{E_0} \frac{dE}{dz} \Delta z_i = -\frac{l_{b,i}}{E_0} k_{RF} e\Delta V_0 \sin \phi_{RF} \equiv -l_{b,i} h_z \qquad (4.24)$$

The quantity:

$$h_z := \frac{1}{E_0} \frac{dE(z)}{dz} = \frac{e\Delta V_0}{E_0} k_{RF} \sin \phi_{RF} \qquad (4.25)$$

is the *linear energy chirp*, relative to the beam mean energy E_0 at the chicane. When the intrinsic beam energy spread is much smaller than the energy spread imparted by the linac, it results $h \approx \frac{\sigma_\delta}{\sigma_{z,i}}$.

The *linear compression factor* is defined as the ratio of initial and final bunch length. It is calculated by inserting Eq. 4.24 into Eq. 4.22:

$$l_{b,f} = l_{b,i} \left(1 - \alpha_c L_1 h_z\right)$$

$$\Rightarrow C := \left| \frac{l_{b,i}}{l_{b,f}} \right| = \frac{1}{|1 - \alpha_c L_1 h_z|}$$

(4.26)

In summary, the curved path in dipole magnets forces particles to shift one respect to another in proportion to their relative energy deviation, and despite their ultra-relativistic velocity. If the product of momentum compaction and relative momentum deviation is large enough, the shift can be comparable to the initial bunch length, which is therefore reduced. In fact, the bunch is shortened if $sign(\alpha_c h_z) > 0$ and still $\alpha_c L_1 h_z < 1$. Since $\alpha_c < 0$ in a 4-dipole chicane (more energetic particles are bent by smaller angles, thus they run shorter path lengths), it must be $h_z < 0$. This means that particles ahead of the on-energy particle ($dz > 0$) must have lower energy ($dE < 0$).

As a quantitative case study, let us consider an electron beam accelerated off-crest by an S-band ($f_{RF} = 3$ GHz) linac of total peak voltage $\Delta V_0 = 500$ MV. The average beam energy at the chicane is specified to be $E_0 = 1$ GeV. The on-energy path length through the chicane is $L_1 = 10$ m, and $\alpha_c = -0.01$. If $C = 10$ is requested, for example, then Eq. 4.26 provides $\phi_{RF} = \arcsin(-0.2865) = -16.6°$ far from the accelerating crest ($h_z = -9 \, \text{m}^{-1}$).

4.2.5 Phase Stability in a Synchrotron

Particles in a high energy synchrotron show longitudinal oscillations in analogy to those in a low energy linac [1]. The driving force is also in this case the RF field. However, since all particles in the synchrotron travel at approximately the same velocity $v_z \approx c$, their relative slippage is not due anymore to the spread of velocities, but to the different curved path lengths induced by a spread in longitudinal momentum, in accordance to Eq. 2.5. Conditions of longitudinal stability will be analysed in the following for constant beam mean energy. In practice, the energy gain per turn provided by the RF cavity is assumed to replenish the energy loss due to radiation emission in dipole magnets. The case of energy ramp is discussed then.

The synchronous particle is, by definition, on-energy, and its revolution frequency ω_s is a sub-harmonic of ω_{RF}. Its RF phase ψ_s determines a constant energy gain per turn $(\delta E)_s$:

$$\omega_{RF} = \frac{d\psi_s}{dt} \equiv h\omega_s = -h\frac{d\theta_s}{dt}$$

$$(\delta E)_s = qV_0 \cos \psi_s$$

(4.27)

and θ_s is the deflection angle along the synchrotron circumference. The coefficient $h \in \mathbb{N}$, called *harmonic number*, is the number of RF cycles per revolution period. It is typically in the range 10^2–10^3 for $f_{RF} = 0.1 - 0.5$ GHz. The minus sign in the upper equation is to show that earlier particles with respect to the synchronous one are at smaller absolute phases.

The reduced variables to describe the longitudinal motion of the generic particle are $\phi = \psi - \psi_s$ and $w = \frac{2\pi \delta E}{\omega_s}$. Their first derivative with respect to time is:

$$
\begin{cases}
\frac{d\phi}{dt} = \frac{d(\psi - \psi_s)}{dt} = -h\frac{(d\theta - d\theta_s)}{dt} = -h\Delta\omega \\
\frac{dw}{dt} \approx \frac{\Delta(\delta E)}{T_0}T_0 = \delta E(\psi) - \delta E(\psi_s) = qV_0(\cos\psi - \cos\psi_s)
\end{cases}
\tag{4.28}
$$

where the time-variation of the energy gain difference is taken over a revolution period T_0.

The derivative of the relative phase in Eq. 4.28 is manipulated by replacing $\Delta\omega$ with η (see Eq. 4.19). Doing so, the longitudinal momentum deviation Δp_z in the definition of the slip factor is approximated with the total momentum deviation ($\Delta p_x, \Delta p_y << \Delta p_z \approx \Delta p$). Then, the total momentum deviation is converted to total energy deviation:

$$
dp = d(\beta\gamma)m_0 c = d\left(\gamma\sqrt{1 - \frac{1}{\gamma^2}}\right)m_0 c = m_0 cd\left(\sqrt{\gamma^2 - 1}\right) =
$$
$$
= m_0 c\frac{d\left(\sqrt{\gamma^2 - 1}\right)}{d\gamma}d\gamma = m_0 c\gamma\frac{\sqrt{1-\beta^2}}{\beta}d\gamma = \frac{m_0 c^2}{\beta c}d\gamma = \frac{dE}{\beta c}
\tag{4.29}
$$

The Top row of Eq. 4.28 becomes:

$$
\frac{d\phi}{dt} = -h\eta\omega_s\frac{\Delta p_z}{p_{z,s}} \approx -\frac{h\eta\omega_s}{p_{z,s}}\frac{\Delta E}{\beta_s c} \equiv \frac{h\eta\Delta E}{R_s p_s} = -\frac{h\eta\omega_s}{2\pi R_s p_{z,s}}w;
$$
$$
\frac{d^2\phi}{dt^2} = -\frac{h\eta\omega_s}{Cp_{z,s}}\frac{dw}{dt} = -\frac{h\eta\omega_s qV_0}{Cp_{z,s}}(\cos\psi - \cos\psi_s)
\tag{4.30}
$$

The "equivalent curvature radius" R_s for the synchronous particle was introduced to make the polygonal path through the synchrotron a circumference of total length $C = \frac{2\pi\beta_s c}{\omega_s} = 2\pi R_s$. It should be noted that, in general, R_s is different from the local curvature radius of the individual dipole magnets.

In the approximation of *small oscillation amplitudes*, i.e., $\psi \approx \psi_s$ and $w \approx w_s$, the second time-derivative of ϕ and w becomes:

$$
\begin{cases}
\frac{d^2\phi}{dt^2} \approx \frac{h\eta\omega_s qV_0}{Cp_{z,s}}\Delta(\cos\psi_s) = -\frac{h\eta\omega_s qV_0}{Cp_{z,s}}\sin\psi_s \cdot \phi \equiv -\Omega_s^2\phi \\
\frac{d^2w}{dt^2} \approx -qV_0\frac{d}{dt}\Delta(\cos\psi_s) = qV_0\sin\psi_s\frac{d\phi}{dt} = -\frac{h\eta\omega_s qV_0}{Cp_{z,s}}\sin\psi_s \cdot w = -\Omega_s^2 w
\end{cases}
\tag{4.31}
$$

We introduced the *synchrotron angular frequency*:

$$
\Omega_s := \frac{2\pi}{T_s} = \sqrt{\frac{qV_0\eta\omega_{RF}\sin\psi_s}{Cp_{z,s}}}
\tag{4.32}
$$

This has the same form of the angular frequency derived in Eq. 4.8 for weakly relativistic motion in a linac, but here replacing $E_{z,0} \to \eta V_0/C$.

In conclusion, when the bunch width in phase and energy is small, i.e., $\Delta z_b \ll \lambda_{RF}$ and $\delta \ll 1$, the longitudinal motion in a synchrotron is stable if $\Omega_s^2 > 0$. In this case, the particle's motion in the longitudinal phase space (ϕ, w) describes an ellipse. Its projection onto the ϕ and the w-axis corresponds to Eq. 4.31. The number of synchrotron oscillations per revolution period is named *synchrotron tune*, $Q_s = \frac{\Omega_s}{\omega_s} = \frac{\Omega_s}{(\beta_s c/R_s)}$. For $\Omega_s \sim$ kHz and $\omega_s \sim$ MHz, one synchrotron oscillation is completed in ~ 1000 turns or so.

4.2.6 Constant of Motion

Equation 4.30 describes particle's motion for arbitrarily large oscillation amplitudes. The driving force can be derived from a scalar potential:

$$\frac{d^2\phi}{dt^2} = F(\phi) = -\frac{dU(\phi)}{d\phi},$$

$$\Rightarrow U(\phi) = -\int_0^\phi F(\phi)d\phi = -\frac{\Omega_s^2}{\sin\psi_s}(\sin\psi - \phi\cos\psi_s) + I_0 \tag{4.33}$$

The constant of integration I_0 can be found as function of the particle's phase ϕ and angular frequency $\dot\phi$, by multiplying both terms of Eq. 4.30 for $\dot\phi$, by integrating them in dt, and finally integrating the r.h.s in $d\phi$:

$$\dot\phi\ddot\phi = \frac{\Omega_s^2}{\sin\psi_s}[\cos(\phi + \psi_s) - \cos\psi_s]\dot\phi;$$

$$\frac{d}{dt}\left(\frac{\dot\phi^2}{2}\right) = \frac{\Omega_s^2}{\sin\psi_s}[\cos(\phi + \psi_s) - \cos\psi_s]\frac{d\phi}{dt}; \tag{4.34}$$

$$\frac{\dot\phi^2}{2} - \frac{\Omega_s^2}{\sin\psi_s}(\sin\psi - \phi\cos\psi_s) = T(\dot\phi) + U(\phi) = I_0$$

The first term of Eq. 4.34-bottom row depends only from the "phase velocity" $\dot\phi$, therefore it can be associated to a kinetic energy $T(\dot\phi)$. The second term, made of an oscillatory and a linear function of ϕ, can be interpreted as a potential energy $U(\phi)$. Consequently, I_0 is the particle's total energy, in proper units of the reduced variables, and it is a constant of motion. This fact should not surprise because the non-interacting particles beam in the accelerator behaves as a conservative system, as long as the particles' total energy does not vary on average over one turn. The individual kinetic and potential contributions to the total energy are, of course, allowed to vary, which makes synchrotron oscillations to happen.

Figure 4.6 offers a graphical representation of the energy gain (Eq. 4.27) and of the RF potential (Eq. 4.33) as function of the particle's phase. The potential is evaluated for three values of the synchronous phase, $\psi_s = \pi/2, 2\pi/3, 5\pi/6$. In proximity of ψ_s, $U(\phi)$ shows a *potential well*, around which the motion remains stable, i.e.,

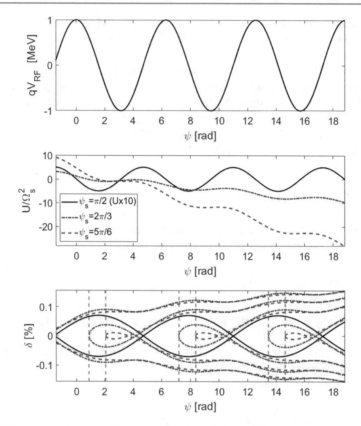

Fig. 4.6 From top to bottom: effective energy gain, normalized RF potential, and separatrix, as function of the particle's phase. U and δ are plotted for 3 values of the synchronous phase. The potential for the stationary bucket ($\psi_s = \pi/2$) is amplified by a factor 10 for better visibility of the potential well. In this figure, $E_0 = 3$ GeV, $T_0 = 1\,\mu s$, $qV_0 = 1$ MeV, $\Omega_s = 3$ kHz

the orbit in the longitudinal phase space is closed (ellipse) because the RF force is restoring. This is the case of the small oscillation amplitudes in Eq. 4.31.

We anticipate that, as shown in the figure, the width of the potential well, both in phase and in amplitude, depends from the value of ψ_s. The absolute level of the well does not change over time for $\psi_s = \pi/2$, which corresponds to null acceleration (see Eq. 4.27). For different values of ψ_s, instead, the level of the potential well varies over consecutive RF cycles, though the beam energy is still approximately constant on average over each individual turn.

4.2.7 RF Acceptance

For sufficiently large oscillation amplitudes in Eq. 4.30, the particle's orbit in the longitudinal phase space can be unbounded. The phase space trajectory which constitutes the boundary between bounded and unbounded motion is called *separatrix*.

Along such special trajectory, the particle turns back in at least one point. This has coordinates $[\psi = 2\pi - \psi_s, \dot{\phi} = 0]$, in correspondence of which the RF force goes to zero (such a phase makes $\ddot{\phi} = 0$ in Eq. 4.30). When the turning point is crossed, the force is not restoring anymore, thus leading to particle loss.

The equation of the seperatrix is found by evaluating I_0 in Eq. 4.34 at the generic particle's phase and velocity, and at the aforementioned turning point:

$$\frac{\dot{\phi}^2}{2} - \frac{\Omega_s^2}{\sin\psi_s}(\sin\psi - \phi\cos\psi_s) = -\frac{\Omega_s^2}{\sin\psi_s}[\sin(2\pi - \psi_s) - 2(\pi - \psi_s)\cos\psi_s];$$

$$\frac{\dot{\phi}^2}{2} = \frac{\Omega_s^2}{\sin\psi_s}[(\sin\psi + \sin\psi_s) - (\psi + \psi_s - 2\pi)\cos\psi_s];$$

$$\frac{\Omega_s^4}{2(qV_0\sin\psi_s)^2}w^2 = \frac{\Omega_s^2}{\sin\psi_s}[(\sin\psi + \sin\psi_s) - (\psi + \psi_s - 2\pi)\cos\psi_s];$$

$$\Rightarrow w(\psi)_{sep} = \pm\frac{qV_0}{\Omega_s}\sqrt{2\sin\psi_s[(\sin\psi + \sin\psi_s) - (\psi + \psi_s - 2\pi)\cos\psi_s]}$$

$$(4.35)$$

where w^2 was introduced in place of $\dot{\phi}^2$ by virtue of Eq. 4.30-top row and of the definition of Ω_s in Eq. 4.32.

Since the bunched beam is typically characterized in terms of relative momentum spread (which is approximately the relative energy spread for ultra-relativistic velocities, see Eq. 4.29), we replace $w(\psi)$ in Eq. 4.35 with $\delta \approx \frac{\delta E}{E_0}$ to find:

$$\delta_{sep}(\psi) = \pm\frac{\omega_s}{2\pi}\frac{w}{E_0} = \pm\frac{\omega_s}{2\pi\Omega_s}\frac{qV_0}{E_0}f(\psi, \psi_s) =$$

$$(4.36)$$

$$= \pm\frac{1}{2\pi Q_s}\frac{qV_0}{E_0}\sqrt{2\sin\psi_s[(\sin\psi + \sin\psi_s) - (\psi + \psi_s - 2\pi)\cos\psi_s]}$$

The region of longitudinal phase space delimited by the separatrix (see Fig. 4.6) is characterized by bounded and therefore stable motion. It is called *RF bucket*. Outside the RF bucket, the motion is unbounded. The maximum extension of the RF bucket in phase and relative energy deviation is said, respectively, *phase acceptance* and *RF energy acceptance* of the accelerator. By virtue of the periodicity of the RF voltage, the number of identical RF buckets is just the harmonic number, $h = \omega_R F/\omega_s$, and the time separation of the buckets is $\Delta t = 1/f_{RF}$.

According to Eq. 4.30, the RF force becomes null also at the synchronous phase, $\ddot{\phi}(\psi_s) = 0$, for which $\dot{\phi}$ reaches a maximum (see Fig. 4.6). The separatrix δ_{sep} in Eq. 4.36 evaluated at $\psi = \psi_s$ corresponds to the *bucket half-height*, or RF energy acceptance δ_{acc}. This is quantified below for an arbitrary synchronous phase ψ_s, by recalling $Q_s = R_s\Omega_s/(\beta_s c)$ and the definition of Ω_s in Eq. 4.32:

$$\delta_{acc}(\psi_s) = \delta_{sep}(\psi = \psi_s) = \pm\frac{\beta_s c}{2\pi R_s\Omega_s}\frac{qV_0}{E_0}\sqrt{(2\sin\psi_s)[2\sin\psi_s - 2(\psi_s - \pi)\cos\psi_s]} =$$

$$= \pm\frac{\beta_s c}{2\pi R_s}\frac{qV_0}{E_0}\sqrt{\frac{2\pi R_s p_{z,s}}{qV_0\eta h\omega_s\sin\psi_s}}\sqrt{2\sin\psi_s}\,G(\psi_s) = \pm\sqrt{\frac{\beta_s}{\pi h\eta}}\sqrt{\frac{qV_0}{E_0}}\,G(\psi_s),$$

$$G(\psi_s) := \sqrt{2[(\psi_s - \pi)\cos\psi_s - \sin\psi_s]}$$

$$(4.37)$$

4.2.8   Stationary Bucket

One could now wonder which is the value of ψ_s that maximizes in absolute sense $G(\psi_s)$, i.e., the RF energy acceptance. First, the range of ψ_s which ensures phase stability has to be identified. This is illustrated in Fig. 4.4: $0 \leq \psi_s \leq \pi/2$ for $\eta < 0$ and $3\pi/2 \leq \psi_s \leq 2\pi$ for $\eta > 0$. Since $G(\psi_s)$ is a monotonic function of ψ_s, it is maximized by $\psi_s = \pi/2$ above transition, and by $\psi_s = 3\pi/2$ below transition. In both cases, $\hat{G} = \sqrt{2}$. Both those phases correspond to null acceleration (see Eq. 4.27), and the RF bucket is said *stationary*.

For values of $\psi_s \neq \pi/2, 3\pi/2$, the absolute level of the potential well changes with the phase as a consequence of some net acceleration over time, the RF bucket area is reduced and made asymmetric in phase (see Fig. 4.6). This suggests that for maximizing the beam injection efficiency from a booster ring, i.e. to maximize the number of particles collected by the storage ring in the RF bucket, the injection should be accomplished at fixed beam energy in the main accelerator (i.e., maximum RF bucket area).

Let us assume a storage ring above transition. The *maximum* RF energy acceptance is found from Eq. 4.36 evaluated at $\psi = \psi_s = \pi/2$, or equivalently from 4.37 evaluated at $\psi_s = \pi/2$:

$$|\hat{\delta}_{acc}| = \frac{1}{\pi Q_s}\frac{qV_0}{E_0} = \sqrt{\frac{2\beta_s}{\pi h|\eta|}}\sqrt{\frac{qV_0}{E_0}} \qquad (4.38)$$

Equation Eq. 4.36 is particularly helpful to calculate the total *bucket area* of the stationary bucket, which therefore results the maximum bucket area among all possible synchronous phases. At first, we evaluate the separatrix equation for $\psi_s = \pi/2$:

$$\delta_{sep}(\psi_s = \pi/2) = \pm\frac{1}{2\pi Q_s}\frac{qV_0}{E_0}\sqrt{2(1+\sin\psi)} = \pm\frac{1}{2\pi Q_s}\frac{qV_0}{E_0}\sqrt{2(1+\cos\phi)} =$$

$$= \pm\frac{1}{\pi Q_s}\frac{qV_0}{E_0}\sqrt{\cos^2\frac{\phi}{2}} = \pm\frac{1}{\pi Q_s}\frac{qV_0}{E_0}\cos\frac{\phi}{2} = \hat{\delta}_{acc}\cos\frac{\phi}{2}$$

$$(4.39)$$

The bucket area can now be calculated:

$$A_{bk} = 2\int_{-\pi}^{\pi}\delta_{sep}(\phi)d\phi = 2|\hat{\delta}_{acc}|\int_{-\pi}^{\pi}\cos\frac{\phi}{2}d\phi = \frac{2}{\pi Q_s}\frac{qV_0}{E_0}\cdot 4 = 8|\hat{\delta}_{acc}| \quad (4.40)$$

The phase acceptance $\delta\phi_{acc}$ can also be calculated as the distance in phase of the two points at which the separatrix crosses the horizontal axis. For the stationary bucket, Eq. 4.39 says that this happens at $[-\pi, 0]$ and $[\pi, 0]$, so that $\delta\phi_{acc} = 2\pi$. Any different synchronous phase from that one of the stationary bucket implies a smaller acceptance (see Fig. 4.6).

4.2.8.1 Discussion: Short Bunches in a Storage Ring
Beam dynamics in a synchrotron is periodic, and particles can be stored for long time. In this case, they emit radiation continuously, and such emission leads to equilibrium

beam sizes independent from the beam parameters at injection. Since the previous analysis applies to the particles' motion in a steady-state condition, we want to use the constant of motion to find the relationship between bunch duration and energy spread at equilibrium, in the approximation of small oscillation amplitudes.

Equation 4.31 says that the generic particle's orbit in the longitudinal phase space $(\phi, \dot{\phi})$ is an ellipse. Let us consider the particle representative of the bunch envelope. The ellipse horizontal and vertical semi-axis has coordinates $[\phi_{max}, 0]$ and $[0, \dot{\phi}_{max}]$. In the approximation of small amplitudes of oscillation ($\phi \approx 0$) in the stationary bucket ($\psi_s = \pi/2$), Eq. 4.34 becomes:

$$\frac{\dot{\phi}^2}{2} - \Omega_s^2 \left[\sin(\phi + \tfrac{\pi}{2}) - \phi \cos \tfrac{\pi}{2}\right] = \frac{\dot{\phi}^2}{2} - \Omega_s^2 \cos \phi \approx \frac{\dot{\phi}^2}{2} - \Omega_s^2 \left(1 - \frac{\phi^2}{2}\right) = const.$$

$$\Rightarrow \dot{\phi}^2 + \Omega_s^2 \phi^2 = I_1 = const.$$

$$(4.41)$$

At the two extremes of the semi-axis, Eq. 4.41 becomes:

$$\begin{cases} \phi = 0 \Rightarrow \dot{\phi} = \dot{\phi}_{max} = \sqrt{I_1} = -\left(\frac{h\eta\omega_s}{p_s C}\right) w_{max} \\ \dot{\phi} = 0 \Rightarrow \Omega_s \phi = \Omega_s \phi_{max} = \sqrt{I_1} = -\left(\frac{h\eta\omega_s}{p_s C}\right) w_{max} \end{cases} \quad (4.42)$$

and we made use of the expression of $\dot{\phi}$ in Eq. 4.30. From the bottom equation it follows:

$$\phi_{max} = -\left(\frac{h\eta\omega_s}{p_s C}\right) \frac{w_{max}}{\Omega_s} = -\frac{h\eta c}{\Omega_s R_s} \frac{w_{max}\omega_s}{2\pi E_0} \approx \left(\frac{h\alpha_c}{Q_s}\right) \delta_{max} \quad (4.43)$$

The very last equality is obtained by recalling the definition of Q_s and the approximation is for particles at ultra-relativistic velocities above transition energy ($\beta_s \to 1$, $\eta \to -\alpha_c$).

The full-width bunch duration is finally estimated:

$$\Delta t_b = \frac{2\phi_{max}}{\omega_{RF}} = (2\delta_{max}) \frac{h\alpha_c \omega_s}{h\omega_s \Omega_s} = (2\delta_{max})\alpha_c \sqrt{\frac{p_s C}{h\omega_s \alpha_c q V_0}} \propto \sqrt{\frac{\alpha_c}{\omega_{RF} V_0}} \quad (4.44)$$

This result suggests that bunch duration in a synchrotron can be shortened by means of an arrangement of the lattice to produce a small momentum compaction—so-called *low-α* mode of operation—and/or a high peak RF voltage at any fixed RF frequency (i.e., a high time-slope of the accelerating voltage).

For typical parameters of a modern electron light source like $V_0 \approx 2$ MV, $E_0 \sim 3$ GeV and $Q_S \sim 0.003$, Eq. 4.38 predicts $\delta_{acc} \approx 7\%$. In reality, nonlinearities associated to higher order terms of the momentum compaction may severely limit the RF energy acceptance, often reduced to 2–4%.

The corresponding electron bunch duration can be estimated by considering that the beam energy spread at equilibrium is typically a small fraction of the RF energy acceptance, and often close to $\delta_{max} = 0.1\%$. We then assume $\omega_s \approx 10$ MHz, $h = 800$ and $\alpha_c = 10^{-3}$. It results $\Delta t_b \approx 23$ ps. Most recent storage rings with 10-fold smaller α_c (so-called "fourth generation") can accommodate 3-fold shorter bunches at low currents. However, RF cavities tuned at a higher harmonic of the fundamental frequency are commonly used to flatten the potential well, hence to elongate the bunch, diluting the charge density, to eventually minimize particles' interaction internal to the bunch. Typical elongations by a factor 3–5 can force the bunch duration to $\sim 60 - 150$ ps fwhm, for ring circumferences in the range 0.2– 2 km, and beam energies in the range 1–6 GeV.

4.2.9 Energy Ramp

We consider a synchrotron in energy ramp mode, where the beam energy is increased turn-by-turn by virtue of "excess" of energy gain in the RF cavity. During the energy ramp, the closed orbit in the accelerator has to be kept (approximately) constant because the local curvature radius r in the dipole magnets is fixed, as dictated by the geometry of the vacuum chamber. By recalling the relationship between longitudinal momentum and curvature radius in the presence of Lorentz's force in Eq.2.5, the aforementioned condition implies a variation of the dipole magnetic field with time, $\dot{p}_z = q\dot{B}_y r$. Hence, the energy gain per turn can be expressed as follows:

$$(\Delta E)_{turn} = (\Delta p_z)_{turn}\frac{c^2}{v_z} = q\dot{B}_y r T_0 c = q\dot{B}_y r 2\pi R_s \equiv qV_0 cos(\psi_s - \psi_0)$$

$$\Rightarrow \psi_s(t) = \psi_0 + arccos\left(2\pi R_s r \frac{\dot{B}_y}{qV_0}\right)$$

$$(4.45)$$

We find that the synchronous phase must vary during the energy ramp, according to the ramp of the magnetic field.

Moreover, by virtue of a non-zero slip factor (whose meaning applies in this case to the motion of the synchronous particle in the presence of variation of the reference energy), we expect that also the revolution frequency varies. This has to be kept synchronous to the radiofrequency, which therefore has to vary as well:

$$f_{RF}(t) = \frac{h\omega_s(t)}{2\pi} = \frac{h}{2\pi R_s}\beta_s(t)c = \frac{h}{2\pi R_s}\frac{p_{z,s}(t)}{\gamma(t)m_0} = \frac{h}{2\pi R_s}\frac{qB_y(t)rc^2}{\sqrt{(p_{z,s}c)^2+(m_0c^2)^2}} =$$

$$= \frac{h}{2\pi R_s}\frac{qB_y(t)rc^2}{\sqrt{(qB_y(t)rc)^2+(m_0c^2)^2}} = \frac{hc}{2\pi R_s}\left(\frac{qr}{m_0c}\right)\frac{B_y(t)}{\sqrt{\left[1+\left(\frac{qr}{m_0c}\right)^2 B_y(t)^2\right]}}$$

$$(4.46)$$

For $p_z c \gg m_0 c^2$, $f_{RF}/h \to f_s$, i.e. the revolution frequency in the ultra-relativistic limit. As an example, Fig. 4.7 shows the variation with time of B_y, Δf_{RF} and ψ_s for protons in energy ramp in the Tevatron collider.

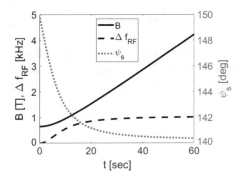

Fig. 4.7 The protons' kinetic energy in the Tevatron collider is assumed to be varied from 150 GeV to 980 GeV in 1 minute. The dipole field increases over time with a parabolic ramp. The synchronous phase and the radiofrequency are varied as well. The following parameters were considered: 6.28 km-long circumference, 10 m-long superconducting dipole magnets, $h = 1110$, and 0.74 deg dipole bending angle (kept constant during the energy ramp)

4.2.9.1 Discussion: Tevatron Proton Collider

The Tevatron is a proton-antiproton synchrotron collider, initially designed for a kinetic energy ramp 10–400 GeV, later on upgraded to 150–980 GeV. It was put in shut down in 2011. The ring circumference is $C = 6.28$ km, the harmonic number is $h = 1110$, and the RF peak voltage $V_0 = 2$ MV. The superconducting dipole magnets are 10 m-long, each dipole bending by 0.74 deg. We assume the energy ramp is concluded in $\Delta t_r = 1$ min. Finally, the gamma-transition is set to 10.

We would like to calculate: (i) the variation of the radio-frequency during the energy ramp in the range 10–400 GeV, and in the range 150–980 GeV, assuming the dipole curvature radius is kept constant, (ii) the average energy gain per turn and the rate of increase of the dipoles' field, for the upgraded energy ramp, and (iii) the synchrotron frequency at the initial and at the final energy, for the upgraded energy ramp.

Δf_{RF} can be calculated by means of Eq. 4.46, where a constant beam orbit implies $r \approx const$. The curvature radius is $r = 10$ m/0.013 rad $= 775$ m. The dipole field at the beginning and at the ed of the energy ramp is calculated through $p_z = eB_y r$. The field ranges [0.05–1.73] T in the low energy ramp, and [0.65–4.22] T in the high energy ramp. The corresponding variation of radiofrequency is $\Delta f_{RF} = -195$ kHz and $\Delta f_{RF} = -1$ kHz, respectively. The main RF is in the range 45–53 MHz.

The average energy gain per turn is calculated from the variation of the longitudinal momentum corresponding to the kinetic energy range 150–980 GeV. Since the protons are ultra-relativistic, we can write $\langle (\Delta E)_{turn} \rangle = \langle \beta c (\Delta p)_{turn} \rangle \approx c \Delta p \frac{T_0}{\Delta t_r} = 0.3$ MeV. The field rate is retrieved from the momentum rate, $(\Delta p)_{turn} \approx er\dot{B}T_0$, which provides $\dot{B} \approx 0.06$ T/s.

Finally, Eq. 4.32 allows us to calculate the synchrotron frequency at the initial and final beam energy. But first, the slip factor and the synchronous phase have to be found. For the former, we assume that the linear optics remains substantially unchanged during the energy ramp, which implies a constant momentum compaction $|\alpha_c| = 1/\gamma_{tr}^2 = 0.01$. Since the γ-factor amounts to $E/(m_p c^2) = [161, 1046]$ at the two kinetic energies of 150 GeV and 980 GeV, the slip factor's absolute value is substantially α_c.

The synchronous phase is retrived from the peak voltage and the energy gain per turn calculated above, which have to satisfy $(\Delta E)_{turn} = eV_0 \cos \psi_s$. Hence, $\psi_s = 1.4$ rad (approximately 81° far from the accelerating crest), and $\frac{\Omega_s}{2\pi} = [0.39, 0.16]$ kHz.

4.2.10 Summary

Single particle longitudinal dynamics in a high energy synchrotron can be summarized as follows.

- When the approximation of small oscillation amplitudes is met, i.e., $\Delta z_b \ll \lambda_{RF}$ and $\delta \ll 1$, the motion is stable for those synchronous phases which make $\Omega_s^2 > 0$ (Eq. 4.32).
- For large amplitudes of oscillation, the motion is stable within the phase space region delimited by the separatrix, namely, the RF bucket (Eq. 4.36). Its half-height is the RF energy acceptance of the accelerator, which depends from the synchronous phase.
- The synchronous phase which maximizes the RF bucket area is that for null acceleration (stationary bucket, Eq. 4.39). The bucket area is linearly proportional to the bucket height. The phase acceptance is also maximum and equal to 2π.
- The RF energy acceptance is weakly dependent but still maximized by a large product $\omega_{RF} V_0$, and small α_c (Eq. 4.38). However, for any equilibrium energy spread, the bunch duration shows the opposite dependence and therefore it is squeezed in proportion to the enlargement of the RF energy acceptance (Eq. 4.44).
- When operated in energy ramp mode, the synchronous phase and the frequency of the RF cavity have to vary with time in order to keep the closed orbit and the energy gain per turn approximately constant, and in accordance to the variation of the dipoles' magnetic field (Eqs. 4.45 and 4.46).

4.3   Transverse Dynamics

4.3.1   Multipolar Field Expansion

The sequence of magnetic elements of an accelerator is called *magnetic lattice* [2,3]. Dipole and quadrupole magnets allow control of the particles' transverse motion at first order in the particle's coordinates. Their principle of operation is sketched in Fig. 4.8. In synchrotrons, other multipole magnets are adopted for control of the dynamics at higher orders. This suggests the need for a description of the magnetic field components through an expansion around the reference orbit.

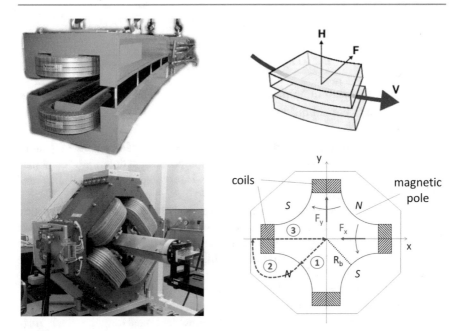

Fig. 4.8 Top: 2 m-long dipole magnet of the Elettra booster ring. Cu-coils surrounding the north and south Fe-pole generate a magnetic field (H) which bends charged particles via Lorentz's force (F). Bottom: quadrupole magnet on a rotating coil measurement bench. On the right, black arrows are magnetic field lines, red arrows are forces exerted on a positive charge entering the plane, and blue lines show a path concatenated to the coils. (Photos credit D. Castronovo, Elettra Sincrotrone Trieste)

Let us consider an electromagnet with Fe-poles at distance R_b from the orbit, and centered on it. R_b is called *bore radius*; in the case of a dipole magnet bending in the horizontal plane, it corresponds to its half-gap. If we describe $B(x, y)$ as an analytical continuous function of the spatial coordinates, we can expand it in Taylor series around the reference orbit. For the vertical field component in proximity of $y = 0$:

$$
\begin{cases}
B_y(x) \approx b_0 + b_1 \left(\frac{x}{R_b}\right) + b_2 \left(\frac{x}{R_b}\right)^2 + \ldots = \sum_0^\infty \frac{R_b^m}{m!} \left(\frac{\partial^m B_y}{\partial x^m}\right) \left(\frac{x}{R_b}\right)^m \\
b_m := \frac{R_b^m}{m!} \left(\frac{\partial^m B_y}{\partial x^m}\right)_{y=0} \qquad m \in \mathbb{N}
\end{cases}
\tag{4.47}
$$

m is the order of expansion, where $m = 0$ is for a dipole, $m = 1$ for a quadrupole, $m = 2$ for a sextupole, etc.

Magnets which generate a Lorentz's force dependent *only* from the coordinate in the force plane ($F_x = F_x(x)$, $F_y = F_y(y)$) are said *normal*. In *skew* magnets, $F_x = F_x(x, y)$, $F_y = F_y(x, y)$. The most general superposition of normal and skew field components, at any order, can be written in a compact complex notation:

$$B_y + i B_x = \sum_0^\infty (b_m + i a_m) \left(\frac{x+iy}{R_b}\right)^m \tag{4.48}$$

where b_m are non-zero in normal magnets and a_m are non-zero in skew magnets. For example, in a normal quadrupole $b_1 = R_b \frac{\partial B_y}{\partial x}$, $a_1 = 0$. In a skew quadrupole, $b_1 = 0$, $a_1 = R_b \frac{\partial B_x}{\partial x}$.

Magnets built on purpose with a single field order are named "separate function" magnets. On the opposite, magnets built with superposed field orders are "combined function" magnets. The most common of this kind are dipole magnets with integrated quadrupole field gradient, hence $B_y = b_0 + b_1 \frac{x}{R_b}$, $B_x = b_1 \frac{y}{R_b}$.

Magnets deviate the particles' trajectory in analogy to lenses in geometric optics. A magnetic lattice made of dipole and quadrupole magnets only implies forces at most linear with the particle's coordinates, and the particle's transverse motion is said *linear optics*. Sextupole, octupole and higher order magnets contribute to the *nonlinear optics*.

Maxwell's equations forbid the abrupt transition from a magnetic field region to a field-free region. Indeed, the region in proximity of the magnets' edges contain residual field components, not described by the aforementioned formulas. Such *fringe fields* typically contribute poorly to the overall focusing property of the magnet. However, they can be important when a large number of magnets is installed in a tight lattice. In this case, interference of fringe fields of adjacent magnets can lead to a noticeable modification of the optics compared to the ideal "hard-edge" field model.

4.3.2 Quadrupole Magnet

A quadrupole magnet is made of four center-symmetric Fe-poles surrounded by coils. These generate the magnetic field, which is deviated by the high relative permeability of the poles towards the center of the magnet, see Fig. 4.8-right plot. For symmetry, the field is null at the exact center of the magnet, while the field lines are more dense in proximity of the poles. Moreover, owing to $\vec{\nabla} \times \vec{B} = 0$ in the region internal to the magnet, a normal quadrupole shows $\frac{\partial B_y}{\partial x} = \frac{\partial B_x}{\partial y}$. This implies that a positive gradient $g = \frac{\partial B_y}{\partial x}$ acting as a restoring or focusing force in the horizontal plane, corresponds to a defocusing force in the vertical plane, and viceversa. Let us verify this with the notation introduced in Eq. 4.48:

$$B_y + i B_x = b_1 \frac{x}{R_b} - a_1 \frac{y}{R_b} + i b_1 \frac{y}{R_b} + i a_1 \frac{x}{R_b} \tag{4.49}$$

$$\Rightarrow \begin{cases} B_y = b_1 \frac{x}{R_b} = \frac{\partial B_y}{\partial x} x \equiv gx \\ \\ B_x = b_1 \frac{y}{R_b} = \frac{\partial B_y}{\partial x} y = \frac{\partial B_x}{\partial y} y = gy \end{cases} \tag{4.50}$$

The Lorentz's force results:

$$\vec{F} = \begin{vmatrix} \hat{x} & \hat{y} & \hat{z} \\ 0 & 0 & v_z \\ gy & gx & 0 \end{vmatrix} = -(qv_z gy)\hat{x} + (qv_z gx)\hat{y} \tag{4.51}$$

In summary, Lorentz's force in a normal quadrupole is linearly proportional to the particle's distance from the magnet center, equal in strength but opposite in sign along the horizontal and the vertical axis. The constant of linear proportionality is the field gradient. At any point of coordinates $|x, y| < R_b$, the field vector can be decomposed in a horizontal and a vertical component.

The field gradient can be expressed as function of R_b and of the Ampere-turns NI by evaluating the magnetic field curvilinear integral along the path in Fig. 4.8 (blue dashed arrows). For simplicity, we assume a relative permeability of the poles much larger than in vacuum ($\mu_r \gg \mu_0$):

$$\oint \vec{H} d\vec{s} = \int_1 \vec{H} d\vec{s} + \int_2 \vec{H} d\vec{s} \int_3 \vec{H} d\vec{s} = \int_0^{R_b} H(r) dr + \int_1^2 \vec{H} d\vec{s} + \int_2^3 H_y dx =$$

$$= \int_0^{R_b} \frac{1}{\mu_0} \frac{\partial B_r}{\partial r} r dr + \int_1^2 \frac{1}{\mu_r} \vec{B} d\vec{s} + 0 \approx \int_0^{R_b} \frac{g}{\mu_0} r dr = \frac{g R_b^2}{2\mu_0} \equiv NI.$$

$$\Rightarrow g = \frac{2\mu_0 NI}{R_b^2}$$

$$\tag{4.52}$$

Hence, the field gradient is made stronger by a smaller bore radius, for any given Ampere-turns. In practice, the size of R_b has to be traded off with the space required to accommodate the beam vacuum chamber.

In the literature, a quadrupole is said to be *focusing* (QF) when it pushes off-axis particles towards the axis in the *horizontal* plane. The opposite is for a *defocusing* quadrupole (QD). Because of its symmetry, a QF can be made QD by simply rotating the magnet by 90°. A normal quadrupole rotated by 45° becomes a skew quadrupole.

A QF becomes a QD for a particle which inverts its direction of motion through the magnet. Since this is equivalent to the change of sign of the particle's charge, some circular colliders have been conceived with two distinct vacuum chambers in the same magnets. The two colliding beams made of particles with opposite charge (e.g., electron-positron, electron-proton, or proton-antiproton colliders), travel on opposite directions. In this way, dipole and quadrupole magnets have exactly the same effect on the two counter-propagating beams, which therefore share the same linear optics.

4.3.3 Strong Focusing

Strong focusing, also named *Alternating-Gradient Focusing*, was first conceived by N. Christofilos in 1949 but not published. In 1952, the strong focusing principle was independently developed by E. Courant, M. S. Livingston, H. Snyder and J. Blewett at Brookhaven National Laboratory, then quickly deployed on the Alternating Gradient Synchrotron at the Brookhaven National Laboratory, USA.

The equations of motion for the generic particle are derived below in the presence of dipoles and quadrupoles only. For this reason, they are said to be in the *linear approximation*. At this stage, the geometry of the accelerator is arbitrary. The derivation is done with the following prescriptions.

- Since we are also interested to the particle's motion on a closed orbit, it is convenient to change the independent variable from t (time) to s, the latter being the longitudinal spatial coordinate along the orbit in the Frenet-Serret reference system. In a synchrotron $s \in [0, C]$. This change of variable implies $\frac{d}{dt} \rightarrow \frac{d}{ds}\frac{ds}{dt} = v_z \frac{d}{ds}$.
- The break of azimuthal symmetry of the magnetic field in a synchrotron in general implies different field values at different points along the accelerator: $B_y = B_y(s)$, $\frac{\partial B_y}{\partial x} = g(s)$.
- The generic particle is assumed to travel close to the reference orbit, $x, y \ll r, R_b$. The local curvature radius r can therefore be expanded at first order in x around the *reference* curvature radius R, or $r = R(1 + x/R)$.
- The generic particle is also assumed to be off-energy, $p_z = p_{z,s}(1 + \delta), \delta \ll 1$. The synchronous particle's longitudinal momentum $p_{z,s}$ is assumed to be constant (locally in a linac, on average over one turn in a synchrotron).

Newton's equation for the horizontal plane is:

$$m\ddot{x}(t) - m\dot{\theta}(t)^2 r = -qv_z B_y;$$

$$\ddot{x}(s) - \frac{1}{v_z^2}\frac{v_z^2}{r(s)} = -\frac{1}{v_z^2}\frac{qv_z B_y(s)}{m};$$

$$\ddot{x}(s) \approx \frac{1}{R(s)}\left[1 - \frac{x}{R(s)}\right] - \frac{qB_y(s)}{p_z};$$

$$\ddot{x}(s) \approx \frac{1}{R(s)}\left[1 - \frac{x}{R(s)}\right] - \frac{qB_0(s)}{p_{z,s}}\left[1 + \frac{g(s)}{B_0(s)}x(s)\right](1 - \delta); \qquad (4.53)$$

$$\ddot{x}(s) \approx \frac{1}{R(s)} - \frac{x}{R(s)^2} - \frac{1}{R(s)} + \frac{\delta}{R(s)} - \left[\frac{qg(s)}{p_{z,s}}\right](1 - \delta)x(s);$$

$$\Rightarrow \ddot{x}(s) + \left[k(s)(1 - \delta) + \frac{1}{R(s)^2}\right]x(s) = \frac{\delta}{R(s)}$$

We have introduced the *normalized quadrupole gradient* $k = \frac{eg}{p_z}$. In practical units:

$$k[m^{-2}] = 0.3\frac{g[T/m]}{p_{z,s}[GeV/c]} \qquad (4.54)$$

We find a similar equation for the vertical plane, for which $R \rightarrow \infty$:

$$\ddot{y}(s) = \frac{qB_x(s)}{p_z};$$

$$\ddot{y}(s) \approx \left[\frac{qg(s)}{p_{z,s}}y(s)\right](1 - \delta); \qquad (4.55)$$

$$\Rightarrow \ddot{y}(s) - k(s)(1 - \delta)y(s) = 0$$

Equations 4.53 and 4.55 are called *Hill's equations*. They describe a *quasi-harmonic* oscillator, i.e., an oscillator in which the angular frequency is a function of the independent variable s. We draw the following observations.

1. Consistently with Maxwell's equations, $k(s)$ has opposite sign in Eqs. 4.53 and 4.55. As previously observed, this means that a QF (QD) is focusing in the x-plane (y-plane) and defocusing in the y-plane (x-plane). This suggests that a sequence of QF and QD could be suitably arranged to guarantee stability along the orbit in both transverse planes.
2. The field gradient can be re-written in terms of the field index introduced for weak focusing in Eq. 2.11:

$$k = \frac{qg}{p_{z,s}} = \frac{q}{q B_0 R} \frac{\partial B_y}{\partial x} = -\frac{1}{R^2} n \qquad (4.56)$$

In this case, the dipolar and the gradient field component are intended to belong to distinct magnets (dipole and quadrupole, respectively). Since commonly $|k| >> \frac{1}{R^2}$, i.e. the focusing force by quadrupoles is much stronger than weak focusing due to the orbit curvature in the dipoles, one has $|n| >> 1$, which implies a large number of oscillations per turn. Such operating mode is named *strong focusing*, as opposite to weak focusing in betatrons and cyclotrons.
3. In the presence of normal magnets only, Hill's Eqs. 4.53 and 4.55 apply to the linearly independent variables $x(s)$, $y(s)$. Their oscillations around the orbit are called *betatron oscillations*.
4. Hill's equations are linear in δ. This leads to a residual focusing term $k\delta << k$ which constitutes a *chromatic* perturbation to the on-energy linear optics. The chromatic term on the r.h.s. of Eq. 4.53 determines a particular solution of the differential equation, due to the orbit distortion of off-energy particles in the magnets.

4.3.3.1 Discussion: FODO Lattice

Let us demonstrate that an ideal magnetostatic force focusing simultaneously in the horizontal and vertical plane cannot exist. This shall bring us to the conclusion that alternated restoring strengths can be implemented, instead, and that under certain circumstances this guarantees stability in both planes along the beam line.

Ideally, we would like to have:

$$F_x = 0, \quad F_y = -k_y y, \quad F_y = -k_z z \qquad (4.57)$$

At the same time, in the region occupied by the beam and internal to the magnet, $\vec{\nabla} \vec{B} = 0$, which implies null divergence of the total Lorentz's force:

$$\frac{\partial F_x}{\partial x} + \frac{\partial F_y}{\partial y} + \frac{\partial F_z}{\partial z} = 0 \quad \Rightarrow \quad k_y = -k_z \qquad (4.58)$$

This proves the thesis.

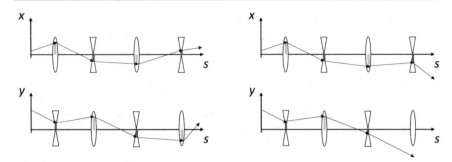

Fig. 4.9 Particle's trajectory in a FODO for a stable (left) and unstable motion (right), in the horizontal (top) and vertical plane (bottom). Focusing quadrupoles in one plane behave as defocusing quadrupoles in the other plane

The principle of alternating strong quadrupole gradients is illustrated in Fig. 4.9. A series of QFs and QDs, denominated FODO lattice in the literature, can guarantee stable motion in *both* transverse planes if a proper choice of the gradients and the distances between the magnets is done. This basically corresponds to a situation in which, in each plane, the particle's distance from the magnetic axis of QFs is systematically larger than that in QDs. Since the quadrupole force is proportional to the particle's distance from the magnet's center (see Eq. 4.51), the restoring force of QFs is always larger than the repulsive one in QDs, and the particle continues oscillating around the reference orbit.

The same lattice, however, can lead to unstable motion, in either one plane or the other or both, if the quadrupoles' gradient and their relative distance is not chosen properly. Either too weak or too strong gradients can lead to instability: the particle starts deviating more and more from the orbit, unitl it gets lost on the vauum chamber.

4.3.4   Principal Trajectories

The solution of Hill's complete equation in Eq. 4.53 is the linear superposition of the solution of the homogeneous equation $x_\beta(s)$ and the particular solution $x_D(s)$. The former is function of initial conditions x_0, x_0', and it is said to describe the *betatron motion*. The latter is proportional to the relative momentum deviation through the momentum-dispersion function, $x_D(s) = D(s)\delta$, and for this reason it is said to describe the *dispersive motion*:

$$\begin{cases} x(s) = x_\beta(s) + x_D(s) = C(s)x_0 + S(s)x_0' + D(s)\delta \\ \\ x'(s) = x_\beta'(s) + x_D'(s) = C'(s)x_0 + S'(s)x_0' + D'(s)\delta \end{cases} \quad (4.59)$$

The derivatives are all intended with respect to the independent variable s. For the sake of brevity, the suffix x of the optical functions C, S, D is suppressed. An identical formalism applies to the vertical plane.

$C(s)$, $S(s)$ are called *principal trajectories*, $D(s)$ is the dispersion function previously introduced in Eq. 4.16. C, S and D are all solutions of Hill's equation, in the homogeneous and complete form, respectively. Equation 4.59 can be re-written in vectorial form by collecting the principal trajectories in a *transfer matrix*:

$$\vec{x}(s) = \begin{pmatrix} x \\ x' \end{pmatrix}_s = \begin{pmatrix} C & S \\ C' & S' \end{pmatrix}_{0 \to s} \begin{pmatrix} x \\ x' \end{pmatrix}_0 = M_{(0 \to s)}\vec{x}_0 \qquad (4.60)$$

The determinant $\det M = CS' - SC' \equiv W(s)$ is the *Wronskian* associated to Hill's second order homogeneous differential equation.

The initial conditions of C, S are chosen as follows: $C(0) = S'(0) = 1$, $C'(0) = S(0) = 0$. Their properties are studied by introducing a "frictional" or "dissipative" term ξ in the homogeneous Hill's equation. Hereafter, we neglect the chromatic focusing error $k\delta$ for simplicity. The dissipative Hill's equations become:

$$\begin{cases} C'' + \xi C' + kC = 0 & \cdot (-S) \\ S'' + \xi S' + kS = 0 & \cdot (C) \end{cases} \qquad (4.61)$$

The two equations are multiplied as indicated in Eq. 4.61 and summed to obtain:

$$(CS'' - SC'') + \xi (CS' - SC') = 0;$$

$$W' + \xi W = 0;$$

$$\Rightarrow W(s) = W(0) \exp\left(-\int_0^s \xi(s')ds'\right)$$

$$\Rightarrow W = 1 \Leftrightarrow \xi(s) = 0 \ \forall s$$

(4.62)

We find that the Wronskian is constant (unitary) under two assumptions:

1. the motion is purely linear in the particle's coordinates (Eq. 4.53),
2. the motion is free of frictional forces (Eq. 4.62).

The physical meaning of the Wronskian is elucidated by considering a particle's trajectory in the transverse phase space (x, x'). An arbitrarily small element of phase space area around the trajectory is, in general, expressed through the cross product $A(s)\hat{k} = d\vec{x}(s) \times d\vec{x}'(s)$, with $d\vec{x}, d\vec{x}'$ the vectors pointing from the representative point of the particle to the vertices of the surface element, see Fig. 4.10.

In general, the two vectors are functions of s, i.e., we assume the existence of a *linear map* $M_{(0 \to s)}$ under which the particle's coordinates and therefore the area transform:

Fig. 4.10 Transformation of an element of phase space area under linear map

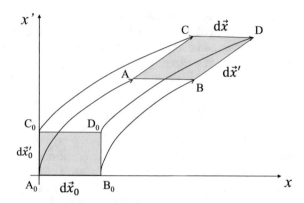

$$\begin{cases} d\vec{x} = d\vec{x}(dx_0, dx_0') \approx \left(\frac{\partial x}{\partial x_0} dx_0, \frac{\partial x}{\partial x_0'} dx_0' \right) \\ \\ d\vec{x}' = d\vec{x}'(dx_0, dx_0') \approx \left(\frac{\partial x'}{\partial x_0} dx_0, \frac{\partial x'}{\partial x_0'} dx_0' \right) \end{cases} \tag{4.63}$$

$$\Rightarrow \vec{A}(s) = \begin{vmatrix} \hat{i} & \hat{j} & \hat{k} \\ \frac{\partial x}{\partial x_0} dx_0 & \frac{\partial x}{\partial x_0'} dx_0' & 0 \\ \frac{\partial x'}{\partial x_0} dx_0 & \frac{\partial x'}{\partial x_0'} dx_0' & 0 \end{vmatrix} =$$

$$= \left(\frac{\partial x}{\partial x_0'} \frac{\partial x'}{\partial x_0} - \frac{\partial x}{\partial x_0'} \frac{\partial x'}{\partial x_0} \right) dx_0 dx_0' \hat{k} = \tag{4.64}$$

$$= (C S' - S' C)\vec{A}(0) = \vec{A}(0)$$

We conclude that the *Jacobian determinant* of the function $\vec{f} = (\vec{x}(x_0, x_0'), \vec{x}'(x_0, x_0'))$ is the Wronskian of the homogeneous Hill's equation. This implies that, under conditions 1. and 2. above, the *phase space area* in proximity of a particle's trajectory is *preserved*, i.e., it is a constant of motion. This is the enunciation of *Liouville's theorem* restricted to linear motion. The theorem will be demonstrated later on in a more general form in the framework of Hamiltonian dynamics.

4.3.5   Transfer Matrices

Hill's equation resembles that of a harmonic oscillator. This suggests solutions for C and S in the form of cos- and sin-like functions, respectively. Their amplitude and phase has to contain information on the magnetic focusing.

Our guess is in Eq. 4.65 below. We demonstrate *a posteriori* that those functions indeed satisfy the homogeneous Eqs. 4.53 and 4.55. We restrict ourselves to the case of *constant focusing* magnets, i.e., $k(s) = const.$ and $R(s) = const.$ inside each element of the accelerator, though k, R can certainly vary from element to element. In other words, k and R are Heaviside step-functions with argument s.

In case a magnetic element has a longitudinal field gradient, we could still apply the assumption of constant focusing to longitudinal slices of the magnets, such that the focusing properties within each slice are approximately constant. The principal trajectories are properly defined within each constant focusing element as follows:

$$\begin{cases} C(s) = \cos\left(\sqrt{K}s\right) \\ S(s) = \frac{1}{\sqrt{K}}\sin\left(\sqrt{K}s\right) \end{cases}, \quad K := k + \frac{1}{R^2} \tag{4.65}$$

We find for example for the $C(s)$ function:

$$C'' + kC = \cos\left(\sqrt{K}s\right)\cdot k + k\cdot\cos\left(\sqrt{K}s\right) = 0 \tag{4.66}$$

The particular solution $D(s)$ can in turn be expressed as function of the principal trajectories $C(s)$, $S(s)$:

$$D(s) = S(s)\int_0^s \frac{C(s')}{R(s')}ds' - C(s)\int_0^s \frac{S(s')}{R(s')}ds' \tag{4.67}$$

and we demonstrate below that such expression satisfies the complete Hill's equation for unitary δ. In the following, we make use of the fact that both C, S satisfy Hill's homogeneous equation, and that $W(C, S)=1$:

$$\begin{aligned} D'' = \frac{d}{ds}D' &= \frac{d}{ds}\left[S'\int\frac{C(s')}{R(s')}ds' + \frac{SC}{R} - C'\int\frac{S(s')}{R(s')}ds' - \frac{CS}{R}\right] = \\ &= S''\int\frac{C(s')}{R(s')}ds' + \frac{S'C}{R} - C''\int\frac{S(s')}{R(s')}ds' - \frac{C'S}{R} = \\ &= S''\int\frac{C(s')}{R(s')}ds' - C''\int\frac{S(s')}{R(s')}ds' + \frac{CS'-SC'}{R} = \\ &= -KS\int\frac{C(s')}{R(s')}ds' + KC\int\frac{S(s')}{R(s')}ds' + \frac{W}{R} = \\ &= -KD + \frac{1}{R} \end{aligned} \tag{4.68}$$

In each transverse plane, the optical functions in Eqs. 4.65 and 4.67 can be used to build a 3×3 *transfer matrix*, so generalizing Eq. 4.60 to the inclusion of dispersive motion, but still constant longitudinal momentum. This way, a sequence of accelerator elements can be described as the ordered product of matrices reflecting the actual sequence of the accelerator components. For example, in the horizontal plane (the suffix x is suppressed for brevity) and with notation $\phi = s\sqrt{K}$:

$$\begin{aligned} \vec{x} = \begin{pmatrix} x \\ x' \\ \delta \end{pmatrix}_s &= \begin{pmatrix} C & S & D \\ C' & S' & D' \\ 0 & 0 & 1 \end{pmatrix}_{0\to s} \vec{x}_0 = \\ &= \begin{pmatrix} \cos\phi & \frac{1}{\sqrt{K}}\sin\phi & \frac{1}{RK}(1-\cos\phi) \\ -\sqrt{K}\sin\phi & \cos\phi & \frac{1}{R\sqrt{K}}\sin\phi \\ 0 & 0 & 1 \end{pmatrix}_{x,s} \vec{x}_0 \equiv M_x\vec{x}_0 \end{aligned} \tag{4.69}$$

For the vertical plane in the absence of vertical dispersion, the generic transfer matrix in Eq. 4.69 is specialized to the case $R \rightarrow \infty$ and $k \rightarrow -k$:

$$\vec{y} = \begin{pmatrix} y \\ y' \\ \delta \end{pmatrix}_s = \begin{pmatrix} \cos\phi & \frac{1}{\sqrt{-k}}\sin\phi & 0 \\ -\sqrt{-k}\sin\phi & \cos\phi & 0 \\ 0 & 0 & 1 \end{pmatrix}_{y,s} \vec{y}_0 \equiv M_y \vec{y}_0 \qquad (4.70)$$

To describe particle's motion in the linear and non-frictional approximation, the product of matrices has to be still a matrix with unitary determinant, as prescribed by Eq. 4.62. Namely, the individual matrices have to belong to an algebric group. This is the case of *symplectic* matrices. We recall that M is a symplectic matrix if it satisfies $M^T G M = G$, with G the anti-symmetric singular matrix

$$G = \begin{pmatrix} 0 & 1 & 0 & \dots & 0 \\ -1 & 0 & 0 & \dots & 0 \\ 0 & 0 & 1 & \dots & 0 \\ 0 & -1 & 0 & \dots & 0 \\ 0 & 0 & 0 & \dots & 1 \\ 0 & 0 & 0 & -1 & 0 \end{pmatrix}_{n \times n} = \begin{pmatrix} O & I_n \\ -I_n & O \end{pmatrix} \qquad (4.71)$$

In summary, the condition det $M = 1$ for each of the transfer matrices describing the accelerator is necessary but no sufficient to describe a phase space area-preserving map. If M is symplectic, instead, the condition det $M = 1$ is automatically satisfied (though not demonstrated here), and the product of N arbitrary symplectic matrices will still give a correct description of the area-preserving map. As long as an accelerator behaves as a linear and conservative system, all its elements have to be represented by symplectic matrices.

4.3.5.1 Discussion: Drift, Dipole, Quadrupole

We want to calculate 3×3 transfer matrices for a drift (straight section in the absence of any external field), a dipole magnet and a quadrupole magnet. The dipole will be approximated to a small bending angle $\theta \ll 1$. The quadrupole's matrix will be calculated in *thin lens* approximation, according to which the quadrupole's length $l_q \rightarrow 0$ but the integrated gradient is non-zero, $kl_q = const$.

The transfer matrix of a drift section is calculated from Eqs. 4.69 and 4.70 by specifying $R \rightarrow \infty$ and $k = 0$:

$$M_{dr,x} = M_{dr,y} = \begin{pmatrix} 1 & L & 0 \\ 0 & 1 & 0 \\ 0 & 0 & 1 \end{pmatrix} \qquad (4.72)$$

The *focal length* is defined by $u' = u_0/f$ ($u = x, y$), hence $f = 1/m_{21}$. As expected, it is infinite in a drift because the particle's angular divergence does not change in that element.

The matrices of a separate function dipole magnet of arc-length l_d and bending angle θ can be calculated from Eqs. 4.69 and 4.70 by specifying $k = 0$. In the horizontal plane, $\phi_x = s/R = l_d/R = \theta$. In the vertical plane, $\theta = 0$ and $R \to \infty$. We find:

$$M_{d,x} = \begin{pmatrix} \cos\theta & R\sin\theta & R(1-\cos\theta) \\ -\frac{1}{R}\sin\theta & \cos\theta & \sin\theta \\ 0 & 0 & 1 \end{pmatrix}$$

$$M_{d,y} = \begin{pmatrix} 1 & l_d & 0 \\ 0 & 1 & 0 \\ 0 & 0 & 1 \end{pmatrix}$$

(4.73)

Since Eq. 4.73 assumes that the field at the entrance and at the exit of the dipole magnet lies on a plane perfectly orthogonal to the direction of the particle's motion, the dipole is classified as "sector". This kind of magnet results a perfect drift in the vertical plane. In the horizontal, it shows a focal length $f = -R/\sin\theta$. For completeness, we report that a dipole magnet with non-zero edge angles, also called "rectangular", shows a similar focusing property, but in the vertical plane only. In both a sector and a rectangular dipole, the focal length goes like $f \propto R^2/l_d \sim (l_d B_y^2)^{-1}$, i.e., the stronger the integrated field is, the shorter the focal length is, i.e., the stronger the focusing provided by the magnet will be.

If we now assume $\theta \ll 1$ and we expand the matrix terms to first order in θ:

$$M_{d,x} \approx \begin{pmatrix} 1 & l_d & \frac{\theta l_d}{2} \\ 0 & 1 & \theta \\ 0 & 0 & 1 \end{pmatrix} \Rightarrow \begin{cases} D_x \approx \frac{\theta l_d}{2} = D'_x \frac{l_d}{2} \\ D'_x \approx \theta \end{cases}$$

(4.74)

For small bending angle, the derivative of the dispersion generated by the dipole magnet is approximately the bending angle, and the dispersion function propagates as if it originated at the middle of the dipole with slope θ.

The transport matrices representative of a quadrupole magnet long l_q are calculated in thin lens approximation. We first specialize Eqs. 4.69 and 4.70 with $R \to \infty$, then take the limit $l_q \to 0$ but for $kl_q = const$:

$$M_{q,x} = \begin{pmatrix} \cos(\sqrt{k}l_q) & \frac{1}{\sqrt{k}}\sin(\sqrt{k}l_q) & 0 \\ -\sqrt{k}\sin(\sqrt{k}l_q) & \cos(\sqrt{k}l_q) & 0 \\ 0 & 0 & 1 \end{pmatrix} \to \begin{pmatrix} 1 & 0 & 0 \\ -kl_q & 1 & 0 \\ 0 & 0 & 1 \end{pmatrix}$$

(4.75)

$$M_{q,y} = M_{q,x}(k \to -k) \to \begin{pmatrix} 1 & 0 & 0 \\ kl_q & 1 & 0 \\ 0 & 0 & 1 \end{pmatrix}$$

As expected, the focal length $|f| = \frac{1}{|kl_q|}$ has opposite sign in the two transverse planes. In fact, the thin lens approximation applies as long as $l_q \ll f$.

4.3.5.2 Discussion: Dog-Leg

Two consecutive dipole magnets, identical but with opposite sign of the curvature radius, can be used to translate the beam with respect to its initial direction of motion. Demonstrate that such a "dog-leg" configuration translates the beam by exactly the amount of dispersion function excited by the lattice, and that the beam direction is not changed because the derivative of the dispersion function at the exit of the line is null.

If $R\theta = L$ is the first sector dipole's length with positive radius and bending angle, then the second dipole has identical geometry but negative curvature radius and negative angle. The 3×3 transfer matrix of the system is:

$$M_x = \begin{pmatrix} 1 & R\sin\theta & R(1-\cos\theta) \\ 0 & 1 & \sin\theta \\ 0 & 0 & 1 \end{pmatrix} \begin{pmatrix} 1 & R\sin\theta & -R(1-\cos\theta) \\ 0 & 1 & -\sin\theta \\ 0 & 0 & 1 \end{pmatrix} =$$

$$= \begin{pmatrix} 1 & 2R\sin\theta & -R\sin^2\theta \\ 0 & 1 & 0 \\ 0 & 0 & 1 \end{pmatrix} \tag{4.76}$$

$$\Rightarrow \begin{cases} D_x = M_{13} = -R\sin^2\theta = -d \\ \\ D'_x = M_{23} = 0 \end{cases} \tag{4.77}$$

4.3.6   Periodic Motion

In this Section, the special case of betatron motion in a periodic lattice, such as in a synchrotron, is considered [4]. Synchrotrons are made of a sequence of N identical "cells", each cell accommodating a series of dipole and quadrupole magnets. The presence of dipoles give them the name of "arcs", or "achromat" if the horizontal dispersion function and its derivative are both zero at the entrance and at the exit of the cell. N is called *superperiod* of the synchrotron.

Each cell is long L_{cell} so that $NL_{cell} = 2\pi R_s = C$. A cell can be represented by a transfer matrix M, as shown in Fig. 4.11. The synchrotron total matrix (in each transverse plane) is $M_t = \prod_{i=1}^{N} M_i = M_i^N$. In general, $M_x \neq M_y$. In an isomagnetic lattice with N_d dipoles per cell, the dipole bending angle is $\theta_d = \frac{2\pi}{N_d}$. For example, Fig. 4.11 sketches a "double-bend" arc lattice with superperiod $N=12$, where each vertex of the polygonal is the center of a dipole magnet. Particles' motion is said to be "stable" if particles can travel on the reference closed orbit, or in proximity of it, for a very large number of turns, ideally $n \to \infty$.

Fig. 4.11 Schematic of a synchrotron with superperiod $N=12$ and double bend arc lattice

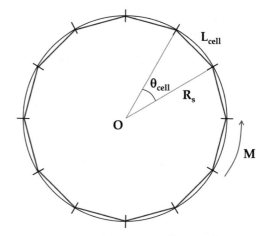

Henceforth, we will adopt the notation $u = x, y$ for all quantities related to the betatron motion, unless differently specified. The periodic homogeneous Hill's equation is:

$$\begin{cases} u''(s) + K(s)u(s) = 0 \\ K(s) = K(s + L_{cell}) = K(s + C) \end{cases} \quad (4.78)$$

The closed form solution of Hill's equation for the whole periodic lattice can be built as a linear superposition of the elements of a basis of the matrix M. In each transverse plane, the basis is given by the eigen-vectors $\vec{u}_j^*$ ($j = 1, 2$) of the matrix. Given λ_j the eigenvalues of M, there exists a diagonal matrix $\Lambda = (\lambda_1, \lambda_2)I$ and an invertible matrix P such that $M = P^{-1}\Lambda P$. The eigenvalues satisfy:

$$M\vec{u}_j^*(s) = \lambda_j \vec{u}_j^* \quad \Rightarrow \quad \vec{u}(s) = \sum_{j=1}^{2} A_j \vec{u}_j^* \quad (4.79)$$

Since the matrix for the whole accelerator still has to contain finite elements after n turns, i.e., the particle's coordinates shall not diverge to infinite for $n \to \infty$, it results:

$$|\lim_{n\to\infty} M^{Nn}\vec{u}(0)| = |\lim_{n\to\infty} M^{Nn} \sum_{j=1}^{2} A_j \vec{u}_{j,0}^*| = |\lim_{n\to\infty} \sum_{j=1}^{2} A_j \lambda_j^{Nn} \vec{u}_{j,0}^*| =$$

$$\leq |A_1 \vec{u}_{1,0}^*| \lim_{n\to\infty} |\lambda_1|^{Nn} + |A_2 \vec{u}_{2,0}^*| \lim_{n\to\infty} |\lambda_2|^{Nn} < \infty$$

$$\Rightarrow |\lambda_j| \leq 1$$

$$(4.80)$$

The eigenvalues of M satisfy the equation $\det(M - \lambda_j I) = 0$. We write M with generic terms and develop the equation accordingly, to retrieve additional properties of the eigenvalues:

$$det \left[\begin{pmatrix} a & b \\ c & d \end{pmatrix} - \begin{pmatrix} \lambda_j & 0 \\ 0 & \lambda_j \end{pmatrix} \right] = det \begin{pmatrix} a - \lambda_j & b \\ c & d - \lambda_j \end{pmatrix} =$$

$$= \lambda_j^2 - (a+d)\lambda_j + (ad - bc) = \lambda_j^2 - Tr(M)\lambda_j + det\, M = \qquad (4.81)$$

$$= \lambda_j^2 - Tr(M)\lambda_j + det\, M = 0,$$

The solutions are:

$$\lambda_{1,2} = \frac{Tr(M) \pm \sqrt{Tr(M)^2 - 4\det M}}{2} \Rightarrow \begin{cases} \lambda_1 \cdot \lambda_2 = \det M = 1 \\ \\ \lambda_1 + \lambda_2 = Tr(M) \end{cases} \qquad (4.82)$$

Equations 4.80 and 4.82 imply $|Tr(M)| = |\lambda_1 + \lambda_2| \leq |\lambda_1| + |\lambda_2| \leq 2$. We demonstrate below that $Tr(M) = \pm 2 \Leftrightarrow \lambda_j = \pm 1$, i.e., $M = \pm I$.

Lemma 1. For $\lambda_1 = x_1 + iy_1$, $\lambda_2 = x_2 + iy_2$, it results:

$$\lambda_1\lambda_2 = (x_1x_2 - y_1y_2) + i(x_1y_2 + x_2y_1) = 1 \Rightarrow \begin{cases} x_1x_2 = 1 + y_1y_2 \\ \\ x_1y_2 = -x_2y_1 \end{cases} \qquad (4.83)$$

Lemma 2. Let us assume generic $\lambda_j \in \mathbb{C}$ and consider the extreme condition $\lambda_1 + \lambda_2 = Tr(M) = \pm 2$. It follows that $y_1 + y_2 = 0$ or $y_1 = -y_2$, and $x_1 + x_2 = \pm 2$. By virtue of Lemma 1, $x_1y_2 = -x_2y_1 \Rightarrow x_1 = -x_2y_1/y_2 = x_2 \Rightarrow x_1 = x_2 = \pm 1 \Rightarrow \lambda_1 = \lambda_2 = \pm 1 \Rightarrow \Lambda = \pm I$ and $M = P^{-1}\Lambda P = \pm I$. Viceversa, if $M = \pm I$ then $Tr(M) = \pm 2$.

When $M = \pm I$, the periodic system is unconstrained (any arbitrary initial condition satisfies the equation of motion) and, owing to Eq. 4.80, any infinitesimal perturbation ϵ to the lattice such that $|\lambda_j + \epsilon| > 1$ will lead to unstable motion. This brings to a tighter *necessary* condition for long-term stability of single particle's motion in the 2-D phase space: $|Tr(M)| < 2$. "Stability" means here bounded values of the particle's coordinates over an arbitrarily large number of passes through a lattice represented by the 2×2 transfer matrix M.

4.3.7   Betatron Function

Floquet's theorem (demonstrated below) applied to a periodic lattice allows us to write the two linearly independent solutions of Hill's equation as periodic functions of s with period C, multiplied by a complex phase:

$$\begin{cases} u_{1,2}(s) = p_{1,2}(s)e^{\pm i(\mu(s) - \mu(0))}, \quad \mu(s) \in \mathbb{R}e\ \forall s \\ \\ p_1(s) = p_2(s) \equiv p(s) = p(s+C)\ \forall s \end{cases} \qquad (4.84)$$

The two amplitudes p_1, p_2 can be made equal without loss of generality because the two exponential functions are already linearly independent. The functions p, μ are intended to be defined in the transverse plane u.

With $u_{1,2}(0) = p_{1,2}(0)$ the solutions at the initial position $s = 0$, and by virtue of the periodicity in Eq. 4.84, the two solutions after one turn result:

$$u_{1,2}(C) = p_{1,2}(C)e^{\pm i(\mu(C)-\mu(0))} = p_{1,2}(0)e^{\pm i(\mu(C)-\mu(0))} =$$

$$= u_{1,2}(0)e^{\pm i(\mu(C)-\mu(0))} \equiv M u_{1,2}(0) \tag{4.85}$$

The very last equality demonstrates that $e^{\pm i(\mu(C)-\mu(0))}$ are the eigenvalues $\lambda_{1,2}$ of M. The real quantity $\mu(C) - \mu(0)$ is called "characteristic exponential coefficient" of the homogeneus Hill's equation. Clearly, those eigenvalues satisfy Eqs. 4.80 and 4.82, and in particular:

$$|Tr(M)| = |\lambda_1 + \lambda_2| = |2\cos\Delta\mu| < 2 \Leftrightarrow \Delta\mu \neq p\pi, p \in \mathbb{N} \tag{4.86}$$

In the literature, u_1, u_2 are written in terms of a positive-definite amplitude $\beta(s) := p^2(s)$ named *betatron function*:

$$\begin{cases} u_{1,2}(s) = \sqrt{\beta(s)}(s)e^{\pm i(\mu(s)-\mu(0))} \\ \beta(s) = \beta(s+C) \end{cases} \tag{4.87}$$

The generic solution of Hill's homogeneous equation is therefore:

$$u(s) = a_1 u_1(s) + a_2 u_2(s) = a_1\sqrt{\beta(s)}(s)e^{i(\mu(s)-\mu_0))} + a_2\sqrt{\beta(s)}(s)e^{-i(\mu(s)-\mu_0)} =$$

$$= \sqrt{2J}\sqrt{\beta(s)}\cos(\mu(s) - \mu_0 + \phi_0), \tag{4.88}$$

where the constants are $\sqrt{J} = \sqrt{2a_1 a_2}$ and $\tan\phi_0 = i\frac{a_1-a_2}{a_1+a_2}$. All quantities are intended to be defined in the u−transverse plane. We draw the following observations.

- The amplitude $\sqrt{2J}$ is a constant of motion, called *single particle invariant*.
- The particle's transverse position along the accelerator is proportional to $\sqrt{\beta_u}$, which contributes to determining the amplitude of oscillation.
- $\Delta\mu(s) = \mu(s) - \mu_0 + \phi_0$ assumes the meaning of a *relative phase advance* of the betatron oscillation.
- Floquet's theorem states the existence a solution of the periodic Hill's equation, i.e., the equation defined for a *periodic lattice*. The solution implies a *periodic betatron function*. Since $\beta(s)$ is an analytic continuous (i.e., differentiable) function of s, its derivative $\beta'(s)$ is also periodic in s.

As a consequence of Floquet's theorem (Eq. 4.84), the betatron phase advance and the betatron function are intrinsically connected. We explicit their relation by substituting the solution $u(s) = p(s)e^{i\mu(s)}$ into Hill's equation:

$$u = pe^{i\mu},$$

$$u' = p'e^{i\mu} + ip\mu'e^{i\mu},$$

$$u'' = \left[p'' + i2p'\mu' + ip\mu'' - p\mu'^2 \right] e^{i\mu}$$

$$\Rightarrow u'' + ku = 0,$$

$$\left(p'' - p\mu'^2 \right) + i \left(2p'\mu' + p\mu'' \right) + kp = 0$$

$$(p'' + kp) - p\mu'^2 + i(2p'\mu' + p\mu'') = 0$$

(4.89)

The real and the imaginary part of the r.h.s. of the last expression have to be individually zero. The latter one gives:

$$\frac{\mu''}{\mu'} = -\frac{2p'}{p};$$

$$\frac{d}{ds} \left(\ln \mu' \right) = -\frac{d}{ds} \left(\ln p^2 \right) + \ln c, \qquad \ln c \equiv 0;$$

$$\int ds(l.h.s.) = \int ds(r.h.s.); \qquad\qquad (4.90)$$

$$\mu' = \frac{1}{p^2};$$

$$\Rightarrow \mu(s) - \mu_0 = \int_0^s \frac{ds'}{\beta(s')}$$

The number of betatron oscillations per turn:

$$Q_u = \frac{\Delta\mu_u}{2\pi} = \oint \frac{ds}{\beta_u(s)} \equiv \frac{2\pi R_s}{\overline{\beta_u}} \qquad\qquad (4.91)$$

is called *betatron tune*. The pair $[Q_x, Q_y]$ is the synchrotron *working point*. The integer part of Q_u is usually in the range 10-100, thus orders of magnitude larger than the synchrotron tune Q_s (see Eq. 4.32).

The last equality on the r.h.s. of Eq. 4.91 defines an *average* betatron function evaluated over the particle's closed orbit. It is important to notice that, in general, $\frac{1}{\overline{\beta_u}} = \langle \frac{1}{\beta_u} \rangle \neq \frac{1}{\langle \beta_u \rangle}$. Figure 4.12 illustrates the generic solution $u(s)$, expressed as function of $\beta_u(s)$ and $\overline{\beta_u}$.

4.3.8   Floquet's Theorem

Floquet's theorem is evoked in Eq. 4.84 for particle's motion in a 2-D phase space (u, u'), with u the particle's position relative to the reference orbit, and $u' = du/ds$

Fig. 4.12 Betatron
oscillation in the x-plane.
Particle's position x(s) is
evaluated for the s-dependent
$\beta_x(s)$ (solid line) and the
average betatron function $\overline{\beta_x}$
(dashed line)

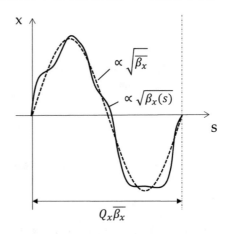

the angular divergence. The most general enunciation assumes a periodic system
with period C:

$$\begin{cases} f'(s) = A(s)f(s) \\ A(s) = A(s+C) \end{cases} \tag{4.92}$$

We now apply Eq. 4.92 to the particle's position and angular divergence,
$f \rightarrow u$, $f' \rightarrow u'$. In our notation, $A(s)$ is still a periodic function, given by the
combination of principal trajectories and arbitrary initial values u_0, u'_0:

$$\begin{cases} u' = Au \\ u'' = Au' + A'u = A(Au) + A'u = (A^2 + A')u \equiv Ku \end{cases} \tag{4.93}$$

The second expression shows that Hill's equation for the betatron motion describes
a special "Floquet's system", where $K(s) = K(s+C)$.

Lemma 1: if $u_1(s)$, $u_2(s)$ are linearly independent solutions of Eq. 4.93, then $\tilde{u}_1 = u_1(s+C)$, $\tilde{u}_2 = u_2(s+C)$ are also linearly independent solutions.

Firstly, we verify that $\tilde{u}_j$, $j = 1, 2$, is a solution of Eq. 4.93 (the subscript is
removed for brevity):

$$\tilde{u}' = A(s+C)\tilde{u} = A(s)\tilde{u} \tag{4.94}$$

by virtue of the periodicity of $A(s)$.

Secondly, we verify the condition of linear independence by demonstrating that the
wronskian of the two solutions is non-zero for at least one value of the s-coordinate
in the range of existence $s \in [0, C]$. To do this, we remind that the vectors $\vec{\tilde{u}}_j$ are
obtained by a single-turn transformation of vectors $\vec{u}_j$ via the transfer matrix M_t:

$$\vec{\tilde{u}} = M_t\vec{u} \quad \Rightarrow \quad \begin{cases} \tilde{u}_j = M_{11}u_j + M_{12}u'_j \\ \tilde{u}'_j = M_{21}u_j + M_{22}u'_j \end{cases} \tag{4.95}$$

The wronskian is:

$$
\begin{vmatrix} \tilde{u}_1 & \tilde{u}_2 \\ \tilde{u}'_1 & \tilde{u}'_2 \end{vmatrix} = = (M_{11}u_1 + M_{12}u'_1)(M_{21}u_2 + M_{22}u'_2) - (M_{11}u_2 + M_{12}u'_2)(M_{21}u_1 + M_{22}u'_1) =
$$
$$
= M_{11}M_{22}u_1u'_2 + M_{12}M_{21}u'_1u_2 - M_{11}M_{22}u_2u'_1 - M_{12}M_{21}u'_2u_1 =
$$
$$
= M_{11}M_{22}(u_1u'_2 - u_2u'_1) + M_{12}M_{21}(u'_1u_2 - u'_2u_1) =
$$
$$
= M_{11}M_{22} - M_{12}M_{21} = 1 \quad \forall s
$$

where we used the fact that u_1, u_2 are linearly independent solutions, and that M_t is symplectic.

Lemma 2: given the "fundamental" matrices $U = \begin{pmatrix} u_1 & 0 \\ 0 & u_2 \end{pmatrix}$ and $\tilde{U} = \begin{pmatrix} \tilde{u}_1 & 0 \\ 0 & \tilde{u}_2 \end{pmatrix}$, there exists a non-singular matrix R such that $\tilde{U} = UR$.
  It has to be:

$$
\tilde{U}_{11} = \tilde{u}_1 = u_1 R_{11}
$$
$$
\tilde{U}_{12} = 0 = u_1 R_{12} \Rightarrow R_{12} = 0
$$
$$
\tilde{U}_{21} = 0 = u_2 R_{21} \Rightarrow R_{21} = 0 \tag{4.96}
$$
$$
\tilde{U}_{22} = \tilde{u}_2 = u_2 R_{22}
$$

Therefore R is diagonal. We also have $\det(\tilde{U}) = \det(U)\det(R) \Rightarrow \det(R) \neq 0$, i.e., R is invertible. Owing to the existence of the logarithm matrix for R, we can write $R = e^{QC}$, where in general $Q_{ij} \in \mathbb{C}$. By virtue of its unitary determinant, it must be:

$$
R(s) = \begin{pmatrix} e^{iq(s)C} & 0 \\ 0 & e^{-iq(s)C} \end{pmatrix}, \quad q(s) \in \mathbb{R}e \tag{4.97}
$$

Lemma 3: given the matrix $P(s) = U(s)e^{-Qs}$, it results $P(s + C) = P(s) \; \forall s$.
  This is demonstrated by calculating $P(s + C)$ and using Lemma 2:

$$
P(s + C) = U(s + C)e^{-Q(s+C)} = \tilde{U}e^{-QC}e^{-Qs} = URe^{-QC}e^{-Qs} =
$$
$$
= UIe^{-Qs} = P(s);
$$

$$
\Rightarrow \begin{cases} U(s) = P(s)e^{Qs} \\ P(s) = P(s + C) \end{cases} \tag{4.98}
$$

Equation 4.98 concludes the demonstration of Floquet's theorem. Indeed, we can define a real function $\mu(s) = q(s)C$ so that:

$$
\begin{cases} U_{11} = u_1(s) = p_1(s)e^{i\mu(s)} \\ U_{22} = u_2(s) = p_2(s)e^{-i\mu(s)} \end{cases} \tag{4.99}
$$

and p_j is periodic with period C.

4.3.9 Courant-Snyder Invariant

A complete description of betatron motion needs the knowledge not only of the particle's position (Eq. 4.88), but also of its angular divergence. This is the angle between the vector of particle's transverse and longitudinal momentum, evaluated at any point along the accelerator. We remind that in the Frenet-Serret coordinate system, the longitudinal versor is tangent to the orbit. Thus, the divergence identifies the particle's instantaneous direction of motion:

$$u'(s) = \frac{p_u}{p_z} = \frac{v_u}{v_z} = \frac{1}{v_z}\frac{du}{dt} = \frac{du}{ds}, \tag{4.100}$$

and by virtue of Eq. 4.88

$$u'(s) = \frac{du}{ds} = \frac{\sqrt{2J}}{2\sqrt{\beta}}\frac{d\beta}{ds}\cos\Delta\mu - \sqrt{2J\beta}\sin\Delta\mu \cdot \frac{d\mu}{ds} =$$
$$= \sqrt{2J}\left[\frac{1}{2\sqrt{\beta}}\frac{d\beta}{ds}\cos\Delta\mu - \frac{1}{\sqrt{\beta}}\sin\Delta\mu\right] =$$
$$= -\sqrt{\frac{2J}{\beta}}\left(\alpha\cos\Delta\mu + \sin\Delta\mu\right), \tag{4.101}$$

$$\alpha := -\frac{\beta'}{2}$$

Equation 4.101 shows that the single particle's betatron oscillations can be described by the so-called *Courant-Snyder* parameters $\alpha_u(s)$, $\beta_u(s)$, or *Twiss functions*. This description is equivalent to that given in terms of principal trajectories in Eqs. 4.69 and 4.70. In fact, the two initial conditions u_0, u_0' are here replaced by the constants $2J$, μ_0, while the linearly independent transfer functions $C_u(s)$, $S_u(s)$ are substituted by the linearly independent functions $\beta_u(s)$, $\alpha_u(s)$.

By definition, the locations at which $\alpha(\bar{s}) = 0$ correspond to a maximum or minimum of $\beta(s)$. Since we can freely define the initial betatron phase $\mu(\bar{s}) = \mu_0 = 0$, we have from Eq. 4.101 that $\alpha(\bar{s}) = 0$ identifies the points along the accelerator in correspondence of which the particle assumes a local maximum or minimum distance from the reference orbit. This is the case, for example, of a particle in the middle of a quadrupole magnet, see Fig. 4.9.

Let us now assume that the full set of functions $u(s)$, $u'(s)$, $\beta(s)$ and $\alpha(s)$ are known at a given s. Since J is a constant of motion, we can retrieve it from the knowledge of the full set of data. After we will have found an expression for it, we will demonstrate that it is indeed a constant of motion under the assumptions 1. and 2. (see Eq. 4.62), which were used to derive the homogeneous Hill's Eqs. 4.53 and 4.55.

At first, we recall the solution of the homogeneous Hill's equation and its first derivative:

$$\begin{cases} u(s) = \sqrt{2J\beta}\cos\Delta\mu \\\\ u'(s) = -\sqrt{\frac{2J}{\beta}}\left(\alpha\cos\Delta\mu + \sin\Delta\mu\right) \end{cases} \tag{4.102}$$

By substituting the first equation into the second one:

$$u' = -\sqrt{\frac{2J}{\beta}}\left(\alpha\frac{u}{\sqrt{2J\beta}} + \sqrt{1-\frac{u^2}{2J\beta}}\right) = -u\frac{\alpha}{\beta} - \sqrt{\frac{2J}{\beta} - \frac{u^2}{\beta^2}};$$

$$\left(u' + u\frac{\alpha}{\beta}\right)^2 + \frac{u^2}{\beta^2} = \frac{2J}{\beta};$$

$$\Rightarrow \begin{cases} 2J = u^2\left(\frac{\alpha^2+1}{\beta}\right) + 2uu'\alpha + \beta u'^2 = \gamma u^2 + 2\alpha uu' + \beta u'^2, \\[2mm] \gamma(s) := \frac{1+\alpha(s)^2}{\beta(s)} \end{cases} \tag{4.103}$$

By virtue of Eq. 4.103, the single particle invariant $2J$ is also called *Courant-Snyder invariant*. The choice of the two linearly independent Courant-Snyder (C-S) parameters among α, β, γ is indeed irrelevant, being the third one dependent from the other two.

To demonstrate that the expression on the r.h.s of Eq. 4.103 is a constant of motion, we consider a generic solution of Hill's homogeneous equation and define a wronskian as follows:

$$\begin{cases} u(s) = a_1 u_1(s) + a_2 u_2(s) \\ u_1 = \sqrt{\beta}e^{i\mu} \end{cases} \Rightarrow V(s) = \begin{vmatrix} u & u' \\ u_1 & u_1' \end{vmatrix} = uu_1' - u'u_1 \tag{4.104}$$

We observe that:

$$\frac{dV}{ds} = u'u_1' + uu_1'' - u''u_1 - u'u_1' = uu_1'' - u''u_1 = u(-ku_1) - (-ku)u_1 =$$
$$= -kuu_1 + kuu_1 = 0$$
$$\Rightarrow V(s) = const. \quad \forall s \tag{4.105}$$

$V(s)$ is expressed in terms of the Courant-Snyder parameters:

$$V(s) = uu_1'(s) - u'u_1(s) = u\left(\frac{1}{2}\frac{\beta'}{\sqrt{\beta}}e^{i\mu} + i\sqrt{\beta}\frac{e^{i\mu}}{\beta}\right) - u'u_1 =$$
$$= u\frac{e^{i\mu}}{\sqrt{\beta}}(-\alpha + i) - u'u_1 = u\sqrt{\beta}e^{i\mu}\left(\frac{i-\alpha}{\beta}\right) - u'u_1 =$$
$$= uu_1\left(\frac{i-\alpha}{\beta}\right) - u'u_1 = u_1\left[u\left(\frac{i-\alpha}{\beta}\right) - u'\right] \tag{4.106}$$

$$\Rightarrow VV^* = u_1u_1^*\left[u^2\left(\frac{1+\alpha^2}{\beta^2}\right) + \frac{2\alpha}{\beta}uu' + u'^2\right] =$$

$$= \gamma u^2 + 2\alpha uu' + \beta u'^2$$

We conclude that since $V(s)$ is a constant of motion, $|V|^2 = VV^*$ is constant as well, and this is exactly the expression of $2J$ in Eq. 4.103.

4.3.10 Phase Space Ellipse

Hill's equation for the betatron motion in a periodic lattice describes a quasi-periodic oscillator. That is, the particle's motion in the phase space (u, u') is bounded and, in the most general case, the orbit at any given s maps an ellipse. The dependence of the C-S parameters from the s-coordinate suggests that, contrary to a pure harmonic oscillation (i.e., constant angular frequency), the ellipse orientation and ellipticity changes at different s.

According to Eq. 4.102, the particle's lateral position oscillates within an envelope of amplitude $\sqrt{2J\beta}$. Local maxima or minima are reached for:

$$x' = 0 \Rightarrow \cos \Delta\mu = -\frac{\sin \Delta\mu}{\alpha} \Rightarrow \cos^2 \Delta\mu = \frac{1}{1+\alpha^2}$$

$$\Rightarrow |x| = \frac{\sqrt{2J\beta}}{\sqrt{1+\alpha^2}} \le \sqrt{2J\beta} \tag{4.107}$$

Fig. 4.9 suggests that this is common to happen in quadrupole magnets.

At all points of the lattice where $\alpha = 0$, the particle's maximum spatial and angular excursion, not to be met simultaneously, are $|\hat{u}| = \sqrt{2J\beta}$ and $|\hat{u}'| = \sqrt{2J/\beta}$, respectively. The C-S invariant introduced in Eq. 4.103 becomes:

$$2J(\alpha = 0) = \gamma u^2 + \beta u'^2 = \frac{u^2}{\beta} + \beta u'^2 = 2J\frac{u^2}{|\hat{u}|^2} + 2J\frac{u'^2}{|\hat{u}'|^2}$$

$$\Rightarrow \frac{u^2}{|\hat{u}|^2} + \frac{u'^2}{|\hat{u}'|^2} = 1 \tag{4.108}$$

$$\Rightarrow |\hat{u}||\hat{u}'| = 2J = \frac{Area}{\pi}$$

Eq. 4.108 describes an up-right phase space ellipse, i.e., the ellipse axes are aligned to the Cartesian axes (u, u'). The area of the ellipse is just the particle's C-S invariant in units of π, i.e., the phase space area enclosed by the particle's orbit is a constant of motion, in the assumption of linear and non-dissipative forces. The phase space ellipse for generic C-S parameters is illustrated in Fig. 4.13. We summarize our findings below.

- Betatron motion of the generic particle is represented by an ellipse in the phase space (u, u'), on which the particle's representative pint lies during the particle's motion in the accelerator. The ellipse area in units of π is the particle's C-S invariant, i.e., it is a constant of motion in the approximation of linear and non-dissipative dynamics.
- The ellipse orientation and ellipticity are uniquely determined by the C-S parameters. By virtue of s-dependent focusing in Hill's equation, the C-S parameters vary with s, and so the ellipse does. In general, we can draw a different ellipse in correspondence of each s along the accelerator. The particle's coordinates at any s belong to the ellipse evaluated at that point.

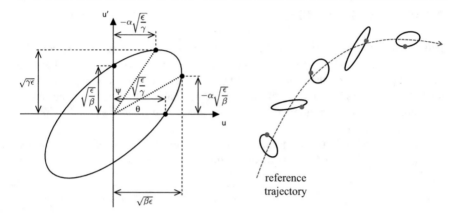

Fig. 4.13 Left: phase space ellipse at a specific s-coordinate in the accelerator, represented in terms of Courant-Snyder parameters and particle's invariant $2J = \epsilon$. One also has $\tan \psi = -\alpha\sqrt{\frac{\epsilon}{\gamma}}\frac{1}{\sqrt{\epsilon\gamma}} = -\frac{\alpha}{\gamma}$, $\tan \theta = -\alpha\sqrt{\frac{\epsilon}{\beta}}\frac{1}{\sqrt{\epsilon\beta}} = -\frac{\alpha}{\beta}$. Right: representative point of a particle (red dot) on the phase space ellipse at different s along the accelerator. In the physical space, the point describes betatron oscillations

- If the motion is periodic, the ellipse remains identical at any given s over consecutive turns (because the C-S parameters are periodic with period equal to the synchrotron circumference). Turn-by-turn, the particle's coordinates at that specific s will map the ellipse as the betatron phase advances.
- The Particle's position is proportional to $\sim\sqrt{\beta}$, its angular divergence to $\sim\frac{1}{\sqrt{\beta}}$. The condition $\alpha = 0$ implies a local maximum or minimum of the betatron function, and the corresponding phase space ellipse is up-right. The individual particle's position and angular divergence do not necessarily reach a maximum there, depending from the betatron phase advance. However, the absolute maximum position and angular divergence become accessible, and equal to $|\hat{u}| = \sqrt{2J\beta}$ and $|\hat{u}'| = \sqrt{2J/\beta}$, respectively.

4.3.11 Floquet's Normalized Coordinates

The phase space representation of particle's betatron motion is expected to reduce to a purely harmonic oscillator if the dependence of the oscillation amplitude from the s-coordinate could be removed. To show this, let us derive Hill's homogeneous equation under a transformation of coordinates normalized to the local betatron function.

The betatron phase advance becomes the independent variable:

$$
\begin{cases} s \rightarrow \Delta\mu = Q\theta \equiv \phi \\[2mm] u \rightarrow w = \frac{u}{\sqrt{\beta}} \end{cases}
\Rightarrow
\begin{cases} w = \sqrt{2J}\cos\phi \\[2mm] w' = \frac{dw}{d\phi} = -\sqrt{2J}\sin\phi \end{cases}
\tag{4.109}
$$

Q is the betatron tune, the pair (J, ϕ) is called *action-angle* variables, and w, w' are the "Floquet's normalized coordinates". An ellipse in the phase space (u, u') becomes a circle of radius $\sqrt{2J}$ in the phase space (w, w'). As the particle advances along the accelerator coordinate s, the representative point lying on the circle moves by an angle $\phi(s)$. As expected, Hill's homogeneous equation becomes that of a pure harmonic oscillator (the angular frequency does not depend from s anymore):

$$w'' = \frac{dw'}{d\phi} = -w \quad \Rightarrow \quad w'' + w = 0 \tag{4.110}$$

It is straightforward to show that, if θ is taken as independent variable instead, then the phase space orbit still maps onto an ellipse, and the angular frequency is just the betatron tune:

$$w''(\theta) = -Q^2 \sqrt{2J} \cos(Q\theta) \quad \Rightarrow \quad w'' + Q^2 w = 0 \tag{4.111}$$

Since $\Delta\mu_t = 2\pi Q$ is the single turn phase advance in a synchrotron, a particle moves in this normalized phase space by a corresponding angle $\theta_t = 2\pi$.

4.3.11.1 Discussion: Periodic Momentum-Dispersion Function

The solution of the Hill's complete equation for the periodic motion, denominated *periodic momentum-dispersion function*, or simply "periodic dispersion", is by definition an eigenvector of the one-turn 3×3 transfer matrix, in analogy to Eq. 4.87 for the betatron motion. Let us find an expression for η, η' as function of the principal trajectories, by exploiting the one-turn matrix properties (trace and determinant).

Being the periodic dispersion an eigen-vector of the one-turn matrix, it must satisfy:

$$\begin{pmatrix} \eta \\ \eta' \\ 1 \end{pmatrix}_s = \begin{pmatrix} C(s) & S(s) & D(s) \\ C'(s) & S'(s) & D'(s) \\ 0 & 0 & 1 \end{pmatrix} \begin{pmatrix} \eta \\ \eta' \\ 1 \end{pmatrix}_0 \tag{4.112}$$

The periodicity over one turn implies $(\eta, \eta', 1)_0^t = (\eta, \eta', 1)_s^t$. We can therefore solve Eq. 4.112 for η and η' at the arbitrary coordinate s. By using the relation $C + S' = Tr(M_t) = 2 \cos \Delta\mu$ and $\det M = CS' - S'C = 1$, we find:

$$\begin{cases} \eta(s) = \dfrac{[1-S'(s)]D(s)+S(s)D'(s)}{2(1-\cos \Delta\mu)} \\[3mm] \eta'(s) = \dfrac{[1-C(s)]D'(s)+C'(s)D(s)}{2(1-\cos \Delta\mu)} \end{cases} \tag{4.113}$$

It emerges that $D(s)$ is the dispersive element of the transfer matrix, while $\eta(s)$ is the periodic solution of Hill's complete equation. When (η, η') are calculated along the beam line starting from initial values $(0, 0)$, it results from Eq. 4.112 that, at the generic coordinate s, $(\eta, \eta')_s = (D, D')_s$. This justifies the naming of dispersion function for $D(s)$ adopted so far, for a beam line in which $D_0 = D'_0 = 0$.

4.3.12 Equivalence of Matrices

Solutions of Hill's homogeneous equation can be expressed equivalently in terms of principal trajectories (see Eq. 4.60) and C-S parameters (see Eq. 4.102). The equivalence implies that principal trajectories mapping the particle's coordinates from one point to another in the accelerator, can be expressed as a combination of C-S parameters evaluated at the beginning and at the end of the line, and viceversa.

The former correspondence is found by recalling u, u' in Eqs. 4.59 and 4.102 (all optical functions refer to the same transverse plane):

$$\begin{cases} u(s) = Cu_0 + Su'_0 = \sqrt{2J\beta}\cos\Delta\mu \\ u'(s) = C'u_0 + S'u'_0 = -\sqrt{\frac{2J}{\beta}}\,(\sin\Delta\mu + \alpha\cos\Delta\mu) \end{cases} \tag{4.114}$$

The Initial conditions $C(0) = S'(0) = 1$, $C'(0) = S(0) = 0$, $\beta(0) = \beta_0$, $\alpha(0) = \alpha_0$, $\Delta\mu(0) = 0$ are used to find:

$$\begin{cases} u(0) = u_0 = \sqrt{2J\beta_0} \\ u'(0) = u'_0 = -\alpha_0\sqrt{\frac{2J}{\beta_0}} \end{cases} \tag{4.115}$$

and therefore:

$$\begin{cases} C = \sqrt{\frac{\beta}{\beta_0}}\cos\Delta\mu + S\frac{\alpha_0}{\beta_0} \\ C' = -\frac{\sin\Delta\mu + \alpha\cos\Delta\mu}{\sqrt{\beta\beta_0}} + S'\frac{\alpha_0}{\beta_0} \end{cases} \tag{4.116}$$

We now impose $CS' - SC' = 1$ to find S':

$$S'\sqrt{\frac{\beta}{\beta_0}}\cos\Delta\mu + SS'\frac{\alpha_0}{\beta_0} + S\left(\frac{\sin\Delta\mu + \alpha\cos\Delta\mu}{\sqrt{\beta\beta_0}}\right) - SS'\frac{\alpha_0}{\beta_0} = 1;$$

$$S' = \frac{1}{\sqrt{\frac{\beta}{\beta_0}}\cos\Delta\mu}\left[1 - S\frac{(\sin\Delta\mu + \alpha\cos\Delta\mu)}{\sqrt{\beta\beta_0}}\right] \tag{4.117}$$

Finally, we remind that S is a solution of Hill's equation, and thereby it can be used to evaluate the C-S invariant at $s = 0$ and at any downstream coordinate. We choose the second point in correspondence of a local maximum of the betatron function ($\alpha = 0$ or $\beta = 1/\gamma$):

$$\beta S'^2 + 2\alpha SS' + \gamma S^2 = \beta_0 S(0)'^2 + 2\alpha_0 S(0)S'(0) + \gamma_0 S(0)^2 = \beta_0;$$

$$\beta S'^2 + \frac{S^2}{\beta} = \beta_0; \tag{4.118}$$

$$\frac{S^2}{\beta\beta_0} = 1 - \frac{\beta_0}{\beta}S'^2$$

We conclude that: (i) since $S(0) = 0$, it has to be $S \sim \sin\Delta\mu$; (ii) it does not depend from α_0, therefore it does not depend from α either; (iii) it has to be a length. These properties are all simultaneously satisfied by $S = \sqrt{\beta\beta_0}\sin\Delta\mu$. This is now plugged into the expressions for C, C' and S' (Eqs. 4.116 and 4.117) to obtain the following correspondence:

$$M = \begin{pmatrix} C & S \\ C' & S' \end{pmatrix} = \begin{pmatrix} \sqrt{\frac{\beta}{\beta_0}}(\cos \Delta\mu + \alpha_0 \sin \Delta\mu) & \sqrt{\beta_0\beta}\sin\Delta\mu \\ \frac{(\alpha - \alpha_0)\cos\Delta\mu - (1 + \alpha\alpha_0)\sin\Delta\mu}{\sqrt{\beta_0\beta}} & \sqrt{\frac{\beta_0}{\beta}}(\cos\Delta\mu - \alpha\sin\Delta\mu) \end{pmatrix}$$

$$(4.119)$$

$$\Rightarrow M_t := M_{turn}(\alpha = 0) = \begin{pmatrix} \cos\Delta\mu_t & \beta\sin\Delta\mu_t \\ -\frac{1}{\beta}\sin\Delta\mu_t & \cos\Delta\mu_t \end{pmatrix} \tag{4.120}$$

The one-turn matrix M_t for a periodic lattice was conveniently chosen in correspondence of a point of local maximum or minimum of the betatron function. The stability condition $|Tr(M_t)| < 2$ is satisfied by $\Delta\mu_t = 2\pi Q_u \neq n\pi, n \in \mathbb{N}$ and Q_u the betatron tune. The condition $Q_u = n/2$ identifies the so-called "integer" and "half-integer" resonance (see later). It is immediate to see from Eq. 4.119 that for *periodic* C-S parameters, we can always write $Tr(M_t) = 2\cos\Delta\mu_t$.

4.3.13 Non-Periodic Motion

We may ask ourselves what is the meaning of the C-S parameters, i.e., a phase-amplitude representation of particle's motion, in a non-periodic lattice, like that one of a single-pass linac.

Naively, since the particle "does not know" whether the lattice it is going to pass through is periodic or not, we might infer that the C-S formalism introduced for periodic motion applies to a non-periodic lattice as well. If we keep the solution of Hill's equation as in Eq. 4.88, we find the differential equation the betatron function has to satisfy in a generic non-periodic system. We adopt the notation $C = \cos\Delta\mu$, $S = \sin\Delta\mu$ for brevity (not to be confused with the principal trajectories), and $\beta(s) = p^2(s)$:

$$u = a\sqrt{\beta}\cos\Delta\mu \equiv apC,$$
$$u' = ap'C - apS\mu',$$
$$u'' = ap''C - ap'S\mu' - ap'S\mu' - ap(C\mu'^2 + S\mu'').$$

$$\mu' = \frac{1}{p^2}, \qquad \mu'' = -\frac{2}{p^3}p'$$

$$(4.121)$$

$$\Rightarrow u'' + Ku = 0;$$
$$ap''C - ap'\frac{S}{p^2} - ap'\frac{S}{p^2} - ap\frac{C}{p^4} + apS^2\frac{p'}{p^3} + aKpC = 0;$$
$$C\left(p'' + Kp - \frac{1}{p^3}\right) - S\left(-\frac{2p'}{p^2} + \frac{2p'}{p^2}\right) = 0;$$

$$\Rightarrow p'' + Kp = \frac{1}{p^3}$$

By replacing the definition $p(s) = \sqrt{\beta(s)}$ into the last expression of Eq. 4.121, we find:

$$\tfrac{1}{2}\beta\beta'' - \tfrac{1}{4}\beta'^2 + K\beta^2 = 1 \tag{4.122}$$

We draw the following observations.

• When $K = 0$, such as in a drift section, the solution of the equation for initial conditions $\beta(0) = \beta_0$, $\beta'(0) = 0$ is $\beta(s) = \beta_0 + \frac{s^2}{\beta_0}$. Namely, the betatron function grows quadratically from the minimum value β_0.

• When $|K| \gg \left[\frac{1}{\beta^2}, \left(\frac{1}{\beta}\frac{d\beta}{ds}\right)^2\right]$, the equation reduces to $\beta'' = -2K\beta$. For example, this is the case of a betatron function far from a minimum, or smooth optics in the presence of moderate strengths. The solution is either oscillatory or exponential depending on the sign of K (for example, passing through a focusing or defocusing quadrupole magnet, respectively).

We conclude that, if Eq. 4.122 has solution, the C-S invariant introduced in Eq. 4.103 is also well-defined in a non-periodic lattice, and subject to the same conditions of linearity and non-frictional forces discussed above.

Single particle's betatron motion in the transverse phase space along a non-periodic lattice is illustrated in Fig. 4.13-right plot. The periodic boundary conditions imposed to β_u, α_u in a synchrotron—if a periodic solution exists—are replaced by arbitrary initial conditions $\beta_{u,0}$, $\alpha_{u,0}$, and $\Delta\mu_{u,0} = 0$. For any given lattice—i.e., a sequence of transfer matrices known in terms of principal trajectories C, S—$\beta_u(s)$, $\alpha_u(s)$ are uniquely defined at any s by virtue of Eq. 4.119. Moreover, for any given particle's invariant J, the particle's coordinates $u(s)$, $u'(s)$ are also uniquely determined by virtue of Eqs. 4.102 and 4.90.

Yet, we have not formally justified the adoption of Floquet's solution for the non-periodic Hill's equation with arbitrary variable coefficient $K(s)$. To do so, we may proceed in two complementary ways.

First, the non-periodic problem can be reduced to a periodic one, by demonstrating that for any choice of β_0, α_0, there exists a lattice making the beam line periodic, which would allow us to invoke Floquet's theorem. This condition is always satisfied by the inverse matrix of the original line, or $MM^{-1} = I$. We note that in this case $Tr(MM^{-1}) = 2$, contrary to the prescription for stability in Eq. 4.80 (we will show that such lattice satisfies the so-called "integer resonance"), and that the initial conditions are unconstrained. The impasse is removed as long as we require the motion to be bounded on a single-pass only, and the initial conditions are arbitrarily set by the user.

Second, we note that any accelerator lattice can be modeled as a series of constant focusing elements, in the sense of Eq. 4.65. This allows the non-periodic Hill's equation with arbitrary variable coefficient $K(s)$ to be reduced locally to an equation with constant coefficient K, which has solution for any K (either oscillatory or hyperbolic, depending on the sign of K).

In conclusion, the equivalence of matrices in Eq. 4.119 is formally preserved for an arbitrary non-periodic lattice as long as a piece-wise constant focusing representation of the beam line is allowed. The description of the motion is exact up to the chosen order of expansion of the matrix elements in the particle's coordinates.

4.3.14 Summary

The analysis of the single particle transverse dynamics has brought to the following findings.

- Alternated strong quadrupole focusing can guarantee stable motion in both transverse planes as predicted by Hill's equations, forcing particles to move in proximity of the reference orbit.
- Particle's motion in the phase space (u, u') is that of a quasi-harmonic oscillator. The Courant-Snyder parameters describe a phase space ellipse, on which the particle's representative point lies. The ellipse area is an invariant of the particle's motion in the assumption of linear, non-dissipative dynamics.
- Linear optics is described through the product of symplectic matrices. These can be written either in terms of the geometric parameters of the accelerator component, i.e., principal trajectories (Eq. 4.65), or of the lattice Courant-Snyder parameters (Eq. 4.119).
- Long-term stability in a periodic system requires that the trace of the 2×2 one-turn transfer matrix be $|Tr(M_t)| < 2$.

4.4   Beam Envelope

4.4.1   Statistical Emittance

Several hundreds' million particles can be stored in a bunch, in linacs as well in synchrotrons. While particle tracking codes are able to model the motion of individual particles, the theory of single particle dynamics can be used to reduce the complexity of the problem by keeping track of the beam envelope, i.e., of the bunch as a whole [5].

It was shown that the C-S parameters determine the shape and the orientation of a particle's phase space ellipse. The ellipse area—the C-S particle's invariant—depends from the particle's initial conditions (Eq. 4.102), which vary from particle to particle. If the motion of all beam particles is described by the same C-S parameters defined by the lattice (Eq. 4.119), then all ellipses are *omothetic*, at any s, and the beam is said to be *matched* to the lattice.

Fig. 4.14 Omothetic phase
space ellipses containing
different percentages of
beam particles

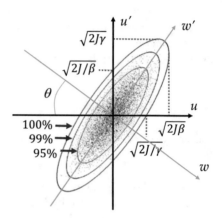

The first consequence of this is that, as long as phase space ellipses can be defined,
i.e., as long as the motion is linear and non-dissipative, particles' orbits in phase space
never overlap. The second implication is that the ensemble of representative points
of the charge distribution can be described by an "envelope ellipse" containing all
or, conventionally, a given percentage of beam particles, as shown in Fig. 4.14. The
phase space area occupied by the discrete distribution of N particles is therefore
approximated to the envelope ellipse area, called *beam emittance*.

Let us assume for simplicity a monochromatic beam (i.e., vanishing energy
spread) in non-dispersive motion. Conventionally, and especially for lepton beams,
the envelope ellipse semi-axes are made to coincide with the Root-Mean-Square
(rms) values $\sigma_u = \sqrt{\langle u^2 \rangle}$, $\sigma_{u'} = \sqrt{\langle u'^2 \rangle}$ of the distribution, where for simplicity we
assume a distribution centered both in u, u', i.e. $\langle u \rangle = \langle u' \rangle = 0$. Owing to the correspondence of ellipse axes and second order momenta of the charge distribution
function, the envelope ellipse phase space area is called *root-mean-square statistical
emittance* or, simply, rms emittance. In the absence of (u, u') correlation, it results
(compare with Eq. 4.108):

$$\epsilon_u = \frac{Area}{\pi} = \sigma_u \sigma_{u'} \tag{4.123}$$

In the most general case $\langle uu' \rangle \neq 0$, the envelope ellipse axes (directions w, w'
in Fig. 4.14) can be rotated with respect to the Cartesian axes by the angle θ which
minimizes the spread in position along the rotated axes. Namely, θ satisfies the least
square condition:

$$\frac{d}{d\theta} \sum_i w_i^2 = \frac{d}{d\theta} \sum_i \left(u_i' \cos \theta - u_i \sin \theta \right)^2 = 0 \tag{4.124}$$

The expansion of the expression above leads to:

$$\frac{d\langle w^2\rangle}{d\theta} = -\frac{2}{N}\sum_i \left(u_i'^2 \sin\theta\cos\theta + u_i u_i' \cos^2\theta - u_i u_i' \sin^2\theta - u_i^2 \sin\theta\cos\theta\right) =$$
$$= -\frac{2}{N}\sum_i \left[u_i'^2 \frac{\sin 2\theta}{2} - u_i u_i'(1 - 2\cos^2\theta) - u_i^2 \frac{\sin 2\theta}{2}\right] =$$
$$= -\frac{1}{N}\sum_i \left[(u_i'^2 - u_i^2)\sin 2\theta + 2u_i u_i' \cos 2\theta\right] =$$
$$= -\frac{1}{N}\sum_i \cos 2\theta \left[(u_i'^2 - u_i^2)\tan 2\theta + 2u_i u_i'\right] = 0$$

$$\Rightarrow \tan(2\theta) = \frac{2\langle uu'\rangle}{\langle u^2\rangle - \langle u'^2\rangle},$$

$$\langle w^2\rangle = \frac{1}{2}\left(\langle u^2\rangle + \langle u'^2\rangle + \frac{2\langle uu'\rangle}{\sin 2\theta}\right),$$

$$\langle w'^2\rangle = \frac{1}{2}\left(\langle u^2\rangle + \langle u'^2\rangle - \frac{2\langle uu'\rangle}{\sin 2\theta}\right)$$

(4.125)

By applying Eq. 4.123 to the new orthogonal axes w, w', and with the help of Eq. 4.125, one finds:

$$\epsilon_u = \sigma_w \sigma_{w'} = \frac{1}{2}\left(\langle u^2\rangle + \langle u'^2\rangle + \frac{2\langle uu'\rangle}{\sin 2\theta}\right)^{1/2}\left(\langle u^2\rangle + \langle u'^2\rangle - \frac{2\langle uu'\rangle}{\sin 2\theta}\right)^{1/2} =$$
$$= \left[(\langle u^2\rangle + \langle u'^2\rangle)^2 - \frac{4\langle uu'\rangle^2}{\sin^2 2\theta}\right]^{1/2} =$$
$$= \left[(\langle u^2\rangle + \langle u'^2\rangle)^2 - 4\langle uu'\rangle^2\left(1 + \cot^2 2\theta\right)\right]^{1/2} =$$
$$= \left\{(\langle u^2\rangle + \langle u'^2\rangle)^2 - 4\langle uu'\rangle^2\left[1 + \frac{(\langle u^2\rangle - \langle u'^2\rangle)^2}{4\langle uu'\rangle^2}\right]\right\}^{1/2} =$$
$$= \frac{1}{2}\left[4\langle u^2\rangle\langle u'^2\rangle - 4\langle uu'\rangle\right]^{1/2} =$$
$$= \sqrt{\langle u^2\rangle\langle u'^2\rangle - \langle uu'\rangle^2}$$

(4.126)

Averages are intended over the particles' ensemble:

$$\langle u^2\rangle = \frac{1}{N}\sum_{i=1}^{N} u_i^2, \qquad \langle u'^2\rangle = \frac{1}{N}\sum_{i=1}^{N} u_i'^2, \qquad \langle uu'\rangle = \frac{1}{N}\sum_{i=1}^{N} u_i u_i' \qquad (4.127)$$

Some additional algebric manipulation of Eq. 4.126 gives:

$$\epsilon_u = \sqrt{\frac{1}{N}\sum_i u_i^2 \frac{1}{N}\sum_i u_i'^2 - \left(\frac{1}{N}\sum_i u_i u_i'\right)^2} =$$

(4.128)

$$= \frac{1}{\sqrt{2}N}\sqrt{\sum_i \sum_j (u_i u_j' - u_j u_i')^2} = \frac{1}{N}\sqrt{2\sum_i \sum_j A_{ij}^2}$$

The rms emittance can therefore be interpreted as the rms value of the phase space areas of triangles (A_{ij}) made of any two points of the distribution and connected to the origin. However, if the transfer map of the accelerator is not linear, triangles do not transform into triangles necessarily. This suggests that the rms emittance is *not* preserved under *nonlinear* transformations.

The percentage of particles contributing to the rms emittance depends from the 2-D distribution function in the phase space. Since in many cases the charge distribution in accelerators is Gaussian, or similar to that, it is instructive to consider a Gaussian distribution hereafter. In such case, it is possible to calculate analytically the percentage of particles contributing to the rms emittance and, in a more general case, to the phase space area extending over an arbitrary number of sigmas in u and u'.

Let us consider the centered 2-D Gaussian distribution in Fig. 4.14; we assume for simplicity $\langle uu' \rangle = 0$:

$$f(u, u') = \frac{1}{2\pi \sigma_u \sigma_{u'}} e^{-\frac{u^2}{2\sigma_u^2}} e^{-\frac{u'^2}{2\sigma_{u'}^2}} \tag{4.129}$$

Projection of the distribution onto the u, u' axis provides, of course, a 1-D Gaussian distribution in the u, u' coordinate, respectively. Projection of the distribution onto the (u, u') plane with a cut-off at k-sigmas both in u, u' generates an ellipse of area $k^2 \sigma_u \sigma_{u'}'$. The ellipse equation is:

$$\frac{u^2}{k^2 \sigma_u^2} + \frac{u'^2}{k^2 \sigma_{u'}^2} = 1 \quad or \quad \frac{u^2}{2\sigma_u^2} + \frac{u'^2}{2\sigma_{u'}^2} = \frac{k^2}{2} \tag{4.130}$$

The relative fraction of particles contained in k-sigmas and normalized to 1 is, by virtue of Eq. 4.130:

$$P = \frac{f(0, 0) - f(x, x'; k)}{\hat{f}(x, x')} = 1 - \frac{f(x, x'; k)}{f(0, 0)} = 1 - e^{-\frac{k^2}{2}} \tag{4.131}$$

The rms emittance ϵ_u corresponds to $k = 1$ and therefore to $P(k = 1) = 39\%$. We find, for example, $P(1.5) = 67.5\%$, $P(2) = 86\%$, $P(2.45) = 95\%$, $P(3) = 98.9\%$, and $P(4) = 99.97\%$.

4.4.2   Transverse Beam Matrix

The equivalence of the envelope ellipse area of a matched beam and the statistical emittance is shown below by recalling the expression of particle's position and angular divergence as solutions of Hill's equation, and by averaging the particle's coordinates over the particles' ensemble. This translates into an average over the particles' invariant J_i and initial betatron phase $\mu_{0,i}$. We find:

$$\langle u^2 \rangle = \frac{1}{N} \sum_{i=1}^{N} 2J_i \beta(s) \cos^2 \Delta\mu_i(s) = \beta(s)\langle J \rangle$$

$$\langle u'^2 \rangle = \frac{1}{N} \sum_{i=1}^{N} \frac{2J_i}{\beta(s)} [\alpha(s)\cos\Delta\mu_i(s) + \sin\Delta\mu_i(s)]^2 = \gamma(s)\langle J \rangle$$

$$\langle uu' \rangle = -\frac{1}{N} \sum_{i=1}^{N} 2J_i \cos\Delta\mu_i(s) [\alpha(s)\cos\Delta\mu_i(s) + \sin\Delta\mu_i(s)] = -\alpha(s)\langle J \rangle$$

$$\Rightarrow \epsilon_u = \sqrt{\langle u^2 \rangle\langle u'^2 \rangle - \langle uu' \rangle^2} = \langle J \rangle\sqrt{\beta\gamma - \alpha^2} = \langle J \rangle$$

$$\Rightarrow \begin{cases} \sigma_u^2(s) = \langle u^2(s) \rangle = \epsilon_u \beta(s) \\[2mm] \sigma_u'^2(s) = \langle u'^2(s) \rangle = \epsilon_u \gamma(s) \\[2mm] \langle uu' \rangle = -\epsilon_u \alpha(s) \end{cases}$$

(4.132)

The condition $\alpha_u = 0$ identifies a local maximum or minimum of the beam's rms size (minimum or maximum of the rms angular divergence). The latter identities can be cast in matrix form:

$$\sigma := \epsilon_u \begin{pmatrix} \beta_u & -\alpha_u \\ -\alpha_u & \gamma_u \end{pmatrix} \equiv \begin{pmatrix} \langle u^2 \rangle & \langle uu' \rangle \\ \langle uu' \rangle & \langle u'^2 \rangle \end{pmatrix}$$

(4.133)

The matrix σ for the u-transverse plane is said (covariant) *beam matrix* and it satisfies $\sqrt{\det \sigma} = \epsilon_u$. It is essential to keep in mind that, while the C-S invariant is the single particle's constant of motion, the emittance is a quantity referring to an ensemble of particles. Indeed, Eq. 4.126 tells us that the emittance of a single particle is zero.

In order to find the transformation rule of σ in the presence of a transfer matrix M, we first consider the generic coordinates (u, u'), and calculate:

$$\vec{u}^T \sigma^{-1} \vec{u} = (u, u')\frac{1}{\epsilon^2}\epsilon \begin{pmatrix} \gamma_u & \alpha_u \\ \alpha_u & \beta_u \end{pmatrix} \begin{pmatrix} u \\ u' \end{pmatrix} = \frac{1}{\epsilon}(u, u') \begin{pmatrix} \gamma_u u + \alpha_u u' \\ \alpha_u u + \beta_u u' \end{pmatrix} =$$

$$= \frac{1}{\epsilon}(\gamma u^2 + 2\alpha uu' + \beta u'^2) = \frac{2J}{\epsilon} = const.$$

(4.134)

Since the single particle's phase space vector transforms as $\vec{u} = M\vec{u}_0$, Eq. 4.134 can be re-written as follows:

$$\vec{u}^T \sigma^{-1} \vec{u} = (M\vec{u}_0)^T \sigma^{-1} M\vec{u}_0 = \vec{u}_0^T M^T \sigma^{-1} M\vec{u}_0 \equiv \vec{u}_0^T \sigma_0^{-1} \vec{u}_0$$

$$\Rightarrow M^T \sigma^{-1} M = \sigma_0^{-1};$$
$$\sigma_0 = \left(M^T \sigma^{-1} M\right)^{-1} = M^{-1}\sigma \left(M^T\right)^{-1};$$
$$M\sigma_0 M^T = MM^{-1}\sigma(M^T)^{-1}M^T;$$

(4.135)

$$\Rightarrow \sigma(s) = M\sigma_0 M^T$$

In practical situations, if the beam is not matched to the lattice, a preliminary manipulation of the rms beam size and divergence is needed in order to adapt the beam's C-S parameters to the design values. This procedure is commonly carried out in the transverse planes with solenoid fields and/or quadrupole magnets, and it is called *optics matching*.

When a non-zero rms energy spread σ_δ is considered, the beam envelope is modified by the presence, if any, of the dispersion function (as for the beam size) and of its first derivative (as for the beam divergence). Owing to the linear superposition of the solutions u_β and $u_D = D_u \delta$ in Eq. 4.59, the most general expression for the rms beam size and angular divergence becomes:

$$\begin{cases} \sigma_u(s) = \sqrt{\langle u_\beta^2(s) \rangle + \langle u_D^2(s) \rangle} = \sqrt{\epsilon_u \beta_u(s) + (D_u(s)\sigma_\delta)^2} \\[4mm] \sigma_{u'}(s) = \sqrt{\langle u_\beta'^2(s) \rangle + \langle u_D'^2(s) \rangle} = \sqrt{\epsilon_u \gamma_u(s) + \left(D_u'(s)\sigma_\delta\right)^2} \end{cases} \tag{4.136}$$

We remind for completeness that, since the dispersion function is also a solution of Hll's complete equation, it is a periodic function when the magnetic lattice is periodic, and so its derivative is. In this case, the periodic dispersion function is noted as $\eta(s) = \eta(s + C)$, $\eta'(s) = \eta'(s + C)$ $\forall s$, see Eq. 4.113.

4.4.3 Transfer of Courant-Snyder Parameters

Equation 4.135 says that if the lattice map is described by principal trajectories, the beam's representative ellipse can be tracked by transporting the beam's C-S parameters through the lattice. The transfer matrix mapping the C-S parameters through the accelerator is found below by using the definition of principal trajectories:

$$\begin{cases} u = Cu_0 + Su_0' & \cdot (S') \\ u' = C'u_0 + Su_0' & \cdot (-S) \end{cases} \Rightarrow (CS' - SC')u_0 = u_0 = (S'u - Su')$$

$$\begin{cases} u = Cu_0 + Su_0' & \cdot (-C') \\ u' = C'u_0 + Su_0' & \cdot (C) \end{cases} \Rightarrow (CS' - SC')u_0' = u_0' = (Cu' - C'u)$$

Then, the C-S invariant is calculated:

$$\gamma_0 u_0^2 + 2\alpha_0 u_0 u_0' + \beta_0 u_0'^2 =$$

$$= \gamma_0 (S'u - Su')^2 + 2\alpha_0 (S'u - Su')(Cu' - C'u) + \beta_0 (Cu' - C'u)^2 =$$

$$= (\gamma_0 S'^2 - 2\alpha_0 C'S' + \beta_0 C'^2)u^2 +$$

$$\quad -2(\gamma_0 SS' - \alpha_0 CS' - \alpha_0 SC' + 2\beta_0 CC')uu' +$$

$$\quad +(\gamma_0 S^2 - 2\alpha_0 CS + \beta_0 C^2)u'^2 \equiv$$

$$\equiv \gamma u^2 + 2\alpha uu' + \beta u'^2$$

Equating member-to-member in the previous expression we find:

$$
\begin{pmatrix} \beta \\ \alpha \\ \gamma \end{pmatrix}_s = \begin{pmatrix} C^2 & -2CS & S^2 \\ -CC' & CS' + SC' & -SS' \\ C'^2 & -2C'S' & S'^2 \end{pmatrix} \begin{pmatrix} \beta \\ \alpha \\ \gamma \end{pmatrix}_0 \tag{4.137}
$$

In reality, the second order momenta of the charge distribution can be retrieved from measurements of the beam size and divergence at a certain location in the accelerator. Equation 4.126 is then used to calculate ϵ_u. The correspondence established by Eq. 4.133 allows one to determine the C-S parameters associated to the charge distribution. If the "beam C-S parameters" are matched to the "lattice C-S parameters" at a specific location, then the beam dynamics can be predicted and controlled from that point on by just looking to the beam envelope. Namely, the beam dynamics is reduced to the evolution of the C-S parameters (in geometric sense, of the beam envelope ellipse) through the lattice according to Eq. 4.137.

4.4.3.1 Discussion: Low-β Insertion

Straight sections, or drifts, are present in linacs and synchrotrons to accommodate, for example, beam diagnostics, insertion devices, detectors, etc. It is sometimes convenient to design the beam optics such that the transverse beam sizes along the drift section are minimized. These sections are therefore named *low-β insertions*. What are the C-S parameters at the entrance of the drift which, for any given drift length L, minimize the betatron function overall? What is the maximum phase advance allowed by such special configuration?

For symmetry, a minimum value of the beam size averaged over the drift implies a beam waist, i.e., a minimum of the betatron function, in the middle of the section. The two transverse planes behave, of course, identically. It is then convenient to define the origin of the s-axis at the middle point of the drift. In each plane, the drift transfer matrix is built with principal trajectories $C = S' = 1$, $C' = 0$, $S = L$ (see Eq. 4.72). Then, the transfer matrix for the C-S parameters with initial conditions $\beta_0 > 0$, $\alpha_0 = 0$ is determined according to Eq. 4.137. The Betatron function propagates from the waist towards the drift ends as follows:

$$
\begin{cases} \beta(s) = C(s)^2 \beta_0 - 2C(s)S(s)\alpha_0 + S(s)^2 \gamma_0 = \beta_0 - 2s\alpha_0 + s^2\gamma_0 = \beta_0 + \frac{s^2}{\beta_0} \\[2mm] \alpha(s) = -CC'\beta_0 + (CS' + SC')\alpha_0 - SS'\gamma_0 = -\frac{s}{\beta_0} \end{cases}
$$

$$\tag{4.138}$$

By construction, $\beta(s)$ is already set to its minimum value β_0 at the midpoint ($s = 0$). Its value is minimized along the whole drift by a suitable choice of β_0, which is found by imposing:

$$
\left(\frac{d\beta(s)}{d\beta_0} \right)_{s=L/2} = 1 - \frac{L^2}{4\beta_0^2} \equiv 0 \ \Rightarrow \ \beta_{0,opt} = \frac{L}{2}. \tag{4.139}
$$

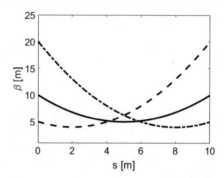

Fig. 4.15 Betatron function along a drift section. The symmetric solution (solid, $\alpha_0 = 1$, $\beta_0 = L$) determines the minimum value of the betatron function averaged over the whole section. All other solutions can lead to even lower local minima, but the lower the minimum is, the faster the rising of the function elsewhere will be, so that the average value is eventually larger (dashed, $\alpha_0 = 0.5$, $\beta_0 = L/2$; dot-dashed, $\alpha_0 = 2$, $\beta_0 = 2L$)

It follows from Eq. 4.138 that the C-S parameters at the edges of the drift are $\beta(L/2) = L$, $\alpha(L/2) = -1$ (the beam is diverging if seen from the midpoint, converging if seen from the opposite direction).

Figure 4.15 shows the betatron function in a low-β insertion for optimal and non-optimal initial C-S parameters. Any deviation from the initial conditions found above leads to a larger average betatron function, although a local minimum can still be obtained.

The betatron phase advance can be calculated from Eq. 4.90:

$$\Delta\mu = \int_{-L/2}^{L/2} \frac{1}{\beta(s)} ds = \int_{-L/2}^{L/2} \frac{1}{\beta_0 + \frac{s^2}{\beta_0}} ds = 2 \arctan\left(\frac{L}{2\beta_0}\right) = \frac{\pi}{2} \qquad (4.140)$$

The phase advance tends to π when $\frac{L}{\beta_0} \to \infty$.

4.4.4 Longitudinal Beam Matrix

Equation 4.31 for the particle's longitudinal oscillations is formally identical to Hills' equation for the betatron motion, Eq. 4.78. In the former case, the oscillations are defined in the longitudinal phase space (z, δ), z and δ being the reduced variables for the particle's longitudinal position and relative longitudinal momentum. When the latter is constant on average over one turn, or approximately constant during adiabatic acceleration, the longitudinal motion is (approximately) purely harmonic.

By virtue of such similarity, the longitudinal motion can be described with formalism identical to that adopted so far for the transverse planes, see Eq. 4.102, with the additional feature of a constant amplitude of oscillation (independent from s). Hence, longitudinal C-S parameters, C-S invariant, statistical emittance, and beam matrix, can be equivalently defined. The single particle's longitudinal invariant J_z is constant as long as the motion can be approximated to a linear and non-dissipative one.

The longitudinal beam matrix Σ transforms as in Eq. 4.135:

$$\Sigma(s) = M\Sigma_0 M^T = \epsilon_z \begin{pmatrix} \beta_z & -\alpha_z \\ -\alpha_z & \gamma_z \end{pmatrix} = \begin{pmatrix} \langle z^2 \rangle & \langle z\delta \rangle \\ \langle z\delta \rangle & \langle \delta^2 \rangle \end{pmatrix} \qquad (4.141)$$

The linear energy correlation along the bunch (linear energy chirp, see Eq. 4.25) and the beam total relative energy spread result:

$$h = \frac{\langle z\delta \rangle}{\langle z^2 \rangle} = -\frac{\alpha_z}{\beta_z}$$
$$(4.142)$$
$$\sigma_\delta = \sqrt{\langle \delta^2 \rangle} = \epsilon_z \gamma_z$$

The Beam's longitudinal emittance is:

$$\epsilon_z = \sqrt{det(\Sigma)} = \sqrt{\langle z^2 \rangle \langle \delta^2 \rangle - \langle z\delta \rangle^2} \rightarrow \sigma_z \sigma_\delta \qquad (4.143)$$

and the limit is for null energy chirp. In this case, the energy spread is said to be *uncorrelated*.

4.4.5 Normalized Emittance

The transverse rms emittance is also named *geometric* (also *natural* in synchrotrons) to recall its geometric meaning in the phase space, $\epsilon_u \sim \sigma_u \sigma_{u'}$. By virtue of Eq. 4.132, it is also a *constant of motion* under the assumption of linear and non-dissipative dynamics.

But, if the beam's mean energy is not constant, ϵ_u is not expected to be constant anymore because a frictional force is introduced, which translates for example into an instantaneous variation of the particle's angular divergence via Eq. 4.100. To quantify the emittance variation with beam's energy, we first consider instantaneous longitudinal acceleration, small energy spread ($\sigma_\delta \ll 1$), and we assume for simplicity $\alpha_u = 0$:

$$\epsilon_u = \sigma_u \sigma_{u'} = \frac{\sigma_u \sigma_{p_u}}{p_{z,s}} = \left(\frac{\sigma_u \sigma_{p_u}}{m_0 c} \right) \frac{1}{\beta_z \gamma} \qquad (4.144)$$

β_z, γ are the common Lorentz factors. It results that the geometric emittance is inversely proportional to the beam's mean energy because pure longitudinal acceleration does affect neither the beam size nor the transverse momentum, but only the longitudinal momentum. In practice, the higher the energy is, the more collimated the beam will be.

To compare the emittance of accelerated beams somehow independently from the final beam's energy, an *energy-normalized emittance* is introduced (simply "normalized emittance" hereafter):

$$\epsilon_{n,u} := \beta_z \gamma \epsilon_u \cong \gamma \epsilon_u \qquad (4.145)$$

and the approximation is for ultra-relativistic particles ($\beta_z \cong \beta \to 1$). It results that $\epsilon_{n,u}$ is also defined in statistical sense, therefore it is constant under the same conditions which apply to ϵ_u.

In analogy to the motion in the transverse planes, the longitudinal rms emittance is $\epsilon_z = \sigma_z \sigma_\delta \propto \frac{1}{\beta_z \gamma}$. Here, we are assuming that particles are in a deep relativistic regime, so that their velocity is approximately c, in spite of their relative momentum deviation, and therefore the bunch length is constant (this is exact in high energy linacs; other effects concur to the determination of emittances in a synchrotron, which will be treated later on). Consequently, the normalized longitudinal emittance is also defined according to Eq. 4.145:

$$\epsilon_{n,z} := \beta_z \gamma \epsilon_z = \beta_z \gamma \cdot \frac{\sigma_z \sigma_{p_z}}{p_{z,s}} = \frac{\sigma_z \sigma_{p_z}}{m_0 c} \approx \frac{\sigma_z \sigma_E}{\beta_z m_0 c^2} \approx \sigma_z \sigma_\gamma \qquad (4.146)$$

The last two approximations use $\Delta p = \Delta E/(\beta c)$, by neglecting the transverse momenta with respect to the longitudinal ($p_x, p_y << p_z \cong p$), and by taking the ultra-relativistic limit $\beta_z \to 1$. Equation 4.146 says that in the presence of linear, ultra-relativisitic and non-dissipative motion, the product of bunch length and *uncorrelated* energy spread is a constant of motion.

4.4.6   Beam Brightness

Many applications of particle accelerators, and especially at high beam energies, require not only a high charge spatial density, but also the capability of the beam to be collimated by strong magnetic field gradients. In other words, the beam envelope should have both small transverse size and angular divergence, which translate into small transverse emittance.

The charge density evaluated over the transverse 4-D phase space and the bunch duration is called 5-D *peak brightness*:

$$B = \frac{Q}{\epsilon_x \epsilon_y \sigma_t} = \frac{I}{\epsilon_x \epsilon_y} \qquad (4.147)$$

Its energy-normalized counterpart is defined in terms of the normalized transverse emittances:

$$B_n = \frac{I}{\epsilon_{x,n} \epsilon_{y,n}} \qquad (4.148)$$

In the ultra-relativistic limit, $B \cong \gamma^2 B_n$.

In high energy light sources, the spectral bandwidth of the emitted radiation can be affected by the beam's relative energy spread σ_δ. For this reason the 6-D beam brightness is conveniently introduced:

$$\mathbb{B} = \frac{I}{\epsilon_x \epsilon_y \sigma_\delta} \tag{4.149}$$

Its energy-normalized counterpart, defined in the ultra-relativistic limit, makes use of the normalized longitudinal emittance in Eq. 4.146:

$$\mathbb{B}_n = \frac{Qc}{\epsilon_{x,n} \epsilon_{y,n} \sigma_z \sigma_\gamma} \tag{4.150}$$

In the ultra-relativisitc limit, $\mathbb{B} \approx \gamma^3 \mathbb{B}_n$.

Light sources emitting coherent radiation can also be interested to high peak currents, because this drives the intensity of the emitted radiation. Incoherent light sources and colliders, instead, are in most cases dealing with the average peak current. This is the total beam charge evaluated over a turn in synchrotrons, the beam duty cycle in linacs. In general, if the beam is made of a train of n_b bunches and the repetition rate of the train production is f_t, the 5-D *average brightness* results:

$$\langle B \rangle = B \cdot \sigma_t \cdot f_t \cdot n_b = \frac{\langle I \rangle}{\epsilon_x \epsilon_y} \tag{4.151}$$

4.4.6.1 Discussion: Normalized Emittance of a Large Energy Spread Beam

Let us demonstrate that for an ultra-relativistic but not monochromatic particle beam, the rms normalized transverse emittance is proportional to the beam momentum spread, and that its expression reduces to Eq. 4.145 for vanishing energy spread.

Equation 4.144 shows that the normalized emittance is defined in terms of second order momenta of the spatial and transverse momentum distribution, and we recall that $u' = p_u/p_z$. We find:

$$\epsilon_{n,u}^2 = \frac{\langle u^2 \rangle \langle p_u^2 \rangle}{m_0^2 c^2} - \frac{\langle up_u \rangle^2}{m_0^2 c^2} = \langle u^2 \rangle \langle \beta_z^2 \gamma^2 u'^2 \rangle - \langle u\beta_z \gamma u' \rangle^2 \tag{4.152}$$

In the simpler case $\langle u' p_z \rangle \approx 0$ we can write:

$$\epsilon_{n,u}^2 \approx \langle u^2 \rangle \langle u'^2 \rangle \langle \beta_z^2 \gamma^2 \rangle - \langle uu' \rangle^2 \langle \beta_z \gamma \rangle^2 \tag{4.153}$$

With the definition of relative momentum spread

$$\sigma_\delta^2 = \frac{\langle p^2 \rangle - \langle p \rangle^2}{\langle p \rangle^2} \approx \frac{\langle \beta_z^2 \gamma^2 \rangle - \langle \beta_z \gamma \rangle^2}{\langle \beta_z \gamma \rangle^2}, \tag{4.154}$$

Eq. 4.153 becomes:

$$\epsilon_{n,u}^2 \approx \langle u^2 \rangle \langle u'^2 \rangle \langle \beta_z \gamma \rangle^2 (1 + \sigma_\delta^2) - \langle uu' \rangle^2 \langle \beta_z \gamma \rangle^2 = \langle \beta_z \gamma \rangle^2 \left(\epsilon_u^2 + \sigma_u^2 \sigma_{u'}^2 \sigma_\delta^2 \right) =$$

$$= \langle \beta_z \gamma \rangle^2 \epsilon_u^2 \left[1 + \left(1 + \frac{\langle uu' \rangle^2}{\epsilon_u^2} \right) \sigma_\delta^2 \right];$$

$$\epsilon_{n,u} = \langle \beta_z \gamma \rangle \epsilon_u \sqrt{1 + \sigma_\delta^2 \left(1 + \alpha_u^2 \right)},$$

$$(4.155)$$

which reduces to Eq. 4.145 for $\sigma_\delta \rightarrow 0$.

Owing to $\alpha_u(s) \neq 0$, the rms transverse normalized emittance evaluated for a not negligible relative energy spread is not constant any longer, even for beam transport at constant mean energy. This has some analogy to the case of an effective emittance in the presence of chromatic motion, where dispersion affects the beam size and divergence. Thereby, the rms emittance calculated from the second order momenta of the distribution (which are now contributed from betatron *and* dispersive motion, see Eq. 4.136) is apparently larger than the emittance in the presence of betatron oscillations only.

In the presence of acceleration, the normalized emittance of a large energy spread beam is not constant even in case of "smooth optics", or $\langle \alpha_u \rangle \approx 0$ along the line, because $\epsilon_{n,u} \sim \gamma \epsilon_u \sigma_\delta \sim 1/\gamma$ (the absolute uncorrelated energy spread is assumed to be approximately constant, see Eq. 4.7 and discussion there). This is consistent with the fact that the statistical emittance in Eq. 4.155 is at third order in the particle coordinates, $\epsilon_{n,u} \sim \sigma_u \sigma_{u'} \sigma_E$. Namely, the single particle's motion is nonlinear, and the beam's optics is affected by chromatic aberrations.

4.4.6.2 Discussion: Ultimate Beam Emittance

What is the smallest transverse emittance of a particle beam?

Since the answer implies the smallest scale (both in space and momentum) of an ensemble of massive particles, it has to have quantistic nature. Heisenberg's Uncertainty Principle applied to the horizontal plane of motion states:

$$\Delta x \Delta p_x \geq \frac{h}{4\pi} \qquad (4.156)$$

where h is the Planck's constant, and the widths are intended to be standard deviations. The beam angular divergence is $\Delta x' = \frac{\Delta p_x}{p_z} = \frac{\Delta p_x}{\beta_z \gamma m_0 c}$, and the equation can be re-written as follows:

$$\epsilon_x = \Delta x \Delta x' = \frac{\Delta x \Delta p_x}{\beta_z \gamma m_0 c} \geq \frac{1}{4\pi \beta_z \gamma} \frac{h}{m_0 c} = \frac{1}{\beta_z \gamma} \frac{\lambda_C}{4\pi}$$

$$(4.157)$$

$$\Rightarrow \epsilon_{n,u} \geq \frac{\lambda_C}{4\pi}$$

The *Compton wavelength* $\lambda_C = \frac{h}{m_0 c}$ is the wavelength of a photon whose energy is equal to the rest energy of the massive particle: $E = h \nu_c = \frac{hc}{\lambda_c} = m_0 c^2$, and

$p = \frac{h}{\lambda_c} = m_0 c = \frac{E}{c}$ (see De Broglie's wave-particle duality in Eq.1.41), where $\lambda_C = 2.426 \cdot 10^{-6}$ μm for an electron. State-of-the-art electron beam normalized emittances have been measured at $\sim 10^{-3}$ μm level for $\sim$fC beam charges, in the keV beam energy range.

References

1. J. Le Duff, Dynamics and acceleration in linear structures, longitudinal beam dynamics in circular accelerators, in *Proceedings of CERN Accelerator School: 5th General Accelerators Physical Course*, CERN 94-01, vol. I, ed. by S. Turner, Geneva, Switzerland (1994), pp. 277–311
2. J. Rossbach, P. Schmuser, Basic course on accelerator optics, in *Proceedings of CERN Accelerator School: 5th General Accelerators Physical Course*, CERN 94-01, vol. I, ed. by S. Turner, Geneva, Switzerland (1994), pp. 17–68
3. M. Sands, *The Physics of Electron Storage Rings: An Introduction*, SLAC-121, UC-28 (ACC) (Stanford Linear Accelerator Center, Menlo Park, CA, USA, 1979), pp. 18–69
4. M. Martini, *An Introduction to Transverse Beam Dynamics in Accelerators*, CERN/PS 96-11 (PA). Lecture at the Joint University Accelerators School, Archamps, France (1996), pp. 1–43, 84–91
5. J. Buon, Beam phase space and emittance, in *Proceedings of CERN Accelerator School: 5th General Accelerators Physical Course*, CERN 94-01, vol. I, ed. by S. Turner, Geneva, Switzerland (1994), pp. 17–88

Hamiltonian Dynamics

5

The single particle's Hamiltonian in the presence of external e.m. field is introduced. It will be used to retrieve Hill's equations of motion, so to underline the approximations required to describe a real accelerator as a Hamiltonian system. The concept of beam emittance in the sense of Liouville's theorem is also presented, and its properties discussed in the wider framework of Poincare'-Cartan invariants. The Hamiltonian formalism is a powerful tool in accelerator physics for the description of particles' motion in the presence of, e.g., nonlinear and coupled dynamics.

5.1 Single Particle Dynamics

5.1.1 Lagrange's Equation

Let us consider a system of N particles characterized at the time t by the generalized coordinates $\vec{q}(t) = (q_1, \ldots, q_N)_t, \dot{\vec{q}}(t) = (\dot{q}_1, \ldots, \dot{q}_N)_t$. For simplicity, we consider a system free from holonomic constraints, such as N particles in free space.

The *Lagrangian* of the system [1,2] is the function $L = L(\vec{q}, \dot{\vec{q}}, t)$ which satisfies the principle of *minimum action* when the system evolves along a phase space trajectory in the time interval $t_2 - t_1$:

$$\delta W = \delta \int_{t_1}^{t_2} L(\vec{q}, \dot{\vec{q}}, t)dt = \int_{t_1}^{t_2} L(\vec{q} + \delta\vec{q}, \dot{\vec{q}} + \delta\dot{\vec{q}}, t)dt - \int_{t_1}^{t_2} L(\vec{q}, \dot{\vec{q}}, t)dt \equiv 0 \tag{5.1}$$

By expanding the Lagrangian at the first order in $q, \dot{q}$ one finds:

$$\delta W = \delta \int_{t_1}^{t_2} \left(\frac{\partial L}{\partial \vec{q}}\delta\vec{q} + \frac{\partial L}{\partial \dot{\vec{q}}}\delta\dot{\vec{q}} \right) dt = \left. \frac{\partial L}{\partial \dot{\vec{q}}}\delta\vec{q} \right|_{t_1}^{t_2} + \int_{t_1}^{t_2} \left[\frac{\partial L}{\partial \vec{q}} - \frac{d}{dt}\left(\frac{\partial L}{\partial \dot{\vec{q}}} \right) \right] = 0 \tag{5.2}$$

© The Author(s), under exclusive license to Springer Nature Switzerland AG 2022
S. Di Mitri, *Fundamentals of Particle Accelerator Physics*, Graduate Texts in Physics,
https://doi.org/10.1007/978-3-031-07662-6_5

and the second term is obtained from integration by parts. By imposing a variation δW with fixed extreme points $\delta \vec{q}(t_1) = \delta \vec{q}(t_2) = 0$, the first term becomes null, while the second term gives a set of N 2nd order differential equations of motion, the so-called *Lagrange's equations*:

$$\frac{d}{dt}\left(\frac{\partial L}{\partial \dot{q}_i}\right) - \frac{\partial L}{\partial q_i} = 0, \quad i = 1, \dots, N \tag{5.3}$$

Experimental observation supported by the assumption of time homogeneity and space isotropy, suggests that the Lagrangian of an ensemble of non-interacting particles in free space is just the total kinetic energy [1]:

$$L = \frac{1}{2}\sum_{i=1}^{N} m_i \dot{q}_i^2 \equiv T \tag{5.4}$$

In the presence of a potential energy of mutual interaction $V(\vec{q})$, the Lagrangian is $L = T - V$, and Lagrange's equations reduce to Newton's equations of motion:

$$\begin{cases} \frac{\partial L}{\partial q_i} = -\frac{\partial V}{\partial q_i} \\ \frac{\partial L}{\partial \dot{q}_i} = m_i \dot{q}_i \end{cases} \Rightarrow m_i \ddot{q}_i = -\frac{\partial V}{\partial q_i} \equiv F(q_i) \tag{5.5}$$

5.1.2 Hamilton's Equations

Each representative point of the particles ensemble can be equivalently described through a new pair of generalized coordinates, whose second component, denominated *generalized momentum*, is a function of $\dot{\vec{q}}$ only:

$$p_i := \frac{\partial L}{\partial \dot{q}_i} \Rightarrow \dot{p}_i = \frac{d}{dt}\frac{\partial L}{\partial \dot{q}_i} = \frac{\partial L}{\partial q_i} = F(q_i) \tag{5.6}$$

It follows that the time-variation of the ith particle's momentum is only due to external forces acting on the particle. If $\vec{F}(\vec{q}) = 0$, $\vec{p}_{tot} = const.$ and the system is *isolated*.

Let us now investigate the properties of L as function of time:

$$\frac{d}{dt}L(\vec{q}, \dot{\vec{q}}; t) = \sum_i \frac{\partial L}{\partial q_i}\frac{dq_i}{dt} + \sum_i \frac{\partial L}{\partial \dot{q}_i}\frac{d\dot{q}_i}{dt} + \frac{\partial L}{\partial t} = \sum_i \left(\frac{\partial L}{\partial q_i}\dot{q}_i + \frac{\partial L}{\partial \dot{q}_i}\ddot{q}_i\right) + \frac{\partial L}{\partial t}$$

$$= \frac{d}{dt}\sum_i \dot{q}_i\frac{\partial L}{\partial \dot{q}_i} + \frac{\partial L}{\partial t}$$

$$\Rightarrow \frac{d}{dt}\left(\sum_i \dot{q}_i\frac{\partial L}{\partial \dot{q}_i} - L\right) \equiv \frac{d}{dt}H(\vec{q}, \vec{p}, t) = -\frac{\partial L}{\partial t}$$

$$\tag{5.7}$$

The function $H(\vec{q}, \vec{p}, t) = \vec{p}\dot{\vec{q}} - L(\vec{q}, \dot{\vec{q}}(\vec{p}), t)$ is the *Hamiltonian* of the system [1,2]. By virtue of Eq. 5.7, if L does not depend explicitly from time, namely $\frac{\partial L}{\partial t} = 0$, the Hamiltonian is a constant of motion, or $H(\vec{q}, \vec{p}, t) = H(\vec{q}, \vec{p}) = const$. If we assume, for example, that this holds for the aforementioned Lagrangian $L = T - V$, we find:

$$\frac{dH}{dt} = \frac{d}{dt}\left(\sum_i m_i \dot{q}_i^2 - T + V\right) = \frac{d}{dt}(T + V) = \frac{dE}{dt} \equiv 0 \qquad (5.8)$$

In conclusion, the Hamiltonian is the total energy and $E = const$.

This result is a special case of *Noether's theorem*, which states that if a group of transformations leaves L invariant in form, then the transformation gives rise to an invariant of the motion, i.e., a quantity which is constant in time. So, if $L = T - V$ does not depend from time, any translation of the t-coordinate leaves L unchanged. The quantity conserved is the total energy. The two quantities (t, E) constitute a Hamiltonian pair in the sense defined below.

From the definition of H in Eq. 5.7, we can calculate the first derivatives of H with respect to its variables:

$$\begin{cases} \dfrac{\partial H}{\partial p_i} = \dot{q}_i \\[2ex] \dfrac{\partial H}{\partial q_i} = -\dfrac{\partial L}{\partial q_i} = -\dfrac{d}{dt}\dfrac{\partial L}{\partial \dot{q}_i} = -\dot{p}_i \end{cases} \qquad (5.9)$$

These expressions are called *Hamilton's equations*, and the generalized coordinates satisfying them are called *canonically conjugated variables* or *pair*. They allow a description of the system dynamics with a set of $2N$ *1st* order differential equations, equivalent to the set of N *2nd* order Lagrange's equations.

A map of the canonically conjugated variables $\vec{q}, \vec{p}$ to the new pair $\vec{Q}(\vec{q}, \vec{p})$, $\vec{P}(\vec{q}, \vec{p})$ is said *canonical transformation* if it preserves Hamilton's equations:

$$\begin{cases} \dfrac{dQ_i}{dt} = \dfrac{\partial H'}{\partial P_i} \\[2ex] \dfrac{dP_i}{dt} = -\dfrac{\partial H'}{\partial Q_i} \end{cases} \qquad (5.10)$$

In general, the Hamiltonian expressed in terms of the new coordinates $H'(\vec{Q}, \vec{P})$ may have a different form than the old one. $\vec{Q}, \vec{P}$ are still canonical coordinates. This allows us to define the Hamiltonian motion as a succession of canonical transformations.

5.1.3 Single Particle Hamiltonian

The Lagrangian of a relativistic particle in free space has to satisfy the principle of minimum action in the laboratory reference frame:

$$W = \int_{t_1}^{t_2} L(t)dt = \int_{\tau_1}^{\tau_2} \gamma L(\tau)d\tau = const.$$

$$\Rightarrow \gamma L(\tau) = inv.$$

(5.11)

where $\tau = dt/\gamma$ is the proper time, hence a Lorentz's invariant.

We require that the Lagrangian be explicitly independent from $\vec{q}$ (invariant for spatial translation) and at most linearly dependent from $\dot{\vec{q}}$. Moreover, the Lagrangian has to have the dimension of energy. The most immediate single particle invariant is the inertial mass $m_0 c^2$, therefore we propose $\gamma L = -m_0 c^2$.

We now consider the presence of an external e.m. force and, in particular, of a scalar electric potential ϕ and a magnetic vector potential $\vec{A}$. In this case, we additionaly prescribe that the Lagrangian be linear in the particle's electric charge, in the potentials, and in the particle's velocity:

$$L = -\frac{m_0 c^2}{\gamma} - e\phi(\vec{u}) + e\vec{v} \cdot \vec{A}$$

(5.12)

The canonical momentum for the ith plane of motion ($i = x, y, z$) is:

$$P_i = \frac{\partial L}{\partial v_i} = \frac{\partial}{\partial v_i}\left(-m_0 c^2 \sqrt{1 - \frac{v_i^2}{c^2}} + e\vec{v}\vec{A}\right) = \gamma m_0 v_i + eA_i = p_i + eA_i$$

$$\Rightarrow v_i = \frac{P_i - eA_i}{\gamma m_0}$$

(5.13)

and $p_i = \gamma m_0 v_i$ is the usual kinetic momentum. The Hamiltonian results:

$$\begin{aligned}
H = \vec{P} \cdot \vec{v} - L &= \frac{\vec{P}(\vec{P} - e\vec{A})}{\gamma m_0} + \frac{m_0 c^2}{\gamma} + e\phi - \frac{e\vec{A}(\vec{P} - e\vec{A})}{\gamma m_0} \\
&= \frac{1}{\gamma m_0}\left[(\vec{P} - e\vec{A})^2 + m_0^2 c^2\right] + e\phi = \frac{(\vec{P} - e\vec{A})^2 + m_0^2 c^2}{(E/c^2)} + e\phi \\
&= c\frac{(\vec{P} - e\vec{A})^2 + m_0^2 c^2}{\sqrt{p^2 + m_0^2 c^2}} + e\phi = c\frac{(\vec{P} - e\vec{A})^2 + m_0^2 c^2}{\sqrt{(\vec{P} - e\vec{A})^2 + m_0^2 c^2}} + e\phi \\
&= c\sqrt{(\vec{P} - e\vec{A})^2 + m_0^2 c^2} + e\phi
\end{aligned}$$

(5.14)

5.1.3.1 Discussion: Lagrangian in Free Space

We want to verify the correctness of our guess for the Lagrangian and the Hamiltonian of a relativistic particle in free space, by demonstrating that the particle's momentum is conserved, and that the Hamiltonian is the particle's total energy. Does the proposed Lagrangian reduce to the well-known classical expression, $L = T$, in the non-relativistic approximation?

Since the Lagrangian proposed in Eq. 5.12 is invariant under spatial translations, the particle's momentum has to be a conserved quantity. We adopt the notation u, v for the particle's position and velocity, in each plane. Lagrange's equation is:

$$\frac{d}{dt} \frac{\partial L}{\partial v} - \frac{\partial L}{\partial u} = 0;$$

$$\frac{d}{dt} \left[-m_0 c^2 \frac{\partial}{\partial v} \sqrt{1 - \frac{v^2}{c^2}} \right] = \frac{d}{dt} \left[m_0 c^2 \frac{\gamma}{2} \frac{2v}{c^2} \right] = m_0 \frac{d(\gamma v)}{dt} = \frac{dp}{dt} = 0 \qquad (5.15)$$

$$\Rightarrow p = const.$$

The Hamiltonian is calculated from the Lagrangian:

$$H = \vec{P} \cdot \vec{v} - L = \gamma m_0 v^2 + \frac{m_0 c^2}{\gamma} = \frac{m_0}{\gamma} (\gamma^2 v^2 + c^2)$$

$$= \frac{m_0 c^2}{\gamma} (1 + \gamma^2 \beta^2) = \frac{m_0 c^2}{\gamma} (1 + \gamma^2 - 1) = \gamma m_0 c^2 \qquad (5.16)$$

$$\Rightarrow H = E$$

It is immediate to see that in the non-relativistic limit:

$$L(\beta \to 0) \approx -m_0 c^2 \left(1 - \frac{\beta^2}{2} \right) = -m_0 c^2 + \frac{1}{2} m_0 v^2 = T - U_0 \qquad (5.17)$$

The mass-energy equivalence introduced by Special Relativity re-defines a baseline of the potential energy of a particle in free space, which amounts to the particle's rest energy.

5.1.4 Hill's Equation

Hill's equations are derived below [3,4] from the single particle Hamiltonian in Eq. 5.14, with two additional prescriptions: (i) acceleration is null or adiabatic, namely $\phi \approx 0$, and (ii) the magnetic field is static and purely transversal, $\vec{A} = (0, 0, A_z)$.

The change of coordinates from a Cartesian system $\vec{u}_c = (x_c, y_c, z_c)$ to a Frenet-Serret system $\vec{u} = (x, y, s)$ is illustrated in Fig. 5.1. It results:

$$\begin{cases} x_c = (R + x) \cos \frac{s}{R} - R \\ y_c = y \\ z_c = (R + x) \sin \frac{s}{R} \end{cases} \qquad (5.18)$$

We now define a generatrix function which will be used to calculate the momenta canonincally conjugated to the particle's spatial coordinates in the Frenet-Serret

Fig. 5.1 Cartesian and
Frenet-Serret reference
systems. Q is the
synchronous particle, P is
the generic particle, R is the
local radius of curvature of
the synchronous particle

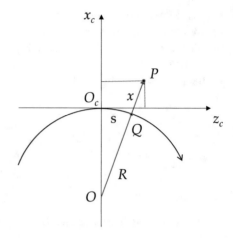

system. If $(p_{x,c},\ p_{y,c},\ p_{z,c},\ )$ are the canonical momenta in the Cartesian system, and
since by definition the generatrix function has to satisfy $\vec{u}_c = -\frac{\partial F}{\partial \vec{p}_{u,c}}$, we propose
$F = -x_c p_{x,c} - y_c p_{y,c} - z_c p_{z,c}$. Then we have:

$$\vec{p} = -\frac{\partial F}{\partial \vec{u}} = -\frac{\partial F}{\partial \vec{u}_c}\frac{\partial \vec{u}_c}{\partial \vec{u}}; \tag{5.19}$$

$$\Rightarrow \begin{cases} p_x = p_{x,c}\cos\frac{s}{R} + p_{z,c}\sin\frac{s}{R} \\ p_y = p_{y,c} \\ p_s = -p_{x,c}\left(1+\frac{x}{R}\right)\sin\frac{s}{R} + p_{z,c}\left(1+\frac{x}{R}\right)\cos\frac{s}{R} \equiv \left(1+\frac{x}{R}\right)p_z \end{cases}$$

Similarly, the magnetic field vector has components:

$$\begin{cases} A_x = A_{x,c}\cos\frac{s}{R} + A_{z,c}\sin\frac{s}{R} \\ A_y = A_{y,c} \\ A_s = -A_{x,c}\left(1+\frac{x}{R}\right)\sin\frac{s}{R} + A_{z,c}\left(1+\frac{x}{R}\right)\cos\frac{s}{R} \equiv \left(1+\frac{x}{R}\right)A_z \end{cases} \tag{5.20}$$

The curvature term x/R in the z-component of the momentum and of the field vector
describes the different orbit travelled by the generic particle in the bending plane by
virtue of its momentum deviation, where in general $\delta = \frac{p_z - p_s}{p_s} \neq 0$.

The Hamiltonian in Eq. 5.14, expressed as function of the canonical pair in the
Frenet-Serret system, becomes:

$$H = c\sqrt{(P_z - eA_z)^2 + (P_y - eA_y)^2 + (P_x - eA_x)^2 + m_0^2 c^2}$$

$$= c\sqrt{\frac{(P_s - eA_s)^2}{\left(1+\frac{x}{R}\right)^2} + (P_y - eA_y)^2 + (P_x - eA_x)^2 + m_0^2 c^2} \tag{5.21}$$

A new Hamiltonian $\mathbb{H} = -P_s(H)$ is now defined, and the independent variable is changed from t to s:

$$\mathbb{H} = -P_s(H) = -\left(1 + \tfrac{x}{R}\right)\sqrt{\left(\tfrac{H}{c}\right)^2 - p_x^2 - p_y^2 - m_0^2 c^2} - eA_s$$
$$= -\left(1 + \tfrac{x}{R}\right) p_z - eA_s \tag{5.22}$$

where we used $H = E$.

Hamilton's equations can now be calculated for a specific form of A_s. This has to generate the well-known dipolar and quadrupolar field components. It is easy to see that the required field vector is:

$$\begin{cases} A_s(x^2, y^2) = -\tfrac{p_s}{e}\left[\tfrac{x}{R} + \left(\tfrac{1}{R^2} - k\right)\tfrac{x^2}{2} - k\tfrac{y^2}{2}\right] \\ \vec{B} = \vec{\nabla} \times \vec{A} \end{cases} \tag{5.23}$$

$$\Rightarrow \begin{cases} B_y = -\tfrac{\partial A_s}{\partial x} = \tfrac{p_s}{eR} + \tfrac{p_s}{eR}\tfrac{x}{R} - \tfrac{p_s}{e}kx = B_{0,y}\left(1 + \tfrac{x}{R}\right) - gx \\ B_x = -\tfrac{\partial A_s}{\partial y} = -\tfrac{p_s}{e}ky = -gy \end{cases} \tag{5.24}$$

Hamilton's equations for the horizontal plane give (with the help of Eqs. 5.22, 5.24):

$$\begin{cases} \tfrac{\partial \mathbb{H}}{\partial x} = -\tfrac{dp_x}{ds} \\ \tfrac{\partial \mathbb{H}}{\partial p_x} = \tfrac{dx}{ds} \end{cases} \tag{5.25}$$

$$\Downarrow$$

$$\ddot{x}(s) = \frac{1}{p_s}\frac{dp_x}{ds} = -\frac{1}{p_s}\frac{\partial \mathbb{H}}{\partial x} = \frac{1}{R}\frac{p_z}{p_s} + \frac{e}{p_s}\frac{\partial A_s}{\partial x}$$

$$= \frac{1}{R}\frac{p_z}{p_s} - \frac{1}{R} - \frac{x}{R^2} + kx = -\left(\frac{1}{R^2} - k\right)x + \frac{\delta}{R}$$

An analogous derivation of the second order differential equation for the vertical plane eventually leads to:

$$\ddot{y}(s) = \frac{1}{p_s}\frac{dp_y}{ds} = -\frac{1}{p_s}\frac{\partial \mathbb{H}}{\partial y} = \frac{e}{p_s c}\frac{\partial A_s}{\partial y} = -ky \tag{5.26}$$

5.2 Liouville's Theorem

5.2.1 Statement

Liouville's theorem [2,3] states that the phase space volume occupied by a Hamiltonian system is preserved by canonical transformations. We present three distinct demonstrations, each of them deepening a different aspect of the theorem.

Lemma 1. *Canonical transformations have unitary determinant and therefore they preserve the area in the canonical phase space.*

Be M a canonical transformation from the pair $(\vec{q}, \vec{p})$ to the pair $(\vec{Q}(\vec{q}, \vec{p}), \vec{P}(\vec{q}, \vec{p}))$. The time derivative of the j-th component of the new spatial coordinate is calculated by making use of the "old" Hamilton's equations:

$$\frac{dQ_j}{dt} = \frac{\partial Q_j}{\partial \vec{q}} \dot{\vec{q}} + \frac{\partial Q_j}{\partial \vec{p}} \dot{\vec{p}} = \frac{\partial Q_j}{\partial \vec{q}} \frac{\partial H}{\partial \vec{p}} - \frac{\partial Q_j}{\partial \vec{p}} \frac{\partial H}{\partial \vec{q}} \tag{5.27}$$

At the same time we have:

$$\frac{\partial H}{\partial P_j} = \frac{\partial H}{\partial \vec{q}} \frac{\partial \vec{q}}{\partial P_j} + \frac{\partial H}{\partial \vec{p}} \frac{\partial \vec{p}}{\partial P_j} \tag{5.28}$$

Since $\vec{Q}, \vec{P}$ are canonical coordinates, i.e. they satisfy Hamilton's equations, we can impose equivalence member-to-member between Eqs. 5.27 and 5.28. The following expressions are found:

$$\left(\frac{\partial Q_j}{\partial q_k} \right)_{\vec{q}, \vec{p}} = \left(\frac{\partial p_k}{\partial P_j} \right)_{\vec{Q}, \vec{P}}, \quad \left(\frac{\partial Q_j}{\partial p_k} \right)_{\vec{q}, \vec{p}} = - \left(\frac{\partial q_k}{\partial P_j} \right)_{\vec{Q}, \vec{P}} \tag{5.29}$$

The phase space area evolves according to the Jacobian determinant J of the transformation, which can now be quantified using the first identity of Eq. 5.29:

$$\int_{V'} d\vec{Q} \times d\vec{P} = J \int_V d\vec{q} \times d\vec{p},$$

$$J = \frac{\partial(\vec{Q}, \vec{P})}{\partial(\vec{q}, \vec{p})} = \frac{\partial(\vec{Q}, \vec{P})}{\partial(\vec{q}, \vec{P})} \frac{\partial(\vec{q}, \vec{P})}{\partial(\vec{q}, \vec{p})} = \frac{\partial \vec{Q}}{\partial \vec{q}} \frac{\partial \vec{P}}{\partial \vec{p}} = \frac{\partial \vec{Q}/\partial \vec{q}}{\partial \vec{p}/\partial \vec{P}} = 1 \tag{5.30}$$

Equation 5.30 demonstrates at once both statements of Lemma 1.

Lemma 2. *Canonical transformations can be represented by square symplectic matrices and therefore they preserve the area in the canonical phase space.*

The first part of the enunciation of Lemma 2 is demonstrated for a 2-D system represented by canonical variables $\vec{v} = (q, p)$. Hamilton's equations are re-written in the compact form $\dot{\vec{v}} = G \vec{\nabla} H$, where G is the singular anti-symmetric square matrix introduced in Eq. 4.71. If M is a canonical transformation explicitly independent from time such that $v(t_2) = M v(t_1)$ or, in short notation, $v_2 = M v_1$, then we also have:

$$\dot{v}_2 = M \dot{v}_1 \Rightarrow \nabla H(v_1) = \nabla H(M^{-1} v_2) = M^t \nabla H(v_2) \tag{5.31}$$

or, more explicitly:

$$\nabla H(v_1) = \begin{pmatrix} \frac{\partial H}{\partial q_1} \\ \frac{\partial H}{\partial p_1} \end{pmatrix} = \begin{pmatrix} \dot{p}_1 \\ -\dot{q}_1 \end{pmatrix} = M^t \begin{pmatrix} \dot{p}_2 \\ -\dot{q}_2 \end{pmatrix} = \begin{pmatrix} m_{11} & m_{21} \\ m_{12} & m_{22} \end{pmatrix} \begin{pmatrix} \dot{p}_2 \\ -\dot{q}_2 \end{pmatrix} \qquad (5.32)$$

$$\Rightarrow \begin{cases} \dot{q}_1 = m_{22}\dot{q}_2 - m_{12}\dot{p}_2 \\ \dot{p}_1 = -m_{21}\dot{q}_2 + m_{11}\dot{p}_2 \end{cases}$$

Equation 5.32 is shown to be true starting from the definition of the map M, i.e., $v_2 = M\dot{v}_1$:

$$\begin{cases} \dot{q}_2 = m_{11}\dot{q}_1 + m_{12}\dot{p}_1 \\ \dot{p}_2 = m_{21}\dot{q}_1 + m_{22}\dot{p}_1 \end{cases} \qquad (5.33)$$

and by solving the two equations for $\dot{q}_1$, $\dot{p}_1$. We find:

$$\begin{cases} \dot{q}_1 = \frac{\dot{q}_2}{m_{11}} - \frac{m_{12}}{m_{11}}\dot{p}_1 \\ \dot{p}_2 = \frac{m_{21}}{m_{11}}\dot{q}_2 - \frac{m_{21}m_{12}}{m_{11}}\dot{p}_1 + m_{22}\dot{p}_1 = \frac{m_{21}}{m_{11}}\dot{q}_2 + \left(\frac{m_{22}m_{11} - m_{21}m_{22}}{m_{11}}\dot{p}_1 \right) = \frac{m_{21}}{m_{11}}\dot{q}_2 + \frac{\dot{p}_1}{m_{11}} \\ \dot{q}_1 = \frac{\dot{q}_2}{m_{11}} - m_{12}\left(\dot{p}_2 - \frac{m_{21}}{m_{11}}\dot{q}_2 \right) = \frac{\dot{q}_2}{m_{11}}(1 - m_{12}m_{21}) - m_{12}\dot{p}_2 = m_{22}\dot{q}_2 - m_{12}\dot{p}_2 \\ \dot{p}_1 = m_{11}\dot{p}_2 - m_{21}\dot{q}_2 \end{cases}$$

which demonstrates Eq. 5.32 or equivalently Eq. 5.31. The demonstration made use twice of the unitary determinant of the canonical transformation, $m_{11}m_{22} - m_{21}m_{22} = 1$ (see Lemma 1).

We now go back to Eq. 5.31 and find:

$$M\dot{v}_1 = MG\nabla H(v_1) = MG\nabla H(M^{-1}v_2) = MGM^t\nabla H(v_2) \equiv \dot{v}_2 = G\nabla H(v_2)$$

$$\Rightarrow MGM^t = G$$

$$(5.34)$$

which is the definition of symplectic matrix for M.

We can show that the phase space area, given by the cross product of two vectors $\vec{v} = (dq_1, dp_1)$ and $\vec{w} = (dq_2, dp_2)$, is preserved by canonical transformations. By virtue of the symplecticity of the transfer matrix:

$$A = \vec{v} \times \vec{w} = \vec{v}^t G \vec{w};$$

$$A' = (M\vec{v})^t G(M\vec{w}) = \vec{v}^t M^t G M \vec{w} = \vec{v}^t G \vec{w} = A,$$

$$(5.35)$$

which concludes the demonstration of Lemma 2.

Lemma 3. *The canonical phase space volume of a Hamiltonian system is a constant of motion.*

We assume that an ensemble of particles occupies, at a given time t, an hyper-volume $V(t)$ in the 6-D canonical phase space $(\vec{q}, \vec{p})$, with $\vec{q} = (q_x, q_y, q_z)$ and $\vec{p} = (p_x, p_y, p_z)$. A surface element $d\vec{S}$ moves with instantaneous velocity $\vec{w}(t) = (\dot{\vec{q}}, \dot{\vec{p}})$. At the time $t + \Delta t$, the particles occupy another volume $V(t + \Delta t)$. The variation of the volume with time is calculated by using the "divergence theorem" (from a closed surface integral to a volume integral), then by applying Hamilton's equations:

$$\frac{dV(t)}{dt} = \oint \vec{w} \cdot d\vec{S} = \int (\vec{\nabla} \cdot \vec{w}) dV = \int \left(\frac{\partial}{\partial \vec{q}} \dot{\vec{q}} + \frac{\partial}{\partial \vec{p}} \dot{\vec{p}} \right) dV =$$

$$= \int \left(\frac{\partial^2 H}{\partial \vec{q} \partial \vec{p}} - \frac{\partial^2 H}{\partial \vec{p} \partial \vec{q}} \right) dV = 0 \tag{5.36}$$

$$\Rightarrow V = \int_{6D} |d\vec{q} \times d\vec{p}| = const.$$

Since the demonstration assumes that the canonical pair $q(t)$, $p(t)$ satisfies Hamilton's equations $\forall t$, the volume is evolving with time according to a canonical transformation of the representative points inside the volume. Thus, canonical transformations preserve the volume.

The following observations can be drawn.

- The demonstration can be applied to a reduced volume dimensionality, like 2-D motion in the transverse plane. In this case, the closed surface (volume) integral becomes a closed line (surface) integral, and the canonical phase space area is preserved.
- The two assumptions of Liouville's theorem—existence of canonically conjugated variables and absence of frictional forces—leave space, for example, to forces nonlinear in the generalized spatial coordinates. For example, high order magnetic fields in an accelerator still preserve the beam's emittance in Liouville sense.
- If the total beam charge is preserved during transport, Liouville's theorem implies that the charge density is also preserved in the 6-D canonical phase space, $\rho(\vec{q}, \vec{p}, t) = \frac{dQ}{d^6 Q/dq^3 dp^3} = const$, and the beam behaves as an incompressible fluid in such space.

5.2.2   Vlasov's Equation

Liouville's theorem expresses the constancy of the phase space density distribution function $\psi(q, p)$ under an Hamiltonian flow, (q, p) being a pair of canonically conjugated variables. The mathematical expression of this statement is called Vlasov's equation:

$$\frac{d\psi(q,p)}{dt} = 0 \quad \Rightarrow \quad \frac{\partial\psi}{\partial t} + \frac{\partial\psi}{\partial q}\dot{q} + \frac{\partial\psi}{\partial p}\dot{p} = 0 \tag{5.37}$$

It can be shown that a distribution function is only function of the Hamiltonian if and only if it is explicitly independent from time, or $\psi(q, p) = \psi(H) \Leftrightarrow \frac{\partial\psi}{\partial t} = 0$.

Vlasov's equation allows us to write:

$$\frac{\partial\psi}{\partial t} = 0 \quad \Rightarrow \quad \frac{\partial\psi}{\partial q}\dot{q} + \frac{\partial\psi}{\partial p}\dot{p} = 0 \tag{5.38}$$

By virtue of Hamilton's equations of motion, Eq. 5.38 becomes:

$$\frac{\partial\psi}{\partial q}\frac{\partial H}{\partial p} - \frac{\partial\psi}{\partial p}\frac{\partial H}{\partial q} = 0 \Rightarrow \begin{cases} \frac{\partial\psi}{\partial q} = \frac{\partial\psi}{\partial p}\frac{\partial p}{\partial H}\frac{\partial H}{\partial q} = \frac{\partial\psi}{\partial H}\frac{\partial H}{\partial q} = -\frac{\partial\psi}{\partial H}\dot{p} \\[2mm] \frac{\partial\psi}{\partial p} = \frac{\partial\psi}{\partial q}\frac{\partial q}{\partial H}\frac{\partial H}{\partial p} = \frac{\partial\psi}{\partial H}\frac{\partial H}{\partial p} = -\frac{\partial\psi}{\partial H}\dot{q} \end{cases} \tag{5.39}$$

which states $\psi = \psi(H)$.

Viceversa, if $\psi = \psi(H)$ we can write:

$$\begin{cases} \frac{\partial\psi}{\partial q} = \frac{\partial\psi}{\partial H}\frac{\partial H}{\partial q} = -\frac{\partial\psi}{\partial H}\dot{p} \\[2mm] \frac{\partial\psi}{\partial p} = \frac{\partial\psi}{\partial H}\frac{\partial H}{\partial p} = \frac{\partial\psi}{\partial H}\dot{q} \end{cases} \Rightarrow \frac{\partial\psi}{\partial q}\dot{q} + \frac{\partial\psi}{\partial p}\dot{p} = -\frac{\partial\psi}{\partial H}\dot{p}\dot{q} + \frac{\partial\psi}{\partial H}\dot{q}\dot{p} = 0 \Rightarrow \frac{\partial\psi}{\partial t} = 0 \tag{5.40}$$

5.2.3 Emittance

The Non-dissipative transverse motion of particles in accelerators is modelled through symplectic matrices. Since these matrices have unitary determinant, the area occupied by beam particles in the *pseudo-canonical* phase space (u, u'), i.e. the geometric emittance, is preserved as long as particles' energy is constant (Eq. 4.64).

The geometric emittance shrinks in proportion to beam's energy when longitudinal acceleration is present (Eq. 4.144). This is a consequence of the fact that the beam's angular divergence couples the transverse and the longitudinal momentum ($u' = p_u/p_z$). The latter one is affected by frictional forces generating either energy increase (RF acceleration) or decrease (radiation emission). Still, the area in the *transverse canonical* phase space (which is assumed to be decoupled from the longitudinal motion, and where frictional forces are absent) has to be preserved (Eqs. 5.30, 5.36).

To see this, let us consider a relativistic beam with Lagrangian $L = T - V$. We choose the generalized spatial coordinate of the i-th particle, $u = x, y$. According to Eq. 5.6, the canonically conjugated momentum is $p_u = m\dot{u} = \beta_u\gamma m_0 c$, i.e., the relativistic transverse momentum. The canonical phase space area is therefore the beam's *normalized* emittance.

In practice, the beam's emittance is calculated starting from the measurement of the second order momenta of the spatial and momentum beam distribution, $u \rightarrow \sigma_u$,

$p_u \to \sigma_{p_u}$. The measured normalized emittance is therefore a statistical emittance (see Eq. 4.144). It results:

$$\epsilon_{c,u} = \sigma_u \sigma_{p_u} = \sigma_u \sigma_{u'} p_z = \beta_z \gamma \epsilon_u \cdot m_0 c = \epsilon_{n,u} \cdot m_0 c \qquad (5.41)$$

Since $m_0 c$ is Lorentz's invariant, we conclude that symplectic matrices, representing canonical transformations, preserve the normalized emittance.

Nevertheless, one important distinction arises between Eqs. 5.36 and 5.41. The former equation applies to a phase space area (hyper-volume) occupied by a *continuous* distribution function in the canonically conjugated variables. The quantity in Eq. 5.36 is therefore an emittance in "Liouville sense". The latter equation, instead, describes the area in the canonical phase space occupied by a *discrete* charge distribution. As pointed out by Eq. 4.128, *nonlinear* motion does *not* preserve the *statistical* emittance, even if defined in terms of canonically conjugated variables.

Similarly, the single particle's longitudinal motion in the approximation of adiabatic acceleration can be described as a Hamiltonian system, whose canonically conjugated variables are (z, p_z). The longitudinal statistical emittance expressed in terms of canonically conjugated variables results:

$$\epsilon_{c,z} = \sigma_z \sigma_{p_z} = \epsilon_{n,z} \cdot m_0 c \qquad (5.42)$$

and $\epsilon_{n,z}$ was defined in Eq. 4.146. All conclusions reached for the transverse plane apply identically to the longitudinal plane.

5.2.4 Acceleration

The 2×2 transfer matrix of an accelerating element is obtained below. A modification to the matrix is proposed, which preserves the normalized emittance. The definition of angular divergence is recalled. We assume that the transverse momentum is not changed by a pure longitudinal electric force $F = e E_z$, such as internally to an RF structure ($u = x, y$):

$$\frac{dp_u}{ds} = \frac{d}{ds}[u'(s)p_z(s)] = 0$$

$$\Rightarrow \quad u'(s)p_z(s) = u'(s)\left[p_{z,0} + \Delta p_z(s)\right] = u'(s)\left[p_{z,0} + \frac{\Delta E(s)}{\beta_z c}\right]$$

$$= u'(s)\left(p_{z,0} + \frac{e E_z s}{\beta_z c}\right) = u_0' p_{z,0} = const.; \qquad (5.43)$$

$$u'(s) = \frac{du}{ds} = \frac{u_0' p_{z,0}}{p_{z,0} + \frac{e E_z s}{\beta_z c}} = u_0' \frac{1}{1+\delta}$$

$$\Rightarrow \quad u(s) = u_0 + u_0' \frac{p_{z,0}\beta_z c}{e E_z} \ln\left(1 + \frac{e E_z s}{\beta_z c p_{z,0}}\right) = u_0 + u_0' \frac{p_{z,0}\beta_z c}{e E_z s} s \ln\left(1 + \frac{e E_z s}{\beta_z c p_{z,0}}\right)$$

$$= u_0 + u_0' \frac{s}{\delta} \ln(1 + \delta)$$

and we used the relation $\Delta p_z = \frac{\Delta E}{\beta_z c}$, with $\delta = \frac{\Delta p_z}{p_z}$.

The transfer matrix for the accelerating element long $s = L$, applied to the vector of coordinates (u, u'), is:

$$M_u^{acc} = \begin{pmatrix} 1 & L\frac{\ln(1+\delta)}{\delta} \\ 0 & \frac{1}{1+\delta} \end{pmatrix} \tag{5.44}$$

The determinant of M_u^{acc} is not 1, therefore it is not symplectic. This is expected because the phase space area in the pseudo-canonical phase space (u, u') is not preserved by acceleration. To recover symplecticity and therefore preservation of the phase space area defined in terms of the canonically conjugated variables (u, p_u), we propose:

$$\tilde{M}_u^{acc} = \begin{pmatrix} 1 & \frac{L}{p_{z,0}}\frac{\ln(1+\delta)}{\delta} \\ 0 & 1 \end{pmatrix}$$

$$\Rightarrow \begin{pmatrix} u \\ p_u \end{pmatrix} = \begin{pmatrix} u_0 + L\frac{p_{u,0}}{p_{z,0}}\frac{\ln(1+\delta)}{\delta} \\ p_{u,0} \end{pmatrix} \rightarrow \begin{pmatrix} u_0 + u_0'L \\ p_{u,0} \end{pmatrix} \tag{5.45}$$

and the limit is for null acceleration, or $\delta \to 0$.

5.3   Poincare'-Cartan Invariants

5.3.1   Phase Space Hypervolumes

The volume preservation stated by Liouville's theorem is in fact a special case of a wider family of invariants, which includes products and sums of n-dimensional phase space hyper-volumes, denominated *Poincare'-Cartan invariants* [5]. For completeness, we list below the invariants without demonstration, in order of increasing dimensionality.

Phase space areas: canonical transformations preserve the sum of canonical phase space areas,

$$\Sigma_{V_2} = \iint_{S_x} dp_x dx + \iint_{S_y} dp_y dy + \iint_{S_z} dp_z dz = const.$$

$$\Rightarrow \epsilon_{n,x} + \epsilon_{n,y} + \epsilon_{n,z} = const. \tag{5.46}$$

and $\epsilon_{n,u}$ ($u = x, y, z$) are 2-D emittances defined in terms of canonically conjugated variables.

Phase space volumes: canonical transformations preserve the sum of canonical phase space hyper-volumes,

$$\Sigma_{V_4} = \int_{V_{xy}} dp_x dx dp_y dy + \int_{V_{yz}} dp_y dy dp_z dz + \int_{V_{xz}} dp_x dx dp_z dz = const.$$

$$\Rightarrow \epsilon_{n,xy} + \epsilon_{n,yz} + \epsilon_{n,xz} = const. \tag{5.47}$$

and $\epsilon_{n,ij}$ $(i, j = x, y, z)$ are 4-D emittances defined in terms of canonically conjugated variables.

2n-dimensional Liouville's theorem: canonical transformations preserve the 2n-dimensional canonical phase space volume,

$$\int_{\Omega'} d^{2n} V' = det(M) \int_{\Omega} d^{2n} V = \int_{\Omega} d^{2n} V$$

$$\Rightarrow \int_S d\vec{p} \times d\vec{q} = \oint_S pdq = const.$$

(5.48)

Equation 5.48 implies that if the motion in the x, y and z-phase space is decoupled, i.e., $H(x, p_x, y, p_y, z, p_z) = H(x, p_x) + H(y, p_y) + H(z, p_z)$, and therefore the 6×6 canonical map M is block-diagonal, the individual phase space area defined in terms of canonically conjugated variables is preserved in each plane. In the presence of coupling between planes of motion, instead, the hyper-volume of the whole coupled phase space is preserved according to Eq. 5.47.

5.3.2 Eigen-Emittance

A particle ensemble in the 6-D canonical phase space $\vec{X} = (x, P_x, y, P_y, z, P_z)$ can be represented as a 6-D rms ellipsoid via the beam matrix (see Eq. 4.133) extended to such high dimension:

$$\sum_{6D} = \langle \vec{X}\vec{X}^t \rangle = \begin{pmatrix} \sigma_{xx} & \sigma_{xy} & \sigma_{xz} \\ \sigma_{yx} & \sigma_{yy} & \sigma_{yz} \\ \sigma_{zx} & \sigma_{zy} & \sigma_{zz} \end{pmatrix}, \quad \sigma_{uw} = \begin{pmatrix} \langle uw \rangle & \langle uP_w \rangle \\ \langle P_u w \rangle & \langle P_u P_w \rangle \end{pmatrix}$$

(5.49)

When off-diagonal terms of $\sum_{6D}$ are not null, Liouville's theorem cannot be applied to the individual sub-spaces, but only to the sum of projected sub-spaces volumes, as shown in Eqs. 5.46, 5.47. Still, it is possible to identify new canonical pairs through which $\sum_{6D}$ can be made diagonal. The transformed phase space areas—denominated *eigen-emittances*—can be individually preserved.

It can be shown that at each s, there exists a symplectic transformation $R(s)$ from the old Cartesian coordinates $\vec{X} = (x, P_x, y, P_y, z, P_z)$ to new coordinates $\vec{Q} = (q_1, R_1, q_2, R_2, q_3, R_3)$, i.e. $\vec{X}(s) = R(s)\vec{Q}(s)$, such that:

$$\sum_{6D} = \langle \vec{X}\vec{X}^t \rangle = \langle R\vec{Q}\vec{Q}^t R^t \rangle = R\langle \vec{Q}\vec{Q}^t \rangle R^{-1} = RDR^{-1}$$

$$D = \langle \vec{Q}\vec{Q}^t \rangle = \begin{pmatrix} \langle q_1^2 \rangle & 0 & \cdots & & & 0 \\ 0 & \langle R_1^2 \rangle & 0 & \cdots & & 0 \\ 0 & 0 & \langle q_2^2 \rangle & 0 & \cdots & 0 \\ 0 & \cdots & 0 & \langle R_2^2 \rangle & 0 & 0 \\ 0 & \cdots & & 0 & \langle q_3^2 \rangle & 0 \\ 0 & \cdots & & & 0 & \langle R_3^2 \rangle \end{pmatrix}$$

(5.50)

The 2-D statistical eigen-emittances are therefore $\epsilon_{n,k} = \sqrt{\langle q_k^2 \rangle \langle R_k^2 \rangle} = \sigma_{q,k} \sigma_{R,k}$, and $k = 1, 2, 3$ identify the *normal modes* of oscillation of the coupled beam. Since $R(s)$ is symplectic, we have $\det(\Sigma_{6D}) = \det(RDR^{-1}) = \det(D) = \sqrt{\epsilon_{n,1} \epsilon_{n,2} \epsilon_{n,3}}$, namely, the 6-D canonical phase space volume is the product of the three 2-D canonical transformed phase space areas.

For constant longitudinal momentum, all constants of motion introduced so far in terms of canonically conjugated variables can be expressed also in terms of pseudo-canonical phase space variables. For the transverse planes, the transverse momentum can be replaced by the angular divergence, and the preservation rules still hold.

5.3.3 Flat and Round Beam

So far, the particle's horizontal and vertical motion were assumed to be decoupled. In reality, they can be coupled because of, e.g., quadrupole roll errors, skew field components, solenoidal fields, etc. Let us consider such transverse coupling, where for simplicity we assume constant longitudinal momentum. Equation 5.46 allows us to write:

$$(\epsilon_x + \epsilon_y)^2 \equiv \epsilon_0^2 = (\epsilon_1 + \epsilon_2)^2 = const. \tag{5.51}$$

Since ϵ_1, ϵ_2 are individually constants of motion, we find:

$$\begin{cases} \epsilon_1 \epsilon_2 = const. \\ \epsilon_1^2 + \epsilon_2^2 = const. \end{cases} \tag{5.52}$$

While the eigen-emittances are defined in such a way that they are preserved along the beam transport, the emittances defined according to the usual Cartesian system of coordinates are not. Their value is actually seen to vary along the accelerator, while their sum is still preserved as in Eq. 5.51.

In synchrotrons, the persistent emission of e.m. radiation in the privileged horizontal bending plane leads to an equilibrium value of ϵ_x, and the vertical emittance results $\epsilon_x = \kappa \epsilon_y$. When $\kappa \approx 0.1 - 1\%$, the coefficient of proportionality is called "weak coupling factor". For betatron functions of same order of magnitude in the two transverse planes, $\sigma_x \gg \sigma_y$, which defines the so-called *flat beam* configuration. By virtue of Eq. 5.51 we find:

$$\epsilon_0 = \epsilon_x + \epsilon_y = (1 + \kappa)\epsilon_x \quad \Rightarrow \quad \begin{cases} \epsilon_x = \frac{1}{1+\kappa} \epsilon_0 \\ \epsilon_y = \frac{\kappa}{1+\kappa} \epsilon_0 \end{cases} \tag{5.53}$$

On the opposite, the *round beam* configuration is obtained in synchrotrons for a coupling factor of $\sim 100\%$, or "full coupling". In this case $\epsilon_x = \epsilon_y = \epsilon_0/2$. It should be noted that the stronger the coupling is, the more inaccurate the standard description of $x-$ and $y-$Courant-Snyder parameters becomes.

At the same time, a particle beam generated with comparable transverse emittances and transported with similar betatron functions in the two planes, can still be defined a "round" beam. This is the typical case of high brightness electron linacs. However, such beam is not necessarily a "coupled" beam. As a matter of fact, coupling could be introduced by solenoidal fields applied to the injection point into the accelerator. Such a round coupled beam is then called "magnetized beam".

5.3.3.1 Discussion: Hamiltonian Flow Through a Magnetic Compressor

Linear theory of magnetic compression in the absence of frictional forces (see Eq. 4.26) allows the bunched beam dynamics to be described as a Hamiltonian flow. Accordingly, the beam longitudinal phase space obeys Liouville's theorem. Let us demonstrate that (i) the minimum bunch length achieved through compression is proportional to the initial uncorrelated energy spread, and (ii) the longitudinal emittance is preserved.

According to the definition of momentum compaction in Eq. 4.15, and with reference to the matrix formalism introduced in Eq. 4.69, the relative longitudinal shift of two ultra-relativistic particles with relative momentum deviation $\delta = dp_z/p_z \approx dE/E$ is:

$$\Delta z = dL = \alpha_c L \delta \equiv R_{56}\delta \quad \Rightarrow \quad R_{56} = \int_0^L \frac{D_x(s')}{R(s')} ds' \tag{5.54}$$

R_{56}, or "longitudinal dispersion", is the linear transfer matrix element which couples the particle's coordinates (z, δ). The beam total energy spread is the sum of correlated and uncorrelated energy spread. At the entrance of the compressor, $\delta_i = \delta_c + \delta_{0,i}$. By recalling the definition of linear energy chirp $h_i = \delta_c/dz_i$ and linear compression factor in Eq. 4.26, the bunch length at the end of the compressor, e.g. represented by the head-tail distance in a two-particle model, results:

$$dz_f = dz_i + \Delta z = dz_i + R_{56}(\delta_c + \delta_{0,i}) = dz_i + R_{56}\left(h_i dz_i + \delta_{0,i}\right)$$

$$= dz_i\left(1 + R_{56}h_i\right) + R_{56}\delta_{0,i} = \frac{dz_i}{C} + R_{56}\delta_{0,i}$$

$$\Rightarrow \sigma_{z,f}^2 = \frac{\sigma_{z,i}^2}{C^2} + R_{56}^2\sigma_{\delta_{0,i}}^2 \tag{5.55}$$

$$\Rightarrow \lim_{C\to\infty} \sigma_{z,f} = R_{56}\sigma_{\delta_{0,i}}$$

Since the beam energy distribution is not modified by the static magnetic fields of the compressor (e.g., a 4-dipole chicane, or an arc made of dipoles and multipoles), the *total* energy spread is preserved:

$$\delta_i = \delta_{0,i} + \delta_c = \delta_{0,i} + h_i dz_i \equiv \delta_{0,f} + h_f dz_f \tag{5.56}$$

By replacing the expression of dz_f in Eq. 5.55 into Eq. 5.56, and using $h_f = Ch_i$ from the definition of linear energy chirp, we get:

$$\delta_i = \delta_{0,f} + h_f \left(\frac{dz_i}{C} + R_{56}\delta_{0,i} \right) = \delta_{0,f} + Ch_i \left(\frac{dz_i}{C} + R_{56}\delta_{0,i} \right)$$

$$= \delta_{0,f} + h_i dz_i + Ch_i R_{56}\delta_{0,i}$$

(5.57)

Finally, by equating the result above with the third term in Eq. 5.56, we find:

$$\delta_{0,f} = \delta_{0,i}(1 - Ch_i R_{56})\delta_{0,i} = C\delta_{0,i}$$

$$\Rightarrow \quad \sigma_{\delta_0,f} = C\sigma_{\delta_0,i}$$

(5.58)

$$\Rightarrow \quad \epsilon_{z,i} = \sigma_{z,i}\sigma_{\delta_0,i} = C\sigma_{z,f} \frac{\sigma_{\delta_0,f}}{C} = \epsilon_{z,f}$$

References

1. L.D. Landau, E.M. Lifshitz, *The Classical Theory of Fields*, 4th edn. (Publication Pergamon Press, New York, 1980), pp. 24–65. ISBN: 9780750627689
2. M. Tabor, *Chaos and Integrability in Nonlinear Dynamics* (Publication by Wiley, New York, 1989), pp. 1–65. ISBN: 978-0-471-82728-3
3. A. Wolski, *The Accelerator Hamiltonian in a Curved Coordinate System, Dynamical Maps for "Linear" Elements* (Lectures given at the University of Liverpool, UK, 2012)
4. R.D. Ruth, Single particle dynamics and nonlinear resonances in circular accelerators, in *SLAC-PUB-3836, Lecture presented at the Join US/CERN School on Particle Accelerators* (Sardinia, Italy, 1985)
5. L.C. Teng, *Concerning n-Dimensional Coupled Motions* (FN-229, National Accelerator Laboratory, 1971), pp. 1–27

Perturbed Linear Optics

6

Four main perturbations to linear optics are introduced in this Chapter: orbit distortion, tune resonances, linear chromaticity, and weak betatron coupling. The former is described in the framework of linear optics, in the presence of magnets' misalignment. When applied to periodic motion, it introduces betatron resonances. Both resonances and linear optics distortion due to particle's momentum deviation—the latter effect is denominated *linear chromaticity*,—force to the adoption of high order magnets to cure, thus to nonlinear optics. Finally, magnets' alignment errors lead to weak coupling of the betatron motion, partly invalidating the uncoupled Hill's equations.

6.1 Orbit Distortion

6.1.1 Single Pass

A thin quadrupole magnet of integrated gradient kl and misaligned by Δu with respect to the beam path, imposes to the beam as a whole an angular deviation, a "kick" in jargon, like if it were a dipole. The scaling down in field order by misaligned magnets is a general effect denominated "feed-down". In a quadrupole, this can also be intended as an additional focusing proportional to the quadrupole-beam relative misalignment, or $\Delta u' = kl\Delta u$.

The generic particle's coordinates transform through the quadrupole according to Eq. 4.75:

© The Author(s), under exclusive license to Springer Nature Switzerland AG 2022   137
S. Di Mitri, *Fundamentals of Particle Accelerator Physics*, Graduate Texts in Physics,
https://doi.org/10.1007/978-3-031-07662-6_6

$$\begin{pmatrix} u_q \\ u'_q \end{pmatrix} = \begin{pmatrix} 1 & 0 \\ kl & 1 \end{pmatrix} \begin{pmatrix} u_0 \\ u'_0 \end{pmatrix} + \begin{pmatrix} 0 \\ kl\Delta u \end{pmatrix}$$

$$\Rightarrow \begin{cases} u_q = u_0 \\ u'_q = (u'_0 + klu_0) + kl\Delta u \equiv u'_\beta + \theta_q \end{cases} \tag{6.1}$$

Since the quadrupole is treated in thin lens approximation, Eq. 6.1 shows that the gradient error does not modify the initial particle's position, but it does the angular divergence.

We now assume a generic, not periodic beam line downstream of the quadrupole magnet. The beam line is ideally aligned to the beam original path. The transfer matrix $M_{1,2}$ has the form of Eq. 4.119. Subscript "1" marks the location of the "perturbation", subscript "2" of the "observation" point downstream. To follow the motion of the beam as a whole downstream of the misaligned quadrupole, M is intended to be applied to the bunch centroid's coordinates:

$$\begin{pmatrix} u \\ u' \end{pmatrix} = M_{1,2} \begin{pmatrix} u_q \\ u'_q \end{pmatrix} = M_{1,2} \begin{pmatrix} u_0 \\ u'_\beta + \theta_q \end{pmatrix}$$

$$\Rightarrow u(s) = \left[m_{11}u_0 + m_{12}(u'_0 + klu_0) \right] + m_{12}\theta_q \equiv u_\beta(u_0, u'_0) + u_p(\theta_q)$$

$$\Rightarrow u_p(s) = kl\Delta u \sqrt{\beta_1 \beta_2(s)} \sin \Delta\mu_{1,2}(s) \tag{6.2}$$

The Beam's position downstream of the kick is proportional to the (square root of) betatron function at the perturbation.

The Beam trajectory distortion induced by misaligned quadrupoles is commonly corrected with a series of small dipole magnets, named "steering" or "corrector" magnets, kicking the beam to restore it on—or in proximity of—the unperturbed trajectory. The Beam's position is recorded at several locations along the accelerator by detectors called "Beam Position Monitors" (BPMs).

The effect of a steering magnet on the beam trajectory is described by an equation of the same form of Eq. 6.2. We then infer that, for any target trajectory to be reached after correction, the kick from a steering magnet is minimized by a large betatron function at the steering magnet location (β_1). At the same time, the sensitivity of the beam position to the steerer is larger when the observation is done at a location of large betatron function (β_2), and for the steerer-to-BPM phase advance $\Delta\mu = (2n + 1)\pi/2, n \in \mathbb{N}$.

6.1.2   Closed Orbit

In a synchrotron, the closed orbit is the reference orbit of the synchronous particle. Its transverse coordinates in the Frenet-Serret reference system are $(u, u') = (0, 0)$ $\forall s$. A closed orbit perturbed by a dipolar kick like that one in Eq. 6.2 still has to satisfy the one-turn periodicity [1]:

$$\begin{pmatrix} u \\ u' \end{pmatrix} = M_t \begin{pmatrix} u \\ u' \end{pmatrix} + \begin{pmatrix} 0 \\ \theta_q \end{pmatrix} \quad \Rightarrow \quad \begin{pmatrix} u \\ u' \end{pmatrix} = (I - M_t)^{-1} \begin{pmatrix} 0 \\ \theta_q \end{pmatrix} \tag{6.3}$$

We choose M_t as in Eq. 4.120, where the one-turn phase advance is $\Delta\mu = 2\pi Q$. We have:

$$(I - M_t)^{-1} = \begin{pmatrix} 1 - \cos 2\pi Q & -\beta \sin 2\pi Q \\ \frac{\sin 2\pi Q}{\beta} & 1 - \cos 2\pi Q \end{pmatrix}^{-1} =$$

$$= \frac{1}{2(1 - \cos 2\pi Q)} \begin{pmatrix} 1 - \cos 2\pi Q & \beta \sin 2\pi Q \\ -\frac{\sin 2\pi Q}{\beta} & 1 - \cos 2\pi Q \end{pmatrix}; \tag{6.4}$$

$$\Rightarrow u(s) = \frac{\theta \beta(s)}{2} \frac{\sin 2\pi Q}{1 - \cos 2\pi Q} = \frac{\theta \beta(s)}{2} \frac{2 \sin \pi Q \cos \pi Q}{2 \sin^2 \pi Q} = \frac{\theta \beta(s) \cos \pi Q}{2 \sin \pi Q}$$

Equation 6.4 describes the closed orbit (c.o.) in the presence of a single kick. Since the kick is assumed to be at the same location s of observation (see Eq. 6.3), we have $\beta_1 = \beta_2 \equiv \beta$. The expression is made more general by considering a series of independent kicks $q = 1, \ldots, N_q$, each kick given in correspondence of the betatron function $\beta(s_q) \equiv \beta_q$. At the generic observation point s_m, we have $\beta(s_m) \equiv \beta_m$, and the betatron phase advance between a kick and the observation point is $\Delta\mu_{q,m} = \mu(s_q) - \mu(s_m) \equiv \mu_q - \mu_m$. If we now assume that the kicks are all independent, the perturbed closed orbit observed at the m-th BPM can be written as linear superposition of all individual perturbations:

$$u_{co}(s) = \sum_{q=1}^{N_q} u_{co,q}(s) = \sum_{q=1}^{N_q} \theta_q \frac{\sqrt{\beta_q \beta_m(s)}}{2 \sin \pi Q} \cos \left(\Delta\mu_{q,m}(s) + \pi Q \right) \tag{6.5}$$

In analogy to the discussion done for the trajectory correction, we draw the following observations.

- When the fractional part of the tune $\to 0$ or 1, the phase advance which minimizes the steerer's strength, i.e., maximizes the orbit sensitivity to steering, is $\Delta\mu_{q,m} = 2n\pi$.
- When the fractional part of the tune ≈ 0.5, the minimum strength (maximum sensitivity) is obtained for $\Delta\mu_{q,m} = (2n + 1)\pi/2$.
- u_{co} oscillates with 2-fold smaller frequency than the betatron oscillation.
- Since u_{co} was derived in the assumption of pure dipole kicks, it does not modify the strong focusing of the lattice, i.e., the lattice C-S parameters. In other words, u_{co} is linearly independent from the solutions of Hill's equation. Moreover, since u_{co} describes the bunch centroid's motion, it corresponds to an identical translation of the orbit for all beam particles. Consequently, the generic particle's position in a synchrotron affected by misaligned quadrupoles and/or dipole field errors, is

the linear superposition of the betatron, dispersive and perturbed orbit motion:

$$
\begin{cases}
u(s) = u_\beta(s) + u_D(s) + u_{co}(s) \\
\\
u'(s) = u'_\beta(s) + u'_D(s) + u'_{co}(s)
\end{cases}
\tag{6.6}
$$

6.1.3   Amplification Factor

When the number of independent dipolar kicks from misaligned quadrupoles is $N_q \gg 1$, and the specific error set Δu_q is unknown, the c.o. distortion can be still estimated assuming a distribution function of the misalignment of each quadrupole magnet having standard deviation σ_q. In most cases, it is reasonable to assume that the misalignment distribution function is the same for all quadrupoles, i.e., σ_q is a statistical uncertainty on the alignment of any quadrupole magnet in the accelerator.

The rms c.o. is first evaluated as due to a single kick (q-th quadrupole magnet) with deviation σ_q (this is defined for a large number of errors seeds, $N_{seed} \gg 1$), and observed at a generic position (m-th monitor):

$$
\sigma_{co}^2(q,m) = \frac{1}{N_{seed}} \sum_{i=1}^{N_{seed}} u_{co,i}^2(s;q,m) =
$$

$$
= \frac{1}{N_{seed}} \sum_i \Delta u_{q,i}^2 \frac{(k_q l_q)^2 \beta_q \beta_m}{(2\sin\pi Q)^2} \cos^2(\Delta\mu_{q,m} + \pi Q) =
\tag{6.7}
$$

$$
= \sigma_q^2 \frac{(k_q l_q)^2 \beta_q \beta_m}{(2\sin\pi Q)^2} \cos^2(\Delta\mu_{q,m} + \pi Q)
$$

Next, we sum over all quadrupoles' kicks, but we still keep a specific observation point:

$$
\sigma_{co}^2(m) = \frac{\beta_m \sigma_q^2}{4\sin^2\pi Q} \sum_{q=1}^{N_q} (k_q l_q)^2 \beta_q \cos^2(\Delta\mu_{q,m} + \pi Q)
\tag{6.8}
$$

Finally, we average over all N_m observation points. Doing so, we approximate $\cos^2$ to its average value over one betatron period (i.e., we assume many betatron oscillations per turn, or $Q \gg 1$):

$$
\sigma_{co}^2 \approx \frac{1}{N_m} \sum_{m=1}^{N_m} \beta_m \frac{\sigma_q^2}{8\sin^2\pi Q} \sum_q (k_q l_q)^2 \beta_q = \frac{\sigma_q^2 \langle \beta_m \rangle}{8\sin^2\pi Q} \sum_q (k_q l_q)^2 \beta_q
\tag{6.9}
$$

By introducing the rms value of the integrated quadrupole gradient $k_q l_q$ but normalized to the local betatron function $\sqrt{\beta_q}$, it results:

$$
\sigma_{kl}^2 := \frac{1}{N_q} \sum_{q=1}^{N_q} (k_q l_q)^2 \beta_q,
\tag{6.10}
$$

and we can re-write the rms c.o. distortion as:

$$\sigma_{co} \approx \frac{\sqrt{\langle \beta_m \rangle N_q}}{2\sqrt{2} \sin \pi Q} \sigma_{kl} \sigma_q \tag{6.11}$$

The ratio $A_u = \frac{\sigma_{co,u}}{\sigma_{q,u}}$ is said *amplification factor* for the u-transverse plane. Roughly speaking, it describes the expected orbit deviation, averaged over the ring lattice, excited by a characteristic quadrupoles' misalignment σ_q. It turns out that, for any given σ_q, $A_u \sim \sqrt{N_q} k_q l_q \langle \beta_u \rangle$.

6.2 Resonances

6.2.1 Resonance Order

Equation 6.11 shows that, owing to the presence of dipolar perturbations, the beam motion becomes unstable, i.e. $u \to \infty$, when the *resonance condition* $Q = r$ (r integer) is satisfied [1,2]. The appearance of tune resonances is an effect intrinsic to strong focusing, and amplified by errors or, as we will see, by nonlinear fields. Indeed, periodicity implies that an error in the magnetic lattice can become a systematic driving force, pushing particles far form the reference orbit, if the particles come back to the error location exactly with the same phase space coordinates. This situation may happen every $pQ = r$ turns, p integer. For the most general case of errors affecting the motion in both transverse planes, the resonance condition reads:

$$pQ_x + qQ_y = r, \quad p, q, r \in \mathbb{N} \tag{6.12}$$

The resonance coefficients depend from the field order associated to the error. The sum $n = |p| + |q|$ is the *resonance order*. According to the notation in Eq. 4.47, a systematic field component of the m-th order gives rise to the (lowest) resonance order $n = m + 1$. Below, we demonstrate that a quadrupole gradient error originates a second order ("half-integer") tune resonance.

The demonstration profits of the graphical representation of the particle's motion in terms of Floquet's coordinates (Eq. 4.109). Let us consider a quadrupole's integrated gradient affected by a small error, $kl + \delta(kl)$ with $\delta(kl) << kl$. For simplicity, we describe the quadrupole in thin lens approximation. Figure 6.1 shows that the particle's position at the location of the quadrupole is not changed, while its divergence is. The error pushes the particle's representative point off the unperturbed orbit of radius a, parallel to the divergence axis w'. For small perturbation $\delta w'$, Δs can be approximated to a segment, and the angle contained by Δs and $\delta w'$ with vertex in P is approximately $Q\theta$. Consequently, the beam *coherent tune-shift* evaluated after one turn ($\theta = 2\pi$) is modified by the small quantity:

$$(\Delta Q)_{turn} = \frac{1}{2\pi} \frac{\Delta s}{a} \approx \frac{1}{2\pi} \frac{\delta w' \cos(Q\theta)}{a} \tag{6.13}$$

Fig. 6.1 Particle's motion in Floquet's phase space, in the presence of quadrupole gradient error $\delta w'$. Q is the betatron tune

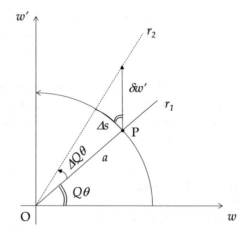

The tune-shift can also be understood by virtue of Eq. 4.91 as a distortion of the betatron function due to the focusing error.

The perturbation $\delta w'$ is made explicit in terms of the gradient error $\delta(kl)$ and of the particle's position inside the magnet, in which we also have $\alpha_u = 0$:

$$\begin{cases} \delta u' = \delta(kl)u \\ w'(\alpha_u = 0) = \sqrt{\beta_u}u' \end{cases} \Rightarrow \delta w' = \sqrt{\beta_u}\delta u' = \sqrt{\beta_u}\delta(kl)u = \sqrt{\beta_u}\delta(kl)\sqrt{\beta_u}w$$

$$(6.14)$$

We observe that $w(P) = a \cos Q\theta$, and by substituting Eq. 6.14 into Eq. 6.13 we obtain:

$$(\Delta Q)_{turn} \approx \tfrac{1}{2\pi}\beta_u\delta(kl)\tfrac{a\cos^2(Q\theta)}{a} = \tfrac{1}{4\pi}\beta_u\delta(kl)\left[1 + \cos(2Q\theta)\right] \qquad (6.15)$$

The tune-shift induced by a quadrupole gradient error is maximum when $(2Q\theta)_{turn} = 4\pi Q = 2\pi r$, i.e., $Q = r/2, r \in \mathbb{N}$.

It can be easily inferred that, since higher order gradient errors are proportional to higher powers of the particle's position in Eq. 6.14, they will lead to higher powers of $\cos(Q\theta)$ in Eq. 6.15. This will eventually generate higher order tune resonances, the resonance order proportional to the magnetic field order. For example, a sextupole magnet generates $3rd$ order resonances, an octupole magnet $4th$ order resonances, etc.

Lower order resonances are generally associated to stronger driving terms. Integer and half-integer resonances are systematic, because dipole and quadrupole magnets are essential components of strong focusing lattices.

In practice, resonances up to $4th$ order are commonly avoided with a suitable choice of the working point. Any residual coherent tune-shift is often minimized by dedicated tune feedback systems through manipulation, for example, of the beam orbit. On top of this, the *incoherent tune-shift*, associated to the individual particle's transverse positions, will tend to limit the oscillation amplitude associated to stable motion. The area in the transverse plane (x, y) in which the particles' motion remains stable for a very large number of turns is said *dynamic aperture*. This can either be larger or smaller than physical restrictions imposed by the vacuum chamber, or *physical aperture*. The minimum of the two areas determines the *transverse acceptance* of the accelerator.

6.2.2 Sum and Difference Resonance

Weak betatron coupling can excite *2nd* order sum and difference resonance, but they have *not* the same impact on the stability of the particle's motion [3]. Weak coupling can be modelled in terms of a perturbative driving force in the homogeneous Hill's equations for the two planes, proportional to the particle's coordinate in the opposite plane. This kind of perturbation can be generated, for example, by quadrupole magnets with a small roll error (typically smaller than 1 mrad or so). By adopting the Floquet's normalized coordinates (see Eq. 4.111):

$$
\begin{cases}
\frac{d^2 w_x}{d\theta^2} + Q_x^2 w_x = \epsilon \cos(m\theta) y \\
\\
\frac{d^2 w_y}{d\theta^2} + Q_y^2 w_y = \epsilon \cos(m\theta) x
\end{cases}
\tag{6.16}
$$

and $m \in \dot{\mathbb{N}}$.

For the perturbative amplitude $\epsilon \ll 1$, the solution of the unperturbed Hill's equation can be substitued in the r.h.s. of Eq. 6.16 to get:

$$
\begin{cases}
\frac{d^2 w_x}{d\theta^2} + Q_x^2 w_x = \epsilon \cos(m\theta) \sqrt{2J_y} \cos(Q_y \theta) = \frac{\epsilon \sqrt{J_y}}{\sqrt{2}} \left[\cos(m + Q_y)\theta + \cos(m - Q_y)\theta \right] \\
\\
\frac{d^2 w_y}{d\theta^2} + Q_y^2 w_y = \epsilon \cos(m\theta) \sqrt{2J_x} \cos(Q_x \theta) = \frac{\epsilon \sqrt{J_x}}{\sqrt{2}} \left[\cos(m + Q_x)\theta + \cos(m - Q_x)\theta \right]
\end{cases}
\tag{6.17}
$$

The driving terms are maximized when the following conditions apply:

$$
\begin{cases}
2\pi(m \pm Q_x) = 2\pi r \\
2\pi(m \pm Q_y) = 2\pi p
\end{cases}
\Rightarrow
\begin{cases}
Q_x = m - r \\
Q_x = r - m \\
Q_y = m - p \\
Q_y = p - m
\end{cases}
\Rightarrow
\begin{cases}
Q_x + Q_y = n \\
|Q_x - Q_y| = n
\end{cases}
\tag{6.18}
$$

and $r, p, n \in \mathbb{N}$.

Stability can be investigated through the analysis of the trace of the one-turn 4×4 matrix perturbed by a quadrupole magnet of focal length f, in the presence of a roll angle $\phi \ll 1$. The perturbed matrix results:

$$\tilde{M}_t = RQR^{-1}Q^{-1}M =$$

$$= \begin{pmatrix} I_2 \cos\phi & I_2 \sin\phi \\ -I_2 \sin\phi & I_2 \cos\phi \end{pmatrix} \begin{pmatrix} F_2 & 0 \\ 0 & D_2 \end{pmatrix} \begin{pmatrix} I_2 \cos\phi & -I_2 \sin\phi \\ I_2 \sin\phi & I_2 \cos\phi \end{pmatrix} \begin{pmatrix} D_2 & 0 \\ 0 & F_2 \end{pmatrix} M = \qquad (6.19)$$

$$= \begin{pmatrix} A & b \\ a & B \end{pmatrix}$$

The one-turn matrix M is written in terms of periodic Courant-Snyder parameters, for the simpler case $\alpha_x = \alpha_y = 0$:

$$M = \begin{pmatrix} M_{2,x} & 0 \\ 0 & M_{2,y} \end{pmatrix} = \begin{pmatrix} \cos\Delta\mu_x & \beta_x \sin\Delta\mu_x & 0 & 0 \\ -\frac{1}{\beta_x}\sin\Delta\mu_x & \cos\Delta\mu_x & 0 & 0 \\ 0 & 0 & \cos\Delta\mu_y & \beta_y \sin\Delta\mu_y \\ 0 & 0 & -\frac{1}{\beta_y}\sin\Delta\mu_y & \cos\Delta\mu_y \end{pmatrix}$$
$$(6.20)$$

Then, the components of $\tilde{M}_t$ result from Eq. 6.19:

$$A = \begin{pmatrix} \cos\Delta\mu_x \cos^2\phi & \beta_x \sin\Delta\mu_x \cos^2\phi \\ -\frac{1}{\beta_x}\sin\Delta\mu_x \cos^2\phi + \cos\Delta\mu_x \frac{2\sin^2\phi}{f} & \cos\Delta\mu_x \cos^2\phi + \beta_x \sin\Delta\mu_x \frac{2\sin^2\phi}{f} \end{pmatrix}$$

$$B = \begin{pmatrix} \cos\Delta\mu_y \cos^2\phi & \beta_y \sin\Delta\mu_y \cos^2\phi \\ -\frac{1}{\beta_y}\sin\Delta\mu_y \cos^2\phi - \cos\Delta\mu_y \frac{2\sin^2\phi}{f} & \cos\Delta\mu_y \cos^2\phi - \beta_y \sin\Delta\mu_y \frac{2\sin^2\phi}{f} \end{pmatrix}$$

$$a = \begin{pmatrix} 0 & 0 \\ \cos\Delta\mu_x \frac{\sin 2\phi}{f} & \beta_x \sin\Delta\mu_x \frac{\sin 2\phi}{f} \end{pmatrix}$$

$$b = \begin{pmatrix} 0 & 0 \\ \cos\Delta\mu_y \frac{\sin 2\phi}{f} & \beta_y \sin\Delta\mu_y \frac{\sin 2\phi}{f} \end{pmatrix}$$
$$(6.21)$$

If λ_j and $\vec{u}_j$ are, respectively, eigenvalue and eigenvector of $\tilde{M}_t$, we can write:

$$(\tilde{M}_t + \tilde{M}_t^{-1})\vec{u}_j = (\lambda_j + \lambda_j^{-1})\vec{u}_j \equiv \kappa_j \vec{u}_j, \quad j = 1,\dots,4 \qquad (6.22)$$

From the calculation of the individual terms in Eq. 6.19 it results:

$$\kappa = \frac{Tr(A+B)}{2} \pm \sqrt{\left(\frac{Tr(A-B)}{2}\right)^2 + |a + b^t|} \qquad (6.23)$$

The radical is made explicit by means of Eq. 6.21, and by noticing that $\sin \Delta\mu_x = \pm \sin \Delta\mu_y$ for the difference and the sum resonance, respectively:

$$
\begin{aligned}
\Delta_\pm &= \left(\frac{Tr(A-B)}{2}\right)^2 + |a + b^t| = \\
&= \frac{\sin^4 \phi}{f^2}(\beta_x \sin \Delta\mu_x - \beta_y \sin \Delta\mu_y)^2 + \frac{\beta_x \beta_y}{f} \sin^2 2\phi \sin \Delta\mu_x \sin \Delta\mu_y = \\
&= \frac{\sin^4 \phi}{f^2}(\beta_x \pm \beta_y)^2 \sin^2 \Delta\mu_x \mp \frac{\beta_x \beta_y}{f} \sin^2 2\phi \sin^2 \Delta\mu_x = \\
&= \frac{\sin^2 \phi \sin^2 \Delta\mu_x}{f^2}\left[(\beta_x^2 + \beta_y^2) \sin^2 \phi \pm 2\beta_x \beta_y (\sin^2 \phi - 2\cos^2 \phi)\right] = \\
&\approx \mp \frac{4\beta_x \beta_y \sin^2 \Delta\mu_x}{f^2}\phi^2 + o(\phi^4)
\end{aligned}
\tag{6.24}
$$

We conclude that for the difference resonance, $\Delta_- > 0$, thereby Eq. 6.23 provides real quantities. Instead, for the sum resonance, $\Delta_+ < 0$, and Eq. 6.23 provides imaginary quantities. We now remind that the eigenvalues of M_t must satisfy $\prod_{j=1}^4 \lambda_j = \det M = 1$ (see Eq. 4.81). So, by virtue of Floquet's theorem we expect $\lambda_{1,2} = p e^{\pm i\mu_j}$, $\lambda_{3,4} = \frac{1}{p} e^{\pm i\mu_j}$. Equation 6.23 allows us to discriminate two situations:

$$
\lambda_j + \frac{1}{\lambda_j} =
\begin{cases}
2\cos \mu_j \in \mathbb{Re} \;\; for \;\; \Delta_- > 0 \;\; \Leftrightarrow \mu \in \mathbb{Re} \\[2mm]
2\cos \mu_j \in \mathbb{Im} \;\; for \;\; \Delta_+ < 0 \;\; \Leftrightarrow \mu \in \mathbb{Im}
\end{cases}
\tag{6.25}
$$

The former case for the difference resonance corresponds to real betatron phase advance, thus stable motion. The latter case of sum resonance corresponds to imaginary betatron phase advance, hence hyperbolic, unstable motion.

6.2.3 Sextupole Resonances and Numerology

Sextupole magnets can induce $3rd$ order resonances. This can be easily inferred by repeating the procedure adopted already to obtain Eq. 6.15, but here considering the quadratic dependence of the magnetic field from the particle's coordinate u. If m is the normalized sextupole's gradient, then we have:

$$
\delta w' = \sqrt{\beta_u} \delta(ml) u^2 = \beta_u^{3/2} \delta(ml) w^2 = \beta_u^{3/2} \delta(ml) a^2 \cos^3(Q\theta)
$$

$$
\Rightarrow (\Delta Q)_{turn} \approx \frac{1}{2\pi} \frac{\delta w' \cos(Q\theta)}{a} = \frac{1}{8\pi} \beta_u^{3/2} \delta(ml) \left[3\cos(Q\theta) + \cos(3Q\theta)\right]
\tag{6.26}
$$

The tune-shift induced by a sextupole gradient error is maximum when $(3Q\theta)_{turn} = 6\pi Q = 2\pi r$, i.e., $Q = r/3, r \in \mathbb{N}$.

It is now apparent that, by virtue of the aforementioned trigonometric relations, and in the approximation of a weak perturbation to the poles geometry ("systematic error"), any magnet of *odd (even) multipole order m* could excite only resonances of *even (odd) resonance order n = m+1* (according to our notation in Eq. 4.47, $m \in \mathbb{N}$).

We also recognize lower order resonances in Eq. 6.26, associated to a feed-down effect of the magnetic field. The feed-down multipole order is $m_l = m - 2p$, $p \in \mathbb{N}$. For example, a decapole magnet ($m = 4$) can behave as a sextupole and a dipole magnet in the sense that it allows systematic resonances of order $n_l = m_l + 1$. When $p = 1$, $m_l = 4 - 2 = 2$ (equivalent sextupole magnet) and therefore $n_l = 2 + 1 = 3$. For $p = 2$, $m_l = 4 - 4 = 0$ (equivalent dipole magnet) and therefore $n_l = 0 + 1 = 1$.

Owing to the finite pole width, higher order components of the multipole magnetic field can be generated, but they are selected by the symmetry of the poles. Consequently, higher order resonances are allowed, so that $n_h = (m + 1)(2r + 1)$, $r \in \mathbb{N}$. For example, a dipole magnet ($m=0$) can also generate $3rd$, $5th$, ... resonance orders, equivalently to a sextupole, a decapole, etc.

In general, both lower and higher order resonances correspond to driving terms in the equations of motion, whose amplitude is smaller than that of the main resonance order $n = m + 1$. This is expected because the magnet is primarily conceived to behave as a multipole magnet of order m (unless combined functions magnets are considered). Random field errors are also commonly present because of manufacturing and assembly inaccuracies. In this case the field or gradient uniformity is perturbed with no symmetry rules, and any resonance order is in principle allowed.

6.3   Linear Chromaticity

6.3.1   Natural Chromaticity

Linear chromaticity is the linear dependence of strong focusing from the particle's longitudinal momentum [1]. It is represented by the normalized gradient "error" $k\delta$ in Eqs. 4.53, 4.55. In a synchrotron, it manifests primarily as a tune shift. The tune shift is intended to be a "coherent" effect when the energy deviation is referring to the synchronous particle but off-energy (e.g., in case of injection energy mismatch into a synchrotron, dipoles field error, energy ramping, etc.). It is "incoherent" when the motion of generic off-energy particles inside a bunch is considered.

The tune shift is quantified below through the matrix formalism. M_t is the one-turn transfer matrix (see Eq. 4.120), M_q and M_e the matrix of the thin lens quadrupole magnet (see Eq. 4.75), associated respectively to the nominal focal length $f^{-1} = kl$ and to the perturbed one, $\tilde{f}^{-1} = (k + \Delta k)l$. The one-turn perturbed transfer matrix is:

$$\tilde{M}_t = M_e M_q^{-1} M_t = \begin{pmatrix} 1 & 0 \\ (k+\Delta k)l & 1 \end{pmatrix} \begin{pmatrix} 1 & 0 \\ -kl & 1 \end{pmatrix} \begin{pmatrix} \cos\Delta\mu_0 & \beta\sin\Delta\mu_0 \\ \frac{1}{\beta}\sin\Delta\mu_0 & \cos\Delta\mu_0 \end{pmatrix} =$$

$$= \begin{pmatrix} \cos\Delta\mu_0 & \beta\sin\Delta\mu_0 \\ \Delta kl\cos\Delta\mu_0 - \frac{1}{\beta}\sin\Delta\mu_0 & \cos\Delta\mu_0 + \beta\Delta kl\sin\Delta\mu_0 \end{pmatrix}$$

$$(6.27)$$

The perturbed phase advance is retrieved from the matrix trace, and evaluated for a small perturbation ($\Delta\mu - \Delta\mu_0 << \Delta\mu_0$):

$$\cos\Delta\mu = \tfrac{1}{2} Tr(\tilde{M}_t) = \cos\Delta\mu_0 + \tfrac{1}{2}\beta\Delta kl\sin\Delta\mu_0;$$
$$\cos\Delta\mu - \cos\Delta\mu_0 \approx -\sin\Delta\mu_0 \cdot \Delta\mu = \tfrac{1}{2}\beta\Delta kl\sin\Delta\mu_0; \qquad (6.28)$$
$$\Rightarrow \Delta Q_u = \tfrac{\Delta\mu_u}{2\pi} = -\tfrac{1}{4\pi}\beta_u\Delta kl$$

When considering the sum of small independent chromatic perturbations distributed along the lattice, $\beta_u\Delta kl \rightarrow \beta_u(s)k(s)\delta ds$, and the linear chromaticity in the u-plane is:

$$\xi_u^{nat} := \tfrac{\Delta Q_u}{\delta} = -\tfrac{1}{4\pi}\oint ds\,\beta_u(s)k(s) \qquad (6.29)$$

ξ_u^{nat} is called *natural chromaticity*. In alternated gradient lattices, $k(s)$ has opposite sign in the x and y plane, at each quadrupole. Nevertheless, with the convention $k > 0$ for focusing magnets, the total chromaticity is always a negative quantity, in both planes, because transverse stability corresponds to an overall focusing effect.

Natural chromaticity in synchrotrons has typical absolute values in the range $\sim 10 - 100$. If not corrected, it can shift the betatron tunes to the proximity of strong resonances. For example, a beam's energy relative mismatch of 0.01% (e.g., 300 keV at the nominal energy of 3 GeV) can lead to a fractional coherent tune shift $\Delta Q \sim 0.005$, and 10-times larger internally to the bunch, where typically $\delta \sim 0.1\%$. A safe control of the working point commonly tolerates residual variations of the tunes $\Delta Q < 0.001$.

6.3.2   Chromaticity Correction

Chromaticity correction cannot be done with additional quadrupole magnets, since they would simply add their contribution to the natural chromaticity further. We then need to recur to higher order field magnets, such as sextupoles, as shown below. Sextupoles act like quadrupoles by means of a feed-down effect, though at the expense of additional nonlinear focusing and therefore higher order resonances.

The magnetic field components in a focusing sextupole magnet are shown in Fig. 6.2, and can be written as follows:

$$\begin{cases} B_y = \tfrac{1}{2}g'(x^2 - y^2) \\ \\ B_x = g'xy \end{cases} \quad , \quad g' = \frac{\partial^2 B_y}{\partial x^2} \qquad (6.30)$$

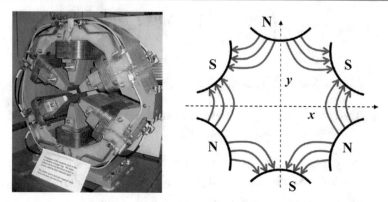

Fig. 6.2 Sextupole magnet of the Australian Synchrotron (left, photo reproduced under Creative Commons Zero license) and sketch of its magnetic field lines

By substituting the solutions of Hill's equations (see Eq. 4.59), we find:

$$\begin{cases} B_y = \frac{1}{2}g'(x_\beta^2 + D_x^2\delta^2 + 2x_\beta D_x\delta - y_\beta^2) \\[2mm] B_x = g'(x_\beta y_\beta + y_\beta D_x\delta) \end{cases} \qquad (6.31)$$

The sextupole field components show a quadratic dependence from the particle's coordinates, $\sim x_\beta^2$, $\sim y_\beta^2$, $\sim \delta^2$, which leads to linear optics distortion denominated *2nd* order *optical aberrations*, geometric and chromatic, respectively. Nevertheless, two additional terms $\propto x_\beta$, y_β can be interpreted as the effect of a quadrupole (linear) gradient, but proportional to the dispersion function at the sextupole's location. Hence, in analogy to the normalized integrated quadrupole gradient kds, we can define a normalized integrated sextupole gradient $m = \frac{eg'}{p_{z,s}}$ such that the sextupole's focal length becomes $\frac{1}{f_{sext}} = mD_x\delta ds$, and ds is the magnet's length.

The one-turn transfer matrix in Eq. 6.27 for the u-plane, but with the addition of a sextupole magnet in the lattice, becomes:

$$M_{t,cor} = S\tilde{M}_t = \begin{pmatrix} 1 & 0 \\ mD_x\delta ds & 1 \end{pmatrix} \begin{pmatrix} \cos\Delta\mu_0 & \beta\sin\Delta\mu_0 \\ \Delta kds\cos\Delta\mu_0 + & \cos\Delta\mu_0 + \\ -\frac{1}{\beta}\sin\Delta\mu_0 & +\beta\Delta kds\sin\Delta\mu_0 \end{pmatrix} \quad (6.32)$$

Its trace gives the sextupole-corrected phase advance:

$$\cos\Delta\mu = \frac{1}{2}Tr(\tilde{M}_{t,cor}) = \cos\Delta\mu_0 + \frac{1}{2}\beta ds\sin\Delta\mu_0(\Delta k + mD_x\delta);$$
$$\cos\Delta\mu - \cos\Delta\mu_0 \approx -\sin\Delta\mu_0 \cdot \Delta\mu = \frac{1}{2}\beta ds\sin\Delta\mu_0(\Delta k + mD_x\delta);$$
$$\Rightarrow \Delta Q_u = \frac{\Delta\mu_u}{2\pi} = -\frac{1}{4\pi}\beta_u(\Delta k + mD_x\delta)ds$$

The linear chromaticity in the two transverse planes is calculated in the presence of the sextupole magnet, properly taking into account the opposite sign of the quadrupole gradient in the two planes:

$$\Rightarrow \begin{cases} \xi_x^{cor} = \frac{\Delta Q_x}{\delta} = -\frac{1}{4\pi} \oint \beta_x(s) \left[k(s) + m(s) D_x(s) \right] ds \\ \\ \xi_y^{cor} = \frac{\Delta Q_y}{\delta} = -\frac{1}{4\pi} \oint \beta_y(s) \left[-k(s) + m(s) D_x(s) \right] ds \end{cases} \tag{6.33}$$

We draw the following conclusions.

1. To zero horizontal and vertical chromaticity simultaneously, we need at least two sextupole "families", providing opposite sign of their gradient. *Focusing* and *defocusing sextupoles* are rotated by 60° around the magnetic axis.
2. To minimize the sextupoles' gradient and therefore correct the chromaticities while reducing the strength of higher order optical aberrations and higher order resonances, sextupoles should be installed in regions of large horizontal dispersion (*chromatic sextupoles*).
3. Internally to dispersive regions, the sextupole gradient is further minimized by installing sextupoles in correspondence of a large betatron function (either horizontal or vertical). Since this has local maxima at quadrupole magnets, chromatic sextupoles should be installed in proximity of quadrupoles.

In practice, chromaticity is corrected to positive values $\sim 1 - 10$ to counteract single and multi-bunch collective effects (see later). A large number of sextupole families is commonly adopted, not only to minimize the individual sextupoles' strength, but also to mutually cancel optical aberrations. Sextupoles devoted to the minimization of aberrations can be installed in non-dispersive regions (*harmonic sextupoles*).

6.3.2.1 Discussion: Chromaticity in Multi-bend Lattices

Synchrotrons are made of arc cells with a number of dipole magnets per arc commonly in the range $N_b = 2 - 7$. The total number of dipole magnets is usually large enough to ensure that the dipole's bending angle is $\theta_b \ll 1$. Lattices with $N_b > 3$ are most recent designs, denominated "multi-bend lattices". We want to evaluate if, generally speaking, the sextupoles' strengths required for chromaticity correction are stronger in a lattice with large or small number of dipoles.

For any gievn natural chromaticity, the strength of chromatic sextupoles is minimized by a large dispersion function at the sextupoles' location, see Eq. 6.33. In the approximation of small bending angles, the dispersion is just proportional to the bending angle (Eq. 4.74). This points out that the larger the angle is, the larger the dispersion function excited in the arc will be.

The dipole bending angle depends in turn from the total number of dipoles, given that the total angle has to be 2π. Thus:

$$m_{sext} \propto \frac{1}{D_x} \propto \frac{1}{\theta_b} = \frac{N_b}{2\pi} \tag{6.34}$$

We thereby expect lower sextupole gradients from a lattice with low number of dipoles per arc. In other words, for few dipoles per arc, each dipole has to bend the beam by a larger angle, which excites in turn a larger dispersion function. This makes the correction of chromaticity by sextupoles more efficient.

References

1. J. Rossbach, P. Schmuser, Basic course on accelerator optics, in *Proceedings of CERN Accelerator School: 5th General Accelerator Physics Course*, CERN 94-01, vol. I, ed. by S. Turner (Geneva, Switzerland, 1994), pp. 69–79
2. E. Wilson, Non-linearities and resonances, in *Proceedings of CERN Accelerator School: 5th General Accelerator Physics Course*, CERN 94-01, vol. I, ed. by S. Turner (Geneva, Switzerland, 1994), pp. 239–252
3. M. Conte, W.W. MacKay, *An Introduction to the Physics of Particle Accelerators*, 2nd edn. (Published by World Scientific, 2008), pp. 127–134, 215–217. ISBN: 978-981-277-960-1

Synchrotron Radiation

7

The main characteristics of synchrotron radiation are recalled, including intensity, spectral, and angular distribution. Starting from the Liénard-Wiechert retarded potentials, the physical meaning of retarded time is discussed. The power radiated by a charged particle in a dipole magnet is calculated, and it is shown to be a Lorentz's invariant. Its formulation is given for the case of linear and centripetal acceleration, representative of, respectively, RF acceleration in a linac, magnetic deflection in a synchrotron.

7.1   Radiated Power

7.1.1   Retarded Potentials

A moving charge q in vacuum generates e.m. field at a distance $\vec{R} = R\hat{n} = \vec{r} - \vec{r}'$, which can be derived from the scalar and vector Liénard-Wiechert "retarded" potentials [1]:

$$
\begin{cases}
\phi(\vec{r}, t) = \frac{1}{4\pi\epsilon_0} \int_V d^3\vec{r}' \frac{\rho(\vec{r}', t')}{|\vec{r}-\vec{r}'|} \\[2mm]
\vec{A}(\vec{r}, t) = \frac{\mu_0}{4\pi} \int_V d^3\vec{r}' \frac{\vec{j}(\vec{r}', t')}{|\vec{r}-\vec{r}'|}
\end{cases},
\tag{7.1}
$$

where $\rho(\vec{r}', t')$ and $\vec{j}(\vec{r}', t')$ are, respectively, the 3-D charge density and current density at the position $\vec{r}'$, and the observation is done at the position $\vec{r}$, see Fig. 7.1.

The source-observer distance is evaluated at the observation time, $\vec{R} = \vec{R}(t)$. The source is evaluated at the retarded time $t' = t - \frac{|\vec{R}(t)|}{c}$, which takes into account the time needed by radiation to travel the distance $\vec{R}$ from the source point to the observation point. In other words, the field perceived at the point P accumulates all

Fig. 7.1 Geometry of a
charge Q moving in the
plane (x, y). P is the
observation point

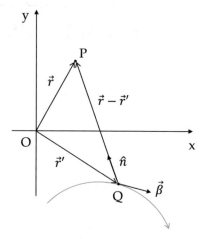

contributions generated by Q, from the past to the observation time, until the very
last instant useful for the light to reach P at the time t. Such instant is indeed the
retarded time t'.

Since the source point Q moves, the distance $\overline{QP}$ changes with time. The distance
at the observation time t is the distance at the time t' plus the space traveled by Q at
velocity $\vec{v}$ in the time interval $t - t'$:

$$|\vec{R}(t)| = |\vec{R}(t')| - \hat{n}\vec{v}(t' - t) = R(t') - \hat{n}\vec{v}\frac{R(t')}{c} = R(t')(1 - \hat{n}\vec{\beta})_{t'} \qquad (7.2)$$

The negative sign of $\hat{n}$ is for a vector directed from Q (source) to P (observer). The
velocity term in Eq. 7.2 can be interpreted as the coefficient to pass from the temporal
coordinate of the retarded time to the observation time:

$$\frac{dt'}{dt} = 1 - \frac{1}{c}\frac{dR(t)}{dt} = 1 - \frac{1}{c}\frac{dR(t')}{dt'}\frac{dt'}{dt}$$

$$\Rightarrow \frac{dt'}{dt} = \frac{1}{1+\frac{1}{c}\frac{dR(t')}{dt'}} = \frac{1}{(1-\hat{n}\vec{\beta})_{t'}} \qquad (7.3)$$

Equation 7.2 allows us to re-write the potentials in Eq. 7.1, and therefore to derive
the fields in the case of static potentials, as function of variables all evaluated at the
retarded time t':

$$\begin{cases} \phi(\vec{r}, t) = \frac{q}{4\pi\epsilon_0}\left[\frac{1}{R(1-\hat{n}\vec{\beta})}\right]_{t'} \\[4mm] \vec{A}(\vec{r}, t) = \frac{q\mu_0}{4\pi}\left[\frac{\vec{\beta}}{R(1-\hat{n}\vec{\beta})}\right]_{t'} \end{cases} \qquad (7.4)$$

$$\Downarrow$$

$$\begin{cases} \vec{E} = -\vec{\nabla}\phi = \frac{q}{4\pi\epsilon_0}\frac{1}{(1-\hat{n}\vec{\beta})_{t'}^3}\left\{\frac{1}{\gamma^2}\frac{(\hat{n}-\vec{\beta})}{R^2} + \frac{\hat{n}\times[(\hat{n}-\vec{\beta})\times\dot{\vec{\beta}}]}{cR}\right\}_{t'} \\[4mm] \vec{B} = \vec{\nabla}\times\vec{A} = \frac{\hat{n}}{c}\times\vec{E} \end{cases} \qquad (7.5)$$

The first term in brackets of the electric field is denominated "near" field. In the non-relativistic limit ($\beta \rightarrow 0$), the field vector aligns to the direction of observation, and it reduces to the classical Coulomb field with radial symmetry. In accelerators, it plays a role, for example, in the interaction of charged particles within the same bunch. This so-called "space charge" force is suppressed at high beam energies, as already derived in Eq. 1.63.

The second term in brackets is denominated "far" field. It is always normal to the direction of observation. This is the radiated field, whose intensity dominates over the near field intensity at large distances from the source, by virtue of the weaker dependence from R.

7.1.2 Larmor's Formula

The e.m. power radiated by a single particle in the non-relativistic limit is calculated by evaluating the volume integral of the Poynting vector (in units of $J/sec/m^2$) associated to the peak value of the far field ($c^2 = \epsilon_0 \mu_0$):

$$\vec{S} = \frac{1}{\mu_0} \left(\vec{E} \times \vec{B} \right) = c^2 \epsilon_0 \vec{E} \times \frac{\hat{n}}{c} \times \vec{E} = c\epsilon_0 |\vec{E}|^2 \hat{n} \tag{7.6}$$

For $\beta \rightarrow 0$:

$$\vec{E} \approx \frac{q}{4\pi\epsilon_0} \left(\frac{\hat{n} \times \hat{n} \times \dot{\vec{\beta}}}{cR} \right)_{t'} = \frac{q}{4\pi\epsilon_0} \left(\frac{\dot{\beta} \sin\theta}{cR} \right)_{t'} \hat{k}$$

$$\Rightarrow P = \int_{\Omega} |\vec{S} \cdot \hat{n}| R^2 d\Omega = c\epsilon_0 \frac{q^2}{16\pi^2\epsilon_0^2} \frac{\dot{\beta}^2}{c^2 R^2} R^2 \int_0^{2\pi} d\phi \int_0^{\pi} d\theta \sin^3\theta = \tag{7.7}$$

$$= \frac{1}{4\pi\epsilon_0} \frac{2}{3} \frac{q^2}{c^3} |\vec{a}|^2 = \frac{1}{4\pi\epsilon_0} \frac{2}{3} \frac{q^2}{m_0^2 c^3} |\frac{d\vec{p}}{dt}|^2$$

The integral assume azimuthal symmetry around the observation radius R. θ is the angle in the plane of motion between the particle's velocity $\vec{\beta}$ and the direction of observation $\hat{n}$. $\Omega = 2\pi \int d\theta \sin^2\theta$ is the solid angle through which the Poynting vector propagates. The whole integration $\int d\phi \int d\theta$ amounts to $8\pi/3$.

Equation 7.7 is called "Larmor's formula". Since it was derived for $\beta \rightarrow 0$, it applies either to the reference frame in which the particle is at rest (but, for example, still moving at relativistic velocity in the laboratory frame), or to a charged particle instantaneously at rest in the laboratory frame. If the particle is constantly at rest, the field perceived at the observation point does not change with time, i.e., the retarded time coincides with the observation time, $t' = t$.

7.1.3   Schwinger's Formula

The relativistic expression of the radiated power was derived by Schwinger, who demonstrated that it is a Lorentz's invariant. This can be understood by realizing that the power is the variation of the particle's energy with time, and that energy and time transform in the same manner under Lorentz transformations.

For the relativistic case, the acceleration squared in Larmor's formula is replaced by the square of the relativistic force divided by the rest mass. In covariant form, this is the scalar self-product of the 4-vector force (see Eq. 1.19), which is indeed a Lorentz's invariant:

$$P = -\frac{1}{4\pi\epsilon_0}\frac{2}{3}\frac{q^2}{c^3}\frac{F^\mu F_\mu}{m_0^2} \tag{7.8}$$

The 4-vector force is defined as the derivative of total energy and momentum with respect to the proper time interval, which is a Lorentz's invariant itself. Hence, the radiated power is demonstrated to be Lorentz's invariant if the time-derivatives preserve their form when energy and momentum are evaluated in a different inertial reference frame.

The demonstration, reported below, is instructive because it offers a practical expression for the calculation of the emitted power for specific configurations of the acceleration vector, as we will see later on. The 4-vector momentum's components are transformed from the laboratory frame ($\frac{E}{c}$, $\vec{p}$) to the moving frame ($\frac{E'}{c}$, $\vec{p}'$) along the x-direction, by virtue of Eqs. 1.16, 1.17 and 1.18:

$$
\begin{aligned}
F^\mu F_\mu &= \frac{dp^\mu}{d\tau}\frac{dp_\mu}{d\tau} = \frac{1}{c^2}\left(\frac{dE}{dt'}\right)^2 - \left(\frac{d\vec{p}}{dt'}\right)^2 = \\
&= \frac{1}{c^2}\left(\gamma\frac{dE'}{dt'} + \gamma v_x\frac{dp'_x}{dt'}\right)^2 - \left(\frac{dp'_y}{dt'}\right)^2 - \left(\frac{dp'_z}{dt'}\right)^2 - \left(\gamma\frac{dp'_x}{dt'} + \gamma\frac{v_x}{c^2}\frac{dE'}{dt'}\right)^2 = \\
&= \left[\frac{\gamma^2}{c^2}\left(\frac{dE'}{dt'}\right)^2 + \gamma^2\frac{v_x^2}{c^2}\left(\frac{dp'_x}{dt'}\right)^2 - \left(\frac{dp'_y}{dt'}\right)^2 - \left(\frac{dp'_z}{dt'}\right)^2 + 2\gamma^2\frac{v_x}{c^2}\frac{dE'}{dt'}\frac{dp'_x}{dt'} + \right. \\
&\qquad \left. -\gamma^2\left(\frac{dp'_x}{dt'}\right)^2 - \gamma^2\frac{v_x^2}{c^4}\left(\frac{dE'}{dt'}\right)^2 - 2\gamma^2\frac{v_x}{c^2}\frac{dE'}{dt'}\frac{dp'_x}{dt'}\right] = \\
&= \frac{1}{c^2}\left(\frac{dE'}{dt'}\right)^2\gamma^2(1-\beta^2) - \left(\frac{dp'_x}{dt'}\right)^2\gamma^2(1-\beta^2) - \left(\frac{dp'_y}{dt'}\right)^2 - \left(\frac{dp'_z}{dt'}\right)^2 = \\
&= \frac{1}{c^2}\left(\frac{dE'}{dt'}\right)^2 - \left(\frac{d\vec{p}'}{dt'}\right)^2
\end{aligned}
\tag{7.9}
$$

Alternatively, the radiated power can be expressed in terms of particle's velocity and acceleration. The ordinary force in the 4-vector force $F^\mu = (\gamma\vec{F}\vec{\beta}, \gamma\vec{F})$ is decomposed in two components, one parallel and one orthogonal to the particle's velocity, $\vec{F} = (F_\parallel, F_\perp) = (\gamma^3 m_0 a_\parallel, \gamma m_0 a_\perp)$. Then, Eq. 7.8 becomes:

$$P = -\frac{1}{4\pi\epsilon_0} \frac{2}{3} \frac{q^2}{m_0^2 c^3} \left[|\gamma(F_\parallel, F_\perp)\vec{\beta}|^2 - |\gamma(F_\parallel, F_\perp)|^2 \right] =$$

$$= -\frac{1}{4\pi\epsilon_0} \frac{2}{3} \frac{q^2}{m_0^2 c^3} \left(\gamma^2 \beta^2 F_\parallel^2 - \gamma^2 F_\parallel^2 - \gamma^2 F_\perp \right) =$$

$$= \frac{1}{4\pi\epsilon_0} \frac{2}{3} \frac{q^2}{m_0^2 c^3} \left[\gamma^2(1-\beta^2) F_\parallel^2 + \gamma^2 F_\perp^2 \right] = \frac{1}{4\pi\epsilon_0} \frac{2}{3} \frac{q^2}{m_0^2 c} (\gamma^6 m_0^2 \ddot{\beta}_\parallel^2 + \gamma^4 m_0^2 \ddot{\beta}_\perp^2) =$$

$$= \frac{1}{4\pi\epsilon_0} \frac{2}{3} \frac{q^2}{c} \gamma^6 (\ddot{\beta}_\parallel^2 + \frac{1}{\gamma^2} \ddot{\beta}_\perp^2 + \ddot{\beta}_\perp^2 - \ddot{\beta}_\perp^2) = \frac{1}{4\pi\epsilon_0} \frac{2}{3} \frac{q^2}{c} \gamma^6 \left[\ddot{\beta}^2 - \ddot{\beta}_\perp^2 \left(1 - \frac{1}{\gamma^2} \right) \right] =$$

$$= \frac{1}{4\pi\epsilon_0} \frac{2}{3} \frac{q^2}{c} \gamma^6 \left[\ddot{\beta}^2 - \ddot{\beta}_\perp^2 \beta^2 \right] = \frac{1}{4\pi\epsilon_0} \frac{2}{3} \frac{q^2}{c} \gamma^6 \left[\ddot{\beta}^2 - \left(\vec{\beta} \times \dot{\vec{\beta}} \right)^2 \right]$$

$$(7.10)$$

As expected, Eq. 7.10 reduces to Eq. 7.7 when $\beta \to 0$.

7.1.4 Radiation Emission in a Linac

Particle acceleration in a RF structure is expected to be accompanied by radiation emission. The radiated power is calculated below in the approximation of pure longitudinal acceleration ($\frac{dp_{x,y}}{dt} = 0$) and "quasi-laminar" motion ($p_{x,y} \ll p_z$). This allows us to write $\frac{dp}{dt} \approx \frac{dp_z}{dt}$, and we recall $dp = \frac{dE}{\beta c}$. The power in covariant form evaluated in the laboratory frame ($dt' = dt/\gamma$) results:

$$P_{lin} = -\frac{1}{4\pi\epsilon_0} \frac{2}{3} \frac{q^2}{m_0^2 c^3} \left[\frac{1}{c^2} \left(\frac{dE}{dt'} \right)^2 - \left(\frac{d\vec{p}}{dt'} \right)^2 \right] =$$

$$\approx \frac{1}{4\pi\epsilon_0} \frac{2}{3} \frac{q^2}{m_0^2 c^3} \gamma^2 \left[\frac{1}{\beta^2 c^2} \left(\frac{dE}{dt} \right)^2 - \frac{1}{c^2} \left(\frac{dE}{dt} \right)^2 \right] =$$

$$= \frac{1}{4\pi\epsilon_0} \frac{2}{3} \frac{q^2}{m_0^2 c^3} \gamma^2 (1-\beta^2) \left(\frac{dE}{\beta c dt} \right)^2 = \frac{1}{4\pi\epsilon_0} \frac{2}{3} \frac{q^2}{m_0^2 c^3} \left(\frac{dE}{ds} \right)^2$$

$$(7.11)$$

$$\Rightarrow \frac{P_{lin}}{\frac{dE}{dt}} = \frac{1}{4\pi\epsilon_0} \frac{2}{3} \frac{q^2}{m_0^2 c^3} \frac{1}{\beta c} \frac{dE}{ds} = \frac{2}{3} \frac{r_0}{\beta} \frac{d\gamma}{ds} \ll 1$$

The last expression quantifies the ratio of power emitted versus power gained by a single particle, by virtue of longitudinal acceleration (r_0 is the particle's classical radius). Since state-of-the-art accelerating gradients are of the order of 100 MV/m, the ratio is smaller than 10^{-13} for ultra-relativistic electrons, and smaller than 10^{-7} for protons. In most practical cases, radiation emission stimulated by RF acceleration can be neglected in the particle's total energy budget.

7.1.5   Radiation Emission in a Synchrotron

Centripetal acceleration in dipole magnets is the main source of radiation emission in a synchrotron. Since the particle's energy, diminished by radiation emission, is replenished by RF acceleration, it is convenient to calculate the radiated power per turn, assuming a constant energy on average in one turn, or $\frac{dE}{dt} \approx 0$. This same condition is assumed for the single pass in a dipole (the amount of radiated power is a small fraction of the beam power). Still, the Lorentz's force determines a variation of the direction of the longitudinal momentum which is at the origin of the radiated power.

The relativistic expression of the Lorentz's force in a dipole magnet of bending radius R is $F_{\perp} = \frac{dp_z}{dt} = \gamma m_0 \frac{v_z^2}{R}$. The radiated power expressed in the laboratory frame results:

$$P_{sr} \approx \frac{1}{4\pi\epsilon_0} \frac{2}{3} \frac{q^2}{m_0^2 c^3} \gamma^2 \left(\frac{d\vec{p}}{dt}\right)^2 = \frac{1}{4\pi\epsilon_0} \frac{2}{3} q^2 c \frac{\beta^4 \gamma^4}{R^2} \tag{7.12}$$

Owing to the strong dependence from the relativistic γ-factor, $P_{sr} \propto E^4/m_0^4$, the emission is enhanced for light particles like leptons, compared to heavier particles like protons, ions, etc., at the *same total energy*. This explains why synchrotron light sources are electron accelerators. Radiation emitted because of centripetal acceleration is called *synchrotron radiation*. For any given total energy, the emission of synchrotron radiation is enhanced by small bending radii, i.e., strong magnetic fields (see Eq. 2.5).

The total energy per turn emitted by a single particle is [2]:

$$U_0 = \oint \frac{P_{sr}}{\beta c} ds = \frac{q^2 \beta^3 \gamma^4}{6\pi\epsilon_0} \oint \frac{ds}{R(s)^2} \rightarrow \frac{q^2}{3\epsilon_0} \frac{\beta^3 \gamma^4}{R} \tag{7.13}$$

and the last expression is for a dense isomagnetic lattice, namely, $C = 2\pi R_{eq} \approx 2\pi R$. In practical units for an ultra-relativistic electron:

$$U_0[keV] = 88.45 \frac{E^4[GeV]}{R[m]} = 26.5 E^3[GeV] B[T] \tag{7.14}$$

This typically amounts to several 100s keV in an electron synchrotron at few GeV's beam energy. In spite of its smallness compared to the nominal beam energy, it would rapidly deplenish the particles' energy after only $\sim 10^4$ turns or so. This motivates the adoption of RF cavities to balance the energy loss. In practice, the total RF peak voltage is at MV level, not only to compensate for the emission of radiation, but also to maximize the RF energy acceptance. Accordingly, the beam's energy is kept constant on average in a turn by running the cavities with a phase close to the zero-crossing.

Since the emission of synchrotron radiation is incoherent, i.e., each particle emits radiation independently from the others in a bunch, the power emitted by the whole bunch in a turn is simply the sum of the emission by N_b particles:

$$P_{sr,b} = N_b \frac{U_0}{T_0} = \frac{U_0 \langle I \rangle}{q} \tag{7.15}$$

where $Q = q N_b$ is the beam's total charge, and $\langle I \rangle$ is the beam average current stored in the accelerator.

7.1.5.1 Discussion: Comparison of Radiated Power

What is the power radiated by a single particle subject to either centripetal or longitudinal acceleration, for equal magnitude of the force in the two cases? Which of the two accelerations maximize the amount of radiated power at ultra-relativistic velocities?

The ratio of radiated power in the two cases is easily calculated from Eq. 7.10:

$$\begin{cases} P_\parallel = -\frac{1}{4\pi\epsilon_0}\frac{2}{3}\frac{q^2}{c^3}\gamma^6|\vec{a}_\parallel|^2 \\ \\ P_\perp = -\frac{1}{4\pi\epsilon_0}\frac{2}{3}\frac{q^2}{c}\gamma^6\dot{\beta}^2(1-\beta^2) = -\frac{1}{4\pi\epsilon_0}\frac{2}{3}\frac{q^2}{c^3}\gamma^4|\vec{a}_\perp|^2 \end{cases} \tag{7.16}$$

We now impose the relativistic force, either parallel or perpendicular to the particle's velocity, to be equal in absolute value: $F_\parallel = \gamma^3 m_0 a_\parallel \equiv F_\perp = \gamma m_0 a_\perp$. Namely, $a_\perp = \gamma^2 a_\parallel$, and we find $\frac{P_\parallel}{P_\perp} = \frac{1}{\gamma^2}$.

The same result is obtained by expressing the radiated power in covariant form:

$$-F^\mu F_\mu = -\frac{\gamma^2}{c^2}(\vec{F}\cdot\vec{u})^2 + \gamma^2|\vec{F}|^2 = \begin{cases} \gamma^2(1-\beta^2)F_\parallel^2 = \gamma^6 m_0^2|\vec{a}_\parallel|^2 \\ \\ \gamma^2 F_\perp^2 = \gamma^4 m_0^2|\vec{a}_\perp|^2 \end{cases} \tag{7.17}$$

$$\Rightarrow \frac{P_\parallel}{P_\perp} = \frac{1}{\gamma^2}$$

In conclusion, when charged particles are at ultra-relativistic velocities, the radiated power is largely more stimulated by centripetal acceleration (e.g., in a dipole magnet) than by longitudinal acceleration (e.g., in a RF structure), for a comparable force in the two situations.

7.2   Angular Distribution

7.2.1   Longitudinal Acceleration

The amount of energy radiated by a single particle in the time interval $dt = t_2 - t_1$, evaluated in the laboratory frame, is written by recalling Eq. 7.7 [1,2]:

$$U = \int_{t_1}^{t_2} P(t)dt = \int_{t_1}^{t_2} dt \int_\Omega d\Omega (|\vec{S}\cdot\hat{n}|R^2)_{t'} = \int_{t_1'}^{t_2'} \int_\Omega d\Omega (|\vec{S}\cdot\hat{n}|R^2)_{t'}\frac{dt}{dt'}dt' =$$

$$= \int_{t_1'}^{t_2'} dt' \int_\Omega d\Omega \frac{dP(t')}{d\Omega} \tag{7.18}$$

As also in Eq. 7.7, an azimuthal symmetry of the emission is assumed, and in this case $d\Omega = 2\pi \sin\theta d\theta$. By virtue of Eq. 7.3, the power angular distribution evaluated at the retarded time is:

$$\frac{dP(t')}{d\Omega} = c\epsilon_0 \left(\frac{q}{4\pi\epsilon_0}\right)^2 \frac{(1-\hat{n}\vec{\beta})_{t'}}{(1-\hat{n}\vec{\beta})_{t'}^6}\frac{\{\hat{n}\times[(\hat{n}-\vec{\beta})\times\dot{\vec{\beta}}]\}_{t'}^2}{c^2 R^2}R^2 =$$

$$= \frac{q^2}{16\pi^2 c\epsilon_0}\frac{1}{(1-\beta\cos\theta)_{t'}^5}\{\hat{n}\times[(\hat{n}-\vec{\beta})\times\dot{\vec{\beta}}]\}_{t'}^2 \tag{7.19}$$

When the acceleration is parallel to the particle's velocity, such as in a RF structure, the cross product in Eq. 7.19 can be re-written as $\hat{n} \times [(\hat{n} - \vec{\beta}) \times \dot{\vec{\beta}}] = \hat{n} \times [(\hat{n} \times \dot{\vec{\beta}}) - (\vec{\beta} \times \dot{\vec{\beta}})] = \hat{n}\dot{\beta} \sin\theta$, and Eq. 7.19 becomes:

$$\frac{dP(t')}{d\Omega} = \frac{q^2}{16\pi^2 c\epsilon_0}\left[\frac{\dot{\beta}^2 \sin^2\theta}{(1-\beta\cos\theta)^5}\right]_{t'}, \tag{7.20}$$

We draw the following observations.

- In the non-relativistic limit $\beta \to 0$, the power angular distribution reduces to the classical dipolar emission, as shown in Fig. 7.2-top left. This has maximum (null) intensity along the direction normal (parallel) to the velocity. The two "lobes" of emission are symmetric in the z-direction. Integrated over $\theta \in [0, \pi]$, the radiated power reduces to Larmor's formula in Eq. 7.7.
- In the ultra-relativistic limit $\beta \to 1$, the maximum of the angular power density is found by imposing $\frac{d}{d\theta}\frac{dP}{d\Omega} \equiv 0$, which translates into $\sin\theta(1 - \cos\theta)^4[2\cos\theta(1 - \beta\cos\theta) - 5\beta\sin^2\theta] = 0$. This gives either $\theta = 0$ (null emission) or, by expanding the square brackets for $\theta \ll 1$ (this assumption will be verified *a posteriori*) and using $\beta = \sqrt{1 - \frac{1}{\gamma^2}} \approx 1 - \frac{1}{2\gamma^2}$, it results $\theta \approx \pm\frac{1}{2\gamma}$.

Particle's motion at large γ's confirms that the emission is at very small angles from the direction of motion. Namely, the two "lobes", characteristic of dipolar emission in the non-relativistic case, are boosted forward and close to the z-axis, in proportion to the beam's energy, as shown in Fig. 7.2-top right. Since the emission is null exactly on-axis, each lobe has to have a characteristic width $\sigma_\theta \approx \frac{1}{\gamma}$.

By virtue of the generic nature of longitudinal acceleration considered so far, the power angular distribution in Eq. 7.20 applies to RF acceleration as well as to longitudinal deceleration, provided for example by a metallic foil on the beam's path. Indeed, this process is used to relate the beam particles' spatio-angular and/or energy distribution to the angular distribution of the emitted radiation. This is named "optical transition radiation" if the frequency range is close to that of visible light.

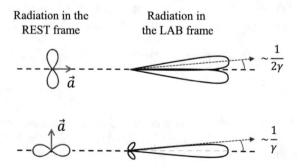

Fig. 7.2 Top view of the radiation power angular distribution for acceleration parallel (top) and normal (bottom) to the charged particle's velocity, for non-relativistic velocity or in the particle's reference frame (left), and for ultra-relativistic velocity in the observer's reference frame (right)

7.2.2 Centripetal Acceleration

When the acceleration is normal to the particle's instantaneous velocity, such as in a dipole magnet, the cross product in Eq. 7.19 becomes $\hat{n} \times [(\hat{n} \times \dot{\vec{\beta}}) - (\vec{\beta} \times \dot{\vec{\beta}})] = \hat{n} \times (\dot{\beta} \cos \theta - \dot{\beta} \beta)\hat{k} = \hat{j} \dot{\beta} (\cos \theta - \beta)$. Equation 7.19 gives [1,2]:

$$\frac{dP(t')}{d\Omega} = \frac{q^2}{16\pi^2 c\epsilon_0} \left[\frac{\dot{\beta}^2 (\cos \theta - \beta)^2}{(1 - \beta \cos \theta)^5} \right]_{t'} \tag{7.21}$$

The term in square brackets can be re-written by expanding the numerator and by isolating the term $(1 - \beta \cos \theta)^2$. This simplifies with the denominator to give:

$$\frac{dP(t')}{d\Omega} = \frac{q^2}{16\pi^2 c\epsilon_0} \left[\frac{\dot{\beta}^2}{(1 - \beta \cos \theta)^3} \right]_{t'} \left[1 - \frac{(1 - \beta^2) \sin^2 \theta}{(1 - \beta \cos \theta)^2} \right]_{t'} \rightarrow \frac{q^2}{16\pi^2 c\epsilon_0} \left[\frac{\dot{\beta}^2}{(1 - \beta \cos \theta)^3} \right]_{t'} \tag{7.22}$$

and the limit is taken for $\beta \rightarrow 1$.

At small angles, the angular power density goes like $\frac{dP}{d\Omega} \propto (1 - \beta \cos \theta)^{-3} \approx (1 - \beta + \beta \frac{\theta^2}{2})^{-3} \approx \frac{8}{\left(\theta^2 + \frac{1}{\gamma^2} \right)^3}$. Radiation emission from centripetal acceleration is maximum on-axis ($\theta = 0$), and it shrinks to 8-fold smaller value at the typical aperture $\theta = \pm \frac{1}{\gamma}$. This is illustrated in Fig. 7.2-bottom plots, for emission in the particle's rest frame and at relativistic motion in the laboratory frame (see also Fig. 1.8).

By virtue of the azimuthal symmetry of emission, an observer at large distance form the source receives light at the characteristic angle $\theta = 1/\gamma$ in the vertical plane. In the horizontal plane, however, the angular collimation characteristic of the instantaneous emission is smeared because particles moving along the circular orbit emit radiation at *any* point tangent to the curved path. For this reason, synchrotron radiation from a dipole magnet is typically selected with slits in the horizontal plane to limit the effective cone of emission to $1/\gamma$ or so (see later Fig. 10.6).

7.3 Spectral Distribution

7.3.1 Critical Frequency

The spectral distribution of synchrotron radiation is derived on the basis of qualitative considerations. Still, all the essential features are kept, and exact results are reported for completeness [2,3]. The spectral bandwidth is first estimated as the inverse of the temporal duration of the radiation pulse emitted along an arc of circumference in a dipole magnet. The time interval is calculated with reference to Fig. 7.3. Since most of the radiation flux is contained in an angular cone of aperture $1/\gamma$, a point-like observer receives light only from a portion of the curved path corresponding to the total angle $2/\gamma$. In particular, the first photon is emitted at point A, and it takes the time $\delta t_1 = 2\frac{\sin(1/\gamma)}{c}$ along the straight line to reach the observer. The last photon is emitted at point B, after the electron has traveled along the curved path in a time interval $\delta t_2 = \frac{2R/\gamma}{\beta c}$. Thus, the duration of the radiation pulse for $\gamma \gg 1$ is:

Fig. 7.3 Top view of synchrotron radiation emission in a dipole magnet, in a characteristic angular cone of amplitude $\theta = 1/\gamma$. The first photon reaching the observer is emitted at point A, the last photon at point B. The bending angle is $2/\gamma$

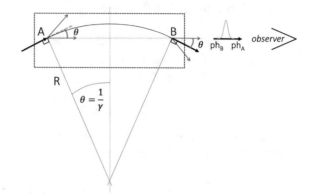

$$\Delta t = \delta t_2 - \delta t_1 = \frac{2R}{c}\left[\frac{1}{\beta\gamma} - \sin\left(\frac{1}{\gamma}\right)\right] \approx \frac{2R}{c\gamma}\left(\frac{1}{\beta} - 1\right) = \frac{2R}{c\gamma}\left(\frac{1}{\sqrt{1 - \frac{1}{\gamma^2}}} - 1\right) \approx$$

$$\approx \frac{2R}{c\gamma}\left(1 + \frac{1}{2\gamma^2} - 1\right) = \frac{R}{c\gamma^3} \sim \frac{T_0}{\gamma^3} \tag{7.23}$$

with T_0 the revolution period.

Since the emission is incoherent (each particle emits independently from all other particles), and since we assume no interference of the emitted waves (radiation emission is continuous along the arc length, and the absence of periodicity of the magnetic field along it does not imply any spectral filtering), the spectral bandwidth is expected to be wide, i.e., $\Delta\omega/\omega_c \approx 1$. The characteristic frequency at which we expect maximum intensity is approximately $\omega_c \approx \Delta\omega \approx \frac{1}{\Delta t} \approx \frac{c\gamma^3}{R}$.

This can also be understood, in the case of periodic motion, considering that the observer (detector) will receive a radiation pulse every time the particle passes in correspondence of the arc length AB [4]. In the frequency domain, the spectral distribution of the pulse is composed of the fundamental (revolution) frequency and of its harmonics. However, a frequency cutoff ω_c is present at high harmonics because of the finite pulse duration, which is of the order of $\omega_c \approx \frac{1}{\Delta t} \approx \frac{c\gamma^3}{R}$.

Thus, we expect the single particle's emission of synchrotron radiation be a discrete spectrum of closely spaced lines up to ω_c. In reality, each spectral line is broadened by the oscillations of many electrons around the main orbit, fluctuations in their kinetic energy, and the statistical nature of emission. All these effects result in a *continuous, broad* spectrum with the aforementioned high frequency cutoff.

In the literature, the "critical" or "cutoff" frequency of synchrotron radiation is defined as $\omega_c = \frac{3}{2}\frac{c\gamma^3}{R}$. This has two peculiar properties: (i) it defines the high frequency cutoff at which the radiation flux starts dropping fast, and (ii) it exactly halves the pulse energy spectral distribution. The corresponding photon energy in practical units is:

$$u_c[keV] = \hbar\omega_c = \frac{3}{4\pi} \frac{hc}{e} \left(\frac{E[GeV]}{m_0 c^2[GeV]}\right)^3 \frac{0.3B[T]}{p_z[GeV/c]} =$$

$$= 2.218 \frac{E^3[GeV]}{R[m]} = 0.665E^2[GeV]B[T] \qquad (7.24)$$

For typical electron beam energies of synchrotron light sources in the range $\sim$ $1 - 10$ GeV and circumferences long $\sim 0.1 - 1$ km, the critical photon energy is in x-rays, and the spectrum extends down to infrared photon energies.

7.3.2   Universal Function

The time integral of the power radiated in a dipole magnet in Eq. 7.19 provides the radiated energy angular distribution. Since the e.m. power is proportional to the squared modulus of the electric field, the Plancherel theorem (the integral of a function's squared modulus is equal to the integral of the squared modulus of its frequency spectrum) allows us to pass from the time domain to the frequency domain of the radiated energy. Moreover, since the frequency of the e.m. field is proportional to the photon energy, the spectral distribution of the radiated energy is obtained eventually [2,3]:

$$\frac{dU}{d\Omega} = \int_{-\infty}^{+\infty} \frac{dP}{d\Omega} dt = c\epsilon_0 \int_{-\infty}^{+\infty} |R\vec{E}(t)|^2 dt = 2c\epsilon_0 \int_0^{+\infty} |R\vec{E}(\omega)|^2 d\omega$$

$$\Rightarrow \frac{dU}{d\omega} = 2c\epsilon_0 \int_\Omega |R\vec{E}(\omega)|^2 d\Omega = \frac{U_0}{\omega_c} S\left(\frac{\omega}{\omega_c}\right), \qquad (7.25)$$

$$S(x) = \frac{9\sqrt{3}}{8\pi} x \int_x^\infty K_{5/3}(x') dx'$$

$K_{5/3}(x)$ is the modified Bessel function and $S(\omega/\omega_c)$ is called *universal function* or *normalized spectrum* of synchrotron radiation. It is calculated by integrating the single particle's radiated energy distribution over the whole solid angle of emission. It is illustrated in Fig. 7.4, normalized to its maximum value 0.57. As mentioned above, it satisfies $\int_0^1 S(u)du = \int_1^\infty S(u)du = 0.5$. We also find that the full-width half-maximum (fwhm) bandwidth is $\Delta\omega \approx \omega_c$, and that, in fact, no radiation is emitted above $\approx 4\omega_c$.

The universal function allows us to calculate the maximum of the power spectral density emitted by a particle beam of total charge $Q_b = N_b q$ in one turn:

$$\left(\frac{dP_{tot}}{d\omega}\right)_{max} [W \cdot sec] = N_b \frac{U_0}{T_0} \frac{\hat{S}}{\omega_c} = 0.57 \frac{N_b P_0}{\omega_c} = 1.5 \cdot 10^{17} E[GeV]\langle I\rangle[mA] \qquad (7.26)$$

and the expression in practical units makes use of Eqs. 7.14 and 7.24.

For completeness, the expression of the characteristic angle of emission from the radiation frequency is reported [6]. Since $\omega_c \sim \gamma^3 \sim \theta^{-3}$, we infer:

$$\frac{\theta_c}{\theta} \approx \left(\frac{\omega_c}{\omega}\right)^{1/3} \quad \Rightarrow \quad \theta_c \approx \theta\left(\frac{\omega_c}{\omega}\right)^{1/3} \approx \frac{1}{\gamma}\left(\frac{\omega_c}{\omega}\right)^{1/3} \qquad (7.27)$$

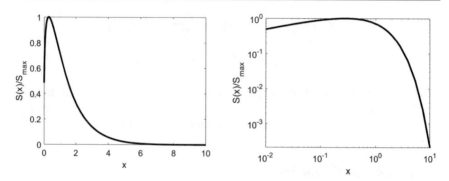

Fig. 7.4 Universal function of synchrotron radiation, normalized to its maximum value $S_{max} = S(0.3) = 0.57$, in linear (left) and log-log scale. $x = \omega/\omega_c$

The characteristic angle θ_c becomes smaller than $1/\gamma$ at $\omega > \omega_c$. The angular-spectral dependence of the universal function shows that synchrotron radiation emission is negligible for $\omega \gg \omega_c$ and $\theta \gg \theta_c$. At very high frequencies, the residual emission is basically on-axis by virtue of extremely small values of θ_c.

In summary:

- the spectral distribution of synchrotron radiation is broadband;
- the spectral intensity falls off exponentially for $\omega > \omega_c$, while it decreases slowly for $\omega < \omega_c$;
- at low frequencies, it is largely independent from the photons' energy, therefore from the electrons' energy, and it is mainly determined by the average current stored in the ring;
- the maximum photon energy is few times the critical energy.

7.3.3 Intensity

The universal function is a powerful tool to retrieve meaningful statistical properties of the photon distribution [2,5]. As an example, the number of photons emitted per second at the energy u by a single electron, on average in one turn, in the unit photon energy interval du, can be calculated as:

$$\frac{dn(\epsilon)}{dt} du = \frac{1}{T_0} \frac{dU}{d\omega} \frac{1}{\hbar\omega} d\omega$$

$$\Rightarrow \dot{n} = \frac{U_0}{T_0} \frac{S}{\omega_c} \frac{1}{\hbar^2\omega} = \frac{U_0}{T_0} \frac{1}{u_c^2} \frac{S(\omega/\omega_c)}{(\omega/\omega_c)} = \frac{P_0}{u_c^2} \frac{S(x)}{x}$$

$$(7.28)$$

This result is used to calculate the number of photons per second or meter, integrated over the whole spectral range:

$$\frac{dN}{dt} = \int_0^\infty \dot{n}(u)du = \frac{15\sqrt{3}}{8}\frac{P_0}{u_c} = \frac{15\sqrt{3}}{8}\frac{26.5E^3[GeV]B[T]}{2.218E^2[GeV]B[T]}\frac{1}{T_0} =$$

$$= \frac{15\sqrt{3}}{8}\frac{2.65}{2.218}E[GeV]\frac{0.3}{2\pi}\frac{B[T]c}{p_z[GeV/c]} \approx 6.2cB[T] \quad [photons/sec] \qquad (7.29)$$

$$\frac{dN}{ds} \approx \frac{dN}{cdt} \approx 6.2B[T] \quad [photons/m]$$

Hence, the total number of photons per second per unit angle in the horizontal plane, summed over the total beam charge, emitted on average in one turn, is:

$$\frac{d^2N_{tot}}{dtd\theta} \approx \frac{dN}{dt}\frac{\langle I\rangle}{e}\frac{R}{c} = 6.2\frac{p_z[GeV/c]}{0.3R}\frac{R}{e}\langle I\rangle = 1.3\cdot 10^{17}E[GeV]\langle I\rangle[A] \quad [\frac{photons}{sec\cdot mrad-hor.}]$$
$$\qquad (7.30)$$

The total spectro-angular intensity can be calculated by differentiating the bottom line of Eq. 7.25 by $d\Omega$, times the number of beam particles N_b:

$$\frac{d^3N_{tot}(\omega)}{dtd\Omega d\omega} = \frac{1}{\hbar\omega}\frac{N_b}{T_0}\frac{d^2U(\omega)}{d\omega d\Omega} = \frac{2c\epsilon_0}{\hbar\omega}\frac{N_b}{T_0}R^2|\vec{E}(\omega)|^2$$

$$\Rightarrow \frac{d^3N_{tot}(\omega)}{dtd\theta d\omega/\omega} = 2.457\cdot 10^{13}E[GeV]\langle I\rangle[A]S(\omega/\omega_c) \quad \left[\frac{photons}{sec\cdot mrad-hor.\cdot 0.1\%bw}\right]$$
$$\qquad (7.31)$$

Equation 7.29 can also be used to quantify the statistical momenta of the photon energy distribution, for example the average and the root mean square photon energy:

$$\langle u\rangle = \frac{1}{N}\int_0^\infty u\dot{n}(u)du = \frac{P}{N} = \frac{8}{15\sqrt{3}}u_c$$

$$\langle u^2\rangle = \frac{1}{N}\int_0^\infty u^2\dot{n}(u)du = \frac{8}{15\sqrt{3}}u_c^2\int_0^\infty xS(s)dx = \frac{11}{27}u_c^2$$
$$\qquad (7.32)$$

7.3.4 Polarization

The electric field vector of synchrotron radiation is seen to oscillate in the bending plane from an observer placed exactly on that plane. The corresponding polarization is therefore linear horizontal. However, the electric field observed off-axis, of not negligible intensity within vertical angles $\pm 1/\gamma$, shows also a vertical component, i.e., the off-axis polarization is elliptical. Since the field vector is seen to rotate, say, clock-wise above the plane and counter-clock-wise below it, the elliptical polarization is, respectively, left- and right-handed.

The contribution of the distinct horizontal and vertical polarization to the single particle radiated power can therefore be discriminated by means of the vertical opening angle. Without a derivation [4,6], we report for completeness the integral of Eq. 7.19 over all horizontal angles of emission, so keeping the dependence of the emitted power from the vertical angle ψ only:

$$\frac{dP}{d\psi} = \int \frac{dP}{d\Omega} \sin\theta d\theta = P_0 \frac{21}{32} \frac{\gamma}{(1+\gamma^2\psi^2)^{5/2}} \left[1 + \frac{5}{7} \frac{\gamma^2\psi^2}{1+\gamma^2\psi^2}\right] \qquad (7.33)$$

The amount of power associated to the horizontal and vertical polarization is, respectively, $P_\sigma = \frac{7}{8}P_0$ and $P_\pi = \frac{1}{8}P_0$.

References

1. J.D. Jackson, *Classical Electrodynamics*, 3rd edn. (Wiley, New York 1999). Chapter 14
2. R.P. Walker, Synchrotron radiation, in *Proceedings of CERN Accelerators School: 5th General Accelerators Physics Course*, CERN 94-01, vol. I, ed. by S. Turner (Geneva, Switzerland, 1994), pp. 437–460
3. K.-J., Kim, Characteristics of synchrotron radiation. AIP Conf. Proc. **184**, 565–632 (1989)
4. G. Margaritondo, A. Primer, Synchrotron radiation. J. Synchr. Rad. **2**, 148–154; vol. 1994, ed. by S. Turner (Geneva, Switzerland, 1995), pp. 437–460
5. A. Balerna, S. Mobilio, *Introduction to Synchrotron Radiation*, in *Synchrotron Radiation*, ed. by S. Mobilio et al., (Springer Berlin Heidelberg, 2015) pp. 3–28 (ed. by S. Turner, Geneva, Switzerland, 1994), pp. 437–460
6. A. Hofmann, Characteristics of synchrotron radiation, in *Proceedings of CERN Accelerator School: Synchrotron Radiation and Free Electron Lasers, CERN 98-04*, ed. by S. Turner (Geneva, Switzerland, 1998), pp. 1–44

Equilibrium Distribution

8

In single-pass or few turns-only recirculating accelerators, the particle beam distribution is largely determined by its configuration at injection, eventually modified at a later stage by RF and magnetic elements. The emission of radiation due to longitudinal acceleration is typically negligible in the particle's energy budget.

On the contrary, the relevant and continuous emission of synchrotron radiation in a storage ring, together with the replenishment of kinetic energy provided by RF cavities, determine a change of the injected 6-D particle distribution over a time scale much longer than the single turn. At the end, the distribution will tend to a Gaussian in each sub-phase space ("equilibrium"), independently from the initial conditions.

This process is commonly described as the sum of two distinct contributions to the particles' invariants, *radiation damping* and *quantum excitation* [1]. The former origins in the *linear* expansion of the radiated energy about the particle's energy, and it results in a damping of the amplitudes of oscillation. The latter emerges when the beam's energy spread due to radiation emission (*second order* momentum) is taken into account. We first consider the impact of these effects on synchrotron oscillations, then we extend the results to the transverse planes.

8.1   Radiation Damping and Quantum Excitation

The Longitudinal motion of a single particle in a storage ring is described by Eq. 4.31, in the assumption that the accelerator behaves as a linear conservative system. This assumption is here revisited. In order to move to homogeneous variables in a normalized longitudinal phase space, Eq. 4.43 is recalled to find:

© The Author(s), under exclusive license to Springer Nature Switzerland AG 2022
S. Di Mitri, *Fundamentals of Particle Accelerator Physics*, Graduate Texts in Physics,
https://doi.org/10.1007/978-3-031-07662-6_8

$$\phi = \omega_{RF}\Delta t = h\omega_s\Delta t \sim \frac{\Delta E}{E_0}\frac{h\alpha_c}{Q_s} = h\omega_s\frac{\Delta E}{E_0}\frac{\alpha_c}{\Omega_s}$$

$$\Rightarrow \Delta t \sim \Delta E\frac{\alpha_c}{E_0\Omega_s}$$

(8.1)

In the following, we will use the notation $\tau \approx \Delta z/c$, $\epsilon = \Delta E$ for the particle's coordinates relative to the synchronous particle.

Let A_ϵ be the single particle's invariant for the longitudinal motion in a conservative system. Its variation with time due to radiation emission is considered. The variation is assumed to be adiabatic, so that the motion can be described as a pure harmonic oscillator on a single turn basis:

$$\begin{cases} \epsilon(t) = A_\epsilon(t)\cos(\Omega_s t + \phi_0) \equiv A_\epsilon(t)\cos\phi \\[2mm] \tau(t) = -\left(\frac{\alpha_c}{E_0\Omega_s}\right)A_\epsilon(t)\sin\phi \end{cases} \Rightarrow \begin{cases} A_\epsilon^2 = \epsilon^2 + \tau^2\left(\frac{E_0\Omega_s}{\alpha_c}\right)^2 \\[2mm] \langle\epsilon^2(t)\rangle_\phi = \frac{A_\epsilon^2(t)}{2} \end{cases}$$

(8.2)

$\langle\epsilon^2(t)\rangle_\phi$ is intended to be averaged over all phases of the bunch particles at a given time t. In this sense, it is just the beam's rms energy spread, and A_ϵ becomes the normalized *rms* amplitude of oscillation.

The Top equation on the r.h.s. of Eq. 8.2 is used to calculate the variation of the squared amplitude, dA_ϵ^2, averaged over all the synchrotron phases. We assume that the instantaneous emission of photons does change the particle's energy by the amount $d\epsilon = -u$, but *not* its phase, or $d\tau = 0$. Then, the radiated energy $u(\epsilon)$ is expanded to first order in the particle's energy:

$$\langle dA_\epsilon^2\rangle_\phi = \langle d\epsilon^2\rangle + \langle d\tau^2\rangle\left(\frac{E_0\Omega_s}{\alpha_c}\right)^2 = 2\langle\epsilon d\epsilon\rangle + \frac{1}{2}\langle(2d\epsilon)d\epsilon\rangle = -2\langle\epsilon u\rangle + \langle u^2\rangle \approx$$

$$\approx -2\langle\epsilon\frac{du}{d\epsilon}\epsilon\rangle + \langle u^2\rangle = -A_\epsilon^2\frac{du}{d\epsilon} + \langle u^2\rangle$$

(8.3)

where the derivative $du/d\epsilon$ is a function of ϵ, characteristic of the photon distribution of synchrotron radiation.

The average growth rate in a turn is additionally averaged over the closed orbit of equivalent radius R:

$$\langle\frac{d}{dt}\langle dA_\epsilon^2\rangle_\phi\rangle_R = \langle\frac{d}{dt}\left(\langle A_\epsilon^2(t)\rangle_\phi - \langle A_\epsilon^2(0)\rangle_\phi\right)\rangle_R = \langle\frac{d}{dt}A_\epsilon^2\rangle_R =$$

(8.4)

$$= -\langle A_\epsilon^2\rangle_R\langle\frac{d}{dt}\frac{du}{d\epsilon}\rangle_R + \langle\frac{d}{dt}\langle u^2\rangle_\phi\rangle_R$$

The physical meaning of the two terms on the r.h.s. of Eq. 8.4 is elucidated below.

8.1.1 Longitudinal Motion

If only the first term on the r.h.s of Eq. 8.4 were taken, we would get:

$$\langle\frac{dA_\epsilon^2}{dt}\rangle_{R,damp} = -\langle A_\epsilon^2\rangle_R\langle\frac{d}{dt}\frac{du}{d\epsilon}\rangle_R$$

(8.5)

We remind that $\frac{du}{dt} = P \propto E^4/R^2 \propto B_y^2 E^2$ is the instantaneous synchrotron radiation power introduced in Eq. 7.12. $P(B_y, E)$ is calculated below at first order in the particle's energy (ϵ), for the off-energy particle travelling on a dispersive orbit in a dipole magnet, i.e., $x \simeq D_x \epsilon/E_0$. This is justified by the fact that, although in general $x = x_\beta + x_D$, we are interested here in the dependence of the radiated power from the off-energy coordinate. Moreover, we assume $\langle x_\beta \rangle = 0$, which allows us to neglect the betatron motion at this stage. Still, to be more general, a quadrupole gradient embedded in the dipole is considered ("combined" dipole magnet).

Because of the energy deviation ϵ, a given bending angle corresponds to a different path length and bending radius for the generic and the synchronous particle, respectively, such that (see Fig. 8.1):

$$\theta = \frac{ds}{R} = \frac{dl}{(R+x)} \Rightarrow dl = \left(1 + \frac{x}{R}\right) ds \tag{8.6}$$

For ultra-relativistic particles $(C = v_z T_0 \simeq c T_0)$ we find:

$$u = \oint \frac{dl}{c} P(B_y, E) = \oint \frac{ds}{c} \left(1 + \frac{x}{R}\right) \left(P_0 + \frac{\partial P}{\partial E} \epsilon + \frac{\partial P}{\partial B_y} \frac{\partial B_y}{\partial x} \frac{\partial x}{\partial E} \epsilon\right) =$$

$$= \oint \frac{ds}{c} \left(1 + \frac{D_x}{R} \frac{\epsilon}{E_0}\right) \left(P_0 + \frac{2P_0}{E_0} \epsilon + \frac{2P_0}{B_y} g D_x \frac{\epsilon}{E_0}\right) =$$

$$\approx \oint \frac{ds}{c} \left(P_0 + P_0 \frac{D_x}{R} \frac{\epsilon}{E_0} + \frac{2P_0}{E_0} \epsilon + 2P_0 k D_x R \frac{\epsilon}{E_0}\right);$$

$$\frac{du}{d\epsilon} = \oint \frac{ds}{c} \left(\frac{2P_0}{E_0} + \frac{P_0}{E_0} \frac{D_x}{R} + 2 \frac{P_0}{E_0} k D_x R\right);$$

$$\frac{d}{dt} \frac{du}{d\epsilon} = \frac{d}{dt} \left(\frac{2U_0}{E_0}\right) + \frac{d}{dt} \oint \frac{ds}{c} \frac{P_0}{E_0} D_x R \left(\frac{1}{R^2} + 2k\right); \tag{8.7}$$

$$\langle \frac{d}{dt} \frac{du}{d\epsilon} \rangle R = \frac{1}{C} \oint \frac{ds}{dt} \frac{2U_0}{E_0} + \frac{1}{C} \oint \frac{ds}{dt} \oint \frac{ds}{c} \frac{P_0}{E_0} D_x R \left(\frac{1}{R^2} + 2k\right) =$$

$$= \frac{2U_0}{T_0 E_0} + \frac{1}{T_0 E_0} \oint \frac{ds}{c} P_0 D_x R \left(\frac{1}{R^2} + 2k\right) =$$

$$= \frac{U_0}{T_0 E_0} \left[2 + \oint ds \left(\frac{P_0 R^2}{U_0 c}\right) \frac{D_x}{R} \left(\frac{1}{R^2} + 2k\right)\right] =$$

$$= \frac{U_0}{T_0 E_0} \left[2 + \frac{\oint ds \frac{D_x}{R} \left(\frac{1}{R^2} + 2k\right)}{\oint \frac{ds}{R^2}}\right] \equiv \frac{U_0}{T_0 E_0} (2 + \mathbb{D})$$

The coefficient $\mathbb{D}$ introduced above depends only from the linear optics of the accelerator. Its physical meaning is elucidated below in the case of, for example, an isomagnetic lattice $(R(s) = R)$ and separate function dipole magnets $(k = 0)$:

$$\mathbb{D} \rightarrow_{(iso)} \frac{1}{C} \oint ds D_x R \left(\frac{1}{R^2} + 2k\right) \rightarrow_{(sep)} \frac{1}{C} \oint \frac{D_x}{R} ds = \alpha_c \tag{8.8}$$

Fig. 8.1 Top view of
reference (ds) and distorted
orbit (dl) in a dipole magnet

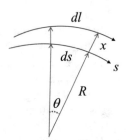

By substituting the result of Eq. 8.7 into Eq. 8.5 we find:

$$\langle \frac{dA_\epsilon^2}{dt} \rangle_{R,damp} = -\langle A_\epsilon^2 \rangle_R \frac{U_0}{T_0 E_0}(2+\mathbb{D}) \equiv -\frac{2}{\tau_\epsilon}\langle A_\epsilon^2 \rangle_R$$

$$\Rightarrow \begin{cases} \langle A_\epsilon^2(t) \rangle_R = \langle A_\epsilon^2(0) \rangle_R e^{-\frac{2t}{\tau_\epsilon}} \\ \\ \alpha_\epsilon = 1/\tau_\epsilon = \frac{U_0}{2T_0 E_0}(2+\mathbb{D}) \equiv \frac{J_\epsilon U_0}{2T_0 E_0} \end{cases} \qquad (8.9)$$

Synchrotron oscillations are damped only if $\mathbb{D} > -2$. α_ϵ is called *damping coefficient* for the longitudinal plane, the characteristic decay time $\tau_\epsilon = 1/\alpha_\epsilon$ is the *longitudinal damping time*, and $J_\epsilon = 2 + \mathbb{D}$ is the *longitudinal partition number*.

Roughly speaking, the damping time corresponds to the number of turns ($1/T_0$) that the particle would require to completely exhaust its energy via radiation emission (U_0/E_0).

If radiation damping were the only process in action, it would shrink the oscillation amplitude, and therefore the beam's longitudinal emittance, to arbitrarily small values over a sufficiently long time. In other words, all representative points in the phase space would collapse to one point, which is in contrast to the experimental observation. This paradox is explained by the presence of the second term on the r.h.s. of Eq. 8.4.

Since each particle emits independently from the others, but photons all belong to the same distribution, averaging over the particles' phase in Eq. 8.3 can be replaced by averaging over the photon energies (Campbell's theorem) [2]. Be $n(u)$ the photon energy distribution of synchrotron radiation introduced in Eq. 7.28 normalized to the total number of emitted photons, $\int_0^\infty n(u)du = N$. Making use of Eqs. 7.29 and 7.32 we find:

$$\langle \frac{dA_\epsilon^2}{dt} \rangle_{R,exc} = \langle \frac{d}{dt} \int_0^\infty u^2 n(u)du \rangle_R = \langle \frac{d}{dt} \int_0^\infty n(u)du \frac{\int_0^\infty u^2 n(u)du}{\int_0^\infty n(u)du} \rangle_R =$$

$$(8.10)$$

$$= \langle \frac{dN}{dt} \langle u^2 \rangle_n \rangle_R = \langle \frac{15\sqrt{3}}{8} \frac{P_0}{u_c} \frac{11}{27} u_c^2 \rangle_R = \frac{55}{24\sqrt{3}} \langle P_0 u_c \rangle_R$$

The balance of quantum excitation in Eq. 8.10 with radiation damping in Eq. 8.9 leads to the existence of an equilibrium value for the oscillation amplitude:

$$\langle \tfrac{dA_\epsilon^2}{dt} \rangle_R = \langle \tfrac{dA_\epsilon^2}{dt} \rangle_{R,damp} + \langle \tfrac{dA_\epsilon^2}{dt} \rangle_{R,exc} = -\tfrac{2}{\tau_\epsilon}\langle A_\epsilon^2 \rangle_R + \tfrac{55}{24\sqrt{3}}\langle P_0 u_c \rangle_R \equiv 0;$$

$$\Rightarrow \frac{\langle A_\epsilon^2 \rangle_R}{2E_0^2} = \left(\frac{\epsilon_{eq}}{E_0}\right)^2 = \sigma_{\delta,eq}^2 = \frac{1}{E_0^2}\frac{55}{96\sqrt{3}}\tau_\epsilon \langle P_0 u_c \rangle_R =$$

$$= \frac{1}{E_0^2}\frac{55}{96\sqrt{3}}\frac{2E_0}{J_\epsilon}\frac{3}{2}\hbar c \gamma^3 \frac{\langle 1/|R^3| \rangle}{\langle 1/R^2 \rangle} = \frac{55}{32\sqrt{3}}\frac{\hbar}{m_0 c}\frac{\gamma^2}{J_\epsilon}\frac{\langle 1/|R^3| \rangle}{\langle 1/R^2 \rangle}$$

(8.11)

$$\Rightarrow \sigma_{\delta,eq}^{iso} = \gamma \sqrt{\frac{C_e}{J_\epsilon |R|}}, \qquad C_e = 3.836 \cdot 10^{-13} m$$

The dependence of $\sigma_{\delta,eq}^2$ from the average radius squared and cubic in Eq. 8.11 comes, respectively, from $\langle P_0 u_c \rangle \sim \langle \frac{1}{R^2} \cdot \frac{1}{R} \rangle$, and $\tau_\epsilon \sim \frac{T_0}{U_0} = \frac{1}{\langle P_0 \rangle} \sim \frac{1}{\langle 1/R^2 \rangle}$. $\sigma_{\delta,eq}^{iso}$ is the rms relative energy spread at equilibrium of an electron beam in an isomagnetic lattice. Since $R \sim E$, one has $\sigma_{\delta,eq} \propto \sqrt{\gamma}$. Such weak dependence explains the observed limited range $\sim 0.06 - 0.12\%$ of the relative energy spread at storage ring light sources spanning $\sim 1 - 8$ GeV beam energies.

The bunch duration at equilibrium is obtained through the usual conversion factor (see Eq. 8.1):

$$\sigma_{t,eq} = \frac{|\alpha_c|}{\Omega_s}\sigma_{\delta,eq}$$

(8.12)

To make this explicit, the expression of the synchrotron frequency in Eq. 4.32 is simplified by introducing the peak accelerating gradient $\dot{V}_0$ in [V/m], and by considering ultra-relativistic electrons (or positrons) above transition energy:

$$\begin{cases} V_0 \sin\psi_s = \frac{dV}{d\psi} = \frac{dV}{cdt}\frac{cdt}{d\psi} = \frac{c\dot{V}_0(s)}{\omega_{RF}} \\ \eta \approx -\alpha_c \\ p_{z,s} \approx \frac{E_0}{c} \end{cases} \Rightarrow \Omega_s \approx \sqrt{\frac{ec\dot{V}_0\alpha_c}{T_0 E_0}}$$

(8.13)

Replacing this into Eq. 8.12, we obtain for an isomagnetic lattice:

$$c\sigma_{t,eq} \approx \alpha_c \gamma \sqrt{\left|\frac{2\pi R_s \gamma m_e c^2}{e\dot{V}_0 \alpha_c}\frac{C_e}{J_\epsilon R}\right|} \approx \sqrt{2\pi C_e}\sqrt{\left|\frac{\alpha_c}{J_\epsilon (e\dot{V}_0/m_e c^2)}\right|}\gamma^{3/2}$$

(8.14)

Typical values of the rms natural bunch duration (e.g., in the presence of a single frequency of the RF system) are of few tens' of picoseconds. Equation 8.14 illustrates in a rigorous manner the dependence of the bunch duration $\sigma_t \propto \sqrt{|\alpha_c/\dot{V}_0|}$, previously inferred in Eq. 4.44.

8.1.2   Horizontal Motion

The equilibrium emittance in the horizontal plane is retrieved by analysing the variation of the Floquet's normalized amplitude of oscillation (Eq. 4.109) in the presence
of radiation emission. The amplitude variation is due to both betatron and dispersive
motion, i.e., $x = x_\beta + x_D$. By introducing the notation $H_x = \gamma_x D_x^2 + 2\alpha_x D_x D_x' + \beta_x D_x'^2$ we find:

$$A_x^2 = \gamma_x x^2 + 2\alpha_x x x' + \beta_x x'^2 = A_{x,\beta}^2 + H_x \left(\frac{\epsilon}{E_0}\right)^2 +$$
$$+ 2\left[\gamma_x x_\beta D_x \tfrac{\epsilon}{E_0} + 2\alpha_x \left(x_\beta D_x \tfrac{\epsilon}{E_0}\right)\left(x_\beta' D_x' \tfrac{\epsilon}{E_0}\right) + \beta_x \left(x_\beta' D_x' \tfrac{\epsilon}{E_0}\right)\right];$$

$$dA_x^2 = dA_{x,\beta}^2 + \tfrac{H_x}{E_0^2} d\epsilon^2 + f\left(x_\beta, x_\beta'\right);$$

$$\langle dA_x^2\rangle_\phi = \langle dA_{x,\beta}^2\rangle_\phi + \tfrac{H_x}{E_0^2}\langle d\epsilon^2\rangle_\phi + f\left(\langle x_\beta\rangle_\phi, \langle x_\beta'\rangle_\phi\right) = \langle dA_{x,\beta}^2\rangle_\phi + \tfrac{H_x}{E_0^2} N_{ph}\langle u^2\rangle_n$$
$$(8.15)$$

We note that, so as J is a definite positive quantity, also H_x is. The very last equality
made use of Eq. 8.10 for the average of the photon energies, and $\langle x_\beta\rangle_\phi = \langle x_\beta'\rangle_\phi = 0$.
As already for the longitudinal motion, also in this case we can discriminate the two
contributions of radiation damping $(dA_{x,\beta}^2)$ and quantum excitation $(\langle u^2\rangle_n)$.

According to Eq. 4.109, the betatron normalized amplitude $dA_{x,\beta}^2$ is:

$$dA_{x,\beta}^2 = dw^2 + dw'^2 = 2w dw + 2w' dw' \qquad (8.16)$$

Since photon emission is assumed to be instantaneous, the variation of the particle's position is null, only the angular divergence is affected. Because of the superposition of betatron and dispersive motion, we find for dw:

$$dx = dx_\beta + dx_D \equiv 0 \Rightarrow dx_\beta = -dx_D = -D_x \tfrac{d\epsilon}{E_0} \Rightarrow dw = \tfrac{D_x}{\sqrt{\beta_x}} \tfrac{u}{E_0} \qquad (8.17)$$

The variation of the normalized angular divergence dw' is retrieved from the
definition of x':

$$\begin{cases} x' = \frac{p_x}{p_z} \approx \frac{p_x}{p} \\ x' = \frac{dx}{ds} = \frac{dx}{dw}\frac{dw}{d\phi}\frac{d\phi}{ds} = \sqrt{\beta_x}w'\frac{1}{\beta_x} = \frac{w'}{\sqrt{\beta_x}} \end{cases} \qquad (8.18)$$

$$\Rightarrow \begin{cases} \frac{dx'}{x'} = \frac{dp}{p} = \frac{d\epsilon}{E_0} \\ \frac{dx'}{x'} = \frac{dw'}{\sqrt{\beta_x}}\frac{1}{x'} = \frac{dw'}{w'} \end{cases} \Rightarrow \frac{dw'}{w'} = -\frac{u}{E_0} \qquad (8.19)$$

The amount of radiated energy $d\epsilon = -u$ is recalled from Eq. 8.7. Here, however, its dependence from the particle's orbit is restricted to the betatron motion:

$$u = \oint \frac{dl}{c} P(B_y) = \oint \frac{ds}{c}\left(1 + \frac{x_\beta}{R}\right)\left(P_0 + \frac{dP}{dB_y}\frac{dB_y}{dx}x_\beta\right) \approx \oint \frac{ds}{c}P_0\left(1 + \frac{x_\beta}{R} + 2kRx_\beta\right)$$
(8.20)

Equations 8.17, 8.19 and 8.20 are plugged into Eq. 8.16, and the orbit-averaged growth rate is calculated:

$$
\begin{aligned}
dA_{x,\beta}^2 &= \oint \frac{ds}{c}\frac{2P_0}{E_0}\left(1 + \frac{x_\beta}{R} + 2kRx_\beta\right)\left(x_\beta\frac{D_x}{\beta_x} - w'^2\right) = \\
&= -\oint \frac{ds}{c}\frac{2P_0}{E_0}\left(w'^2 + w'^2\frac{x_\beta}{R} + 2w'^2kRx_\beta - D_x\frac{x_\beta}{\beta_x} - D_x\frac{x_\beta^2}{\beta_x R} - 2D_x x_\beta^2\frac{kR}{\beta_x}\right); \\
\langle dA_{x,\beta}^2\rangle_\phi &= -\oint \frac{ds}{c}\frac{2P_0}{E_0}\left(\langle w'^2\rangle_\phi - D_x\frac{\langle x_\beta^2\rangle_\phi}{\beta_x R} - 2D_x kR\frac{\langle x_\beta^2\rangle_\phi}{\beta_x}\right) = \\
&= -\oint \frac{ds}{c}\frac{2P_0}{E_0}\left(\langle w^2\rangle_\phi - \frac{D_x}{R}\langle w^2\rangle_\phi - 2D_x kR\langle w^2\rangle_\phi\right) = \\
&= -\oint \frac{ds}{c}A_{x,\beta}^2\frac{P_0}{E_0}\left[1 - D_x R\left(\frac{1}{R^2} + 2k\right)\right]; \\
\frac{d}{dt}\langle dA_{x,\beta}^2\rangle_\phi &= -\frac{d}{dt}\left(\frac{U_0}{E_0}A_{x,\beta}^2\right) + \frac{d}{dt}\oint \frac{ds}{c}A_{x,\beta}^2\frac{P_0}{E_0}D_x R\left(\frac{1}{R^2} + 2k\right); \\
\langle\frac{d}{dt}\langle dA_{x,\beta}^2\rangle_\phi\rangle_R &= -\frac{1}{c}\oint \frac{ds}{dt}\frac{U_0}{E_0}A_{x,\beta}^2 + \frac{1}{c}\oint \frac{ds}{dt}\oint \frac{ds}{c}A_{x,\beta}^2\frac{P_0}{E_0}D_x R\left(\frac{1}{R^2} + 2k\right) = \\
&= -\langle A_{x,\beta}^2\rangle_R\frac{U_0}{T_0 E_0}\left[1 - \oint ds\left(\frac{P_0 R^2}{U_0 c}\right)\frac{D_x}{R}\left(\frac{1}{R^2} + 2k\right)\right] = \\
&= -\langle A_{x,\beta}^2\rangle_R\frac{U_0}{T_0 E_0}\left[1 - \frac{\oint \frac{D_x}{R}\left(\frac{1}{R^2} + 2k\right)ds}{\oint \frac{ds}{R^2}}\right] \equiv \\
&\equiv -\langle A_{x,\beta}^2\rangle_R\frac{U_0}{T_0 E_0}(1 - \mathbb{D}) \equiv -\frac{2}{\tau_x}\langle A_{x,\beta}^2\rangle_R
\end{aligned}
$$
(8.21)

The coefficient $\tau_x = 1/\alpha_x$ is the *horizontal damping time*, $\alpha_x = \frac{U_0}{2T_0 E_0}(1 - \mathbb{D}) = \frac{J_x U_0}{2T_0 E_0}$ is the damping coefficient for the horizontal plane, and J_x the horizontal partition number. Damping is effective as long as $\alpha_x > 0$ or $\mathbb{D} < 1$.

Equation 8.21 is substituted into Eq. 8.15, averaged over the reference orbit, and forced to equilibrium:

$$\langle\frac{d}{dt}\langle dA_x^2\rangle_\phi\rangle_R = \langle\frac{d}{dt}\langle dA_{x,\beta}^2\rangle_\phi\rangle_R + \langle\frac{H_x}{E_0^2}\frac{dN_{ph}}{dt}\langle u^2\rangle_n\rangle_R =$$

$$= -2\frac{\langle A_{x,\beta}^2\rangle_R}{\tau_x} + \frac{55}{24\sqrt{3}}\frac{\langle H_x P_0 u_c\rangle_R}{E_0^2} \equiv 0;$$

$$\Rightarrow \langle A_{x,\beta}^2\rangle_{R,eq} = \frac{55}{48\sqrt{3}}\frac{\tau_x}{E_0^2}\langle H_x P_0 u_c\rangle_R = \frac{55}{48\sqrt{3}}\frac{1}{E_0^2}\frac{2E_0}{\langle P\rangle_R}\frac{3}{2}\hbar c\gamma^3\langle H_x P_0/R\rangle_R = \quad (8.22)$$

$$= \frac{55}{16\sqrt{3}}\frac{\hbar}{m_e c}\frac{\gamma^2}{J_x}\frac{\langle H_x/|R^3|\rangle_R}{\langle 1/R^2\rangle_R};$$

$$\Rightarrow \epsilon_{x,eq} = \frac{\langle A_{x,\beta}^2\rangle_{R,eq}}{2} = C_e\frac{\gamma^2}{J_x}\frac{\langle H_x/|R^3|\rangle_R}{\langle 1/R^2\rangle_R} \rightarrow_{(iso)} C_e\frac{\gamma^2}{J_x}\frac{\langle H_x\rangle_R}{|R|}$$

The limit taken for an isomagnetic lattice puts in evidence the action of the H_x-function. The beam's emittance is generated by the spread of particles' betatron orbits in the transverse phase space, driven by the radiating process. Since the local bump of the betatron amplitude (at the instant of emission) is equal (in absolute value) to the dispersive bump, the average dispersion function and its derivative eventually contribute to determining the particle distribution at equilibrium.

8.1.3   Vertical Motion

The absence of vertical dispersion would suggest null quantum excitation and therefore null vertical emittance for sufficiently long time, by virtue of the only surviving term of radiation damping in Eq. 8.15. In reality, a non-zero vertical beam size is always observed at equilibrium because of the change of the particle's vertical momentum in response of photons emitted off the bending plane, within the characteristic angular divergence $y' = \theta_y \approx \frac{1}{\gamma}$ of synchrotron radiation.

Owing to the absence of vertical dispersion in ideal configurations, $dA_y^2 = dA_{y,\beta}^2$. This is expanded up to second order in the radiated energy, in the assumption that the instantaneous photon emission does not change the particle's position, but only its angular divergence:

$$dA_{y,\beta}^2 = d(w^2 + w'^2) = 2wdw + 2w'dw' + \tfrac{1}{2}2(dw')^2 = 2w'^2\tfrac{dw'}{w'} + (dw')^2 \tag{8.23}$$

By using Eqs. 8.18 and 8.19:

$$dA_{y,\beta}^2 = -2w'^2\tfrac{u}{E_0} + w'^2\tfrac{u^2}{E_0^2} = -2w'^2\tfrac{u}{E_0} + \beta_y\theta_y^2\tfrac{u^2}{E_0^2} \tag{8.24}$$

The variation of the squared amplitude is first averaged over all the betatron phases, then the rate of its growth is averaged over the orbit, and eventually forced to zero at equilibrium:

$$\frac{d}{dt}\langle dA_{y,\beta}^2\rangle_\phi = -2\langle w'^2\rangle_\phi\frac{P_0}{E_0} + \frac{\beta_y\theta_y^2}{E_0^2}\frac{d}{dt}\langle u^2\rangle_\phi$$

$$= -A_{y,\beta}^2\frac{P_0}{E_0} + \frac{\beta_y\theta_y^2}{E_0^2}\frac{dN_{ph}}{dt}\langle u^2\rangle_n;$$

$$\langle\frac{d}{dt}\langle dA_{y,\beta}^2\rangle_\phi\rangle_R = -\langle A_{y,\beta}^2\rangle_R\frac{P_0}{E_0} + \frac{55}{24\sqrt{3}}\frac{1}{E_0^2}\langle\beta_y\theta_y^2 P_0 u_c\rangle_R =$$

$$= -\frac{2\langle A_{y,\beta}^2\rangle_R}{\tau_y} + \frac{55}{24\sqrt{3}}\frac{1}{E_0^2}\langle\beta_y\theta_y^2 P_0 u_c\rangle_R \equiv 0;$$

$$\Rightarrow \langle A_{y,\beta}^2\rangle_{R,eq} = \frac{55}{48\sqrt{3}}\frac{2E_0}{J_y\langle P\rangle_R}\frac{1}{E_0^2}\frac{3}{2}\hbar c\gamma^3\frac{1}{2\gamma^2}\langle\beta_y P_0 u_c\rangle_R$$

$$= \frac{55}{32\sqrt{3}}\frac{\hbar c}{m_e c}\frac{1}{J_y}\frac{\langle\beta_y/|R^3|\rangle_R}{\langle 1/R^2\rangle_R};$$

$$\Rightarrow \epsilon_{y,eq} = \frac{\langle A_{y,\beta}^2\rangle_{R,eq}}{2} = \frac{C_e}{2J_y}\frac{\langle\beta_y/|R^3|\rangle_R}{\langle 1/R^2\rangle_R} \to_{(iso)} \frac{C_e}{2J_y}\frac{\langle\beta_y\rangle_R}{|R|} \tag{8.25}$$

The limit is taken for an isomagnetic lattice. The damping coefficient for the vertical plane is $\alpha_y = 1/\tau_y = \frac{J_y U_0}{2T_0 E_0}$, and τ_y is the *vertical damping time*. In the absence of vertical dispersion, the vertical partition number is $J_y = 1$.

Typical values of the average betatron function and bending radius in electron synchrotrons would lead to $\epsilon_{y,eq} \leq 0.1$ pm. In reality, the vertical emittance at equilibrium is contributed by betatron coupling and spurious vertical dispersion ($\langle H_y \rangle_R \neq 0$), for example due to misaligned magnetic elements, skew quadrupole magnets, etc. Both betatron coupling and spurious vertical dispersion depend from the horizontal motion, and therefore the vertical emittance is proportional to the horizontal one via the so-called coupling coefficient, $\epsilon_y = \kappa \epsilon_x$.

8.1.4   Robinson's Theorem

Robinson's theorem [1,2] states that *the sum of the 3 partition numbers is constant, whatever the magnetic lattice and the RF parameters are.*

To demonstrate the theorem, we consider the 6×6 transfer matrix of a synchrotron, $M(s) = M(s + C)$, which applies to the vector $\vec{u} = (x, x', y, y', t, \delta)^t = (\vec{u}_1, \vec{u}_2, \vec{u}_3)^t$. By virtue of Floquet's theorem (see e.g. Eq. 4.99), the eigenvalues of the matrix can be written in the form:

$$M\vec{u}^* = e^{-(\vec{\alpha} \pm i\vec{\beta})T_0}\vec{u}^*$$

$$\Rightarrow \det M = \prod_{j=1}^3 e^{-(\alpha_j \pm i\beta_j)T_0} = e^{-\sum_j 2\alpha_j T_0} \approx 1 - \sum_j 2\alpha_j T_0 \tag{8.26}$$

where $j = 1, 2, 3$ for the three plans of motion, and the very last equality is for $|\alpha_j T_0| \ll 1$, which will be verified a posteriori.

The matrix $M_{ds} = M(s + ds)$ describes the instantaneous emission of radiation through an infinitesimal element ds of the orbit:

$$M_{ds} = I + \delta M \approx \begin{pmatrix} 1 + m_1 & \cdots & \cdots & \cdots \\ \cdots & 1 + m_2 & \cdots & \cdots \\ \cdots & \cdots & \cdots & 1 + m_6 \end{pmatrix} \tag{8.27}$$

For infinitesimal instantaneous perturbation to the particle's motion via emission of photon energy u, the change in angular divergence is the one calculated in Eq. 8.18, which does not depend upon other coordinates:

$$\frac{dx'}{x'} = \frac{dy'}{y'} = -\frac{u}{E_0} = -\frac{\delta\epsilon_{RF}}{E_0} \quad \Rightarrow m_2 = m_4 = -\frac{\delta\epsilon_{RF}}{E_0} \tag{8.28}$$

The quantity $\delta\epsilon_{RF}$ is the amount of energy restored by RF cavities to keep the particle's energy constant on average in one turn. Its absolute value is equal to the radiated energy, and the negative sign indicates that the divergence lowers as the particle is accelerated ($\delta\epsilon_{RF} > 0$) by the RF field.

Since the rate of emission is quadratic in the particle's total energy (see e.g. Eq. 7.12), the change in relative energy deviation is:

$$\frac{d\delta}{\delta} = \frac{d\delta}{dt}\frac{dt}{\delta} = \frac{d(dE)/dt}{dE/dt} = \frac{dP}{P} = -2\frac{dE}{E_0} = -2\frac{u}{E_0} \quad \Rightarrow m_6 = -2\frac{u}{E_0} \qquad (8.29)$$

Equations 8.28, 8.29 demonstrate that M_{ds} is diagonal. Moreover, since the photon emission does not change the particle's position in the 3 planes of motion, $du_1 = du_3 = du_5 = 0 \Rightarrow m_1 = m_3 = m_5 = 0$, and m_2, m_4, m_6 are the only non-zero terms.

Finally, $\det M$ is calculated with the help of Eqs. 8.28 and 8.29 at the first order in u:

$$\det M_{ds} \cong \prod_{i=1}^{6}(1 + m_i) \cong 1 + m_2 + m_4 + m_6 + o(m_i m_k) = 1 - 2\frac{\delta\epsilon_{RF}}{E_0} - 2\frac{u}{E_0};$$

$$\det M = \det\left(\prod_{ds} M_{ds}\right) = \prod_{ds}(\det M_{ds}) = \prod_{ds}\left[1 - \frac{2}{E_0}(\epsilon_{RF} + u)\right] = 1 - \frac{4U_0}{E_0}$$
$$(8.30)$$

The equality of $\det M$ in Eqs. 8.26 and 8.30 demonstrates that

$$\sum_{j=1}^{3}\alpha_j = \sum_{j=1}^{3}J_j\frac{U_0}{2T_0 E_0} = \frac{2U_0}{T_0 E_0} \quad \Rightarrow \quad \sum_{j=1}^{3}J_j = 4 \qquad (8.31)$$

We draw the following considerations.

- The initial assumption $|\alpha_j T_0| \sim U_0/E_0 \ll 1$ is met in any practical case.
- The theorem holds as long as the external fields are known a priori, i.e., no beam-induced fields are considered.
- The theorem still applies to the case of linear coupling between horizontal and vertical plane (in this case, the normal mode emittances reach equilibrium), as well as to vertical dipole magnets (vertical dispersion).
- Only the choice $J_i > 0$ leads to stable motion. The characteristic time scale to reach equilibrium is the damping time. In particular, by virtue of Eqs. 8.9, 8.21 and 8.25, and when horizontal dispersion only is considered, the beam distribution reaches an equilibrium in all the 3 planes of motion only if $-2 < \mathbb{D} < 1$.

Robinson's theorem can be put in a Lorentz's invariant form by noticing that, according to Eqs. 8.9, 8.21 and 8.25, $2\alpha_{tot} = \sum_i 2\alpha_i$ is the rate of reduction of the 6-D phase space volume. Being it the inverse of a characteristic time, its Lorentz's invariant form is found by passing from the time interval in the laboratory frame to the proper time interval:

$$2\alpha_\tau = \frac{dt}{d\tau}2\alpha_{tot} = 2\gamma\sum_{i=1}^{3}\alpha_i = \frac{2E_0}{m_0 c^2}\frac{2P_0}{E_0} = \frac{4P_0}{m_0 c^2} \qquad (8.32)$$

This says that the characteristic rate of phase space reduction due to radiation damping is 4 times the ratio of the power radiated in one turn and the particle's rest energy. Not surprisingly, α_τ is the ratio of two Lorentz's invariants.

8.1.5 Radiation Integrals

The partition numbers can be cast in the form of integrals of functions—so-called *radiation integrals*—which depend only from the accelerator optics. In case of combined dipole magnets, the radiation integrals are:

$$I_2 = \oint \frac{ds}{R^2}, \quad I_3 = \oint \frac{ds}{|R^3|}, \quad I_4 = \oint ds \, \frac{D_x}{R} \left(\frac{1}{R^2} + 2k \right), \quad I_5 = \oint ds \, \frac{H_x}{|R^3|}$$

$$\Rightarrow J_x = 1 - \frac{I_4}{I_2}, \quad J_y = 1, \quad J_\epsilon = 2 + \frac{I_4}{I_2}$$

$$\tag{8.33}$$

The damping times become:

$$\tau_x = \frac{3T_0}{R\gamma^3} \frac{1}{I_2 - I_4}, \quad \tau_y = \frac{3T_0}{R\gamma^3} \frac{1}{I_2}, \quad \tau_\epsilon = \frac{3T_0}{R\gamma^3} \frac{1}{2I_2 + I_4} \tag{8.34}$$

For completeness, we report some other relevant quantities at equilibrium, for ultra-relativistic electrons:

$$\epsilon_x = C_e \gamma^2 \frac{I_5}{I_2 - I_4} = C_e \frac{\gamma^2}{J_x} \frac{I_5}{I_2}$$

$$\sigma_\delta^2 = C_e \gamma^2 \frac{I_3}{2I_2 + I_4} = C_e \frac{\gamma^2}{J_\epsilon} \frac{I_3}{I_2} \tag{8.35}$$

$$U_0 = \frac{\epsilon_c^2}{6\pi\epsilon_0} \gamma^4 I_2$$

Synchrotrons based on separate function magnets $(k = 0)$, or with small quadrupole gradients compared to the dipoles' weak focusing $k \ll R^{-2}$, show $J_x \approx 1$, $J_y = 1$ and $J_\epsilon \approx 2$. Therefore, they naturally provide damping in all the 3 planes of motion.

In general, the partition numbers can be tuned through diverse techniques, among which the most common are recalled below.

- Gradient ("Robinson") wiggler magnet. This is a few-poles wiggler magnet aiming at reducing $J_\epsilon \to 2$ while increasing $J_x \to 1$. Equation 8.33 shows that this can be obtained by increasing I_2. But, at the same time, I_4 has to be kept small, which implies $2D_x k/R < 0$ or equivalently $\frac{D_x}{B_y} \frac{dB_y}{dx} < 0$. Namely, in each wiggler pole, the dipolar field and the field gradient have to show opposite sign. A symmetric distribution of the magnetic components allows the dispersion to be closed at the end, and the initial direction of motion to be preserved.
- Dipole ("damping") wiggler magnet. One or multiple wigglers with pure dipolar field are installed in the ring to stimulate additional emission of radiation. This basically shortens the damping times in all planes, but depending upon the wiggler field, if installed in a dispersive region or not, it can either enlarge or reduce the emittances at equilibrium.

- Variation of the RF frequency. A small variation of the main RF above transition energy determines a variation of the orbit length according to $\frac{df}{f} = -\frac{dL}{L} = -\alpha_c \frac{dE}{E_0}$. The orbit shift inside quadrupole magnets can be such that the emission of synchrotron radiation in those magnets is enhanced (feed-down dipole effect), contributing to additional damping.

8.1.5.1 Discussion: To Be or Not to Be at Equilibrium?

What is the typical damping time of an electron beam in a 3 GeV, 0.6 km-long storage ring, in the presence of synchrotron radiation emission from 0.6 T dipole magnets? What is the damping time of a proton beam at the same total energy, assuming the same dipole magnet curvature radius? Estimate the damping time of the proton beam stored in the 27 km-long circumference of LHC, at 7 TeV total energy and assuming a dipole field of 20 T.

Since the damping times are of the order of the time needed for the particle to exhaust its total energy by radiation emission, their approximate value is $T_0 E_0 / U_0$, where for the electrons $T_0 = 0.6$ km/c $= 2$ μs and $U_0 = 429$ keV from Eq. 7.14. We find $\tau_e \approx 14$ ms.

Since $\tau \approx T_0 E_0 / U_0 \approx E_0 / P_0$, Eq. 7.12 allows us to write $\tau \sim \frac{E_0 R^2}{\beta^4 \gamma^4}$. For same total energy and curvature radius of the dipole magnets, the ratio of proton and electron beam damping time is:

$$\frac{\tau_p}{\tau_e} = \frac{\beta_e^4 \gamma_e^4}{\beta_p^4 \gamma_p^4} = \left(\frac{\gamma_e^2 - 1}{\gamma_p^2 - 1} \right)^2 \approx 10^{13} \quad \Rightarrow \quad \tau_p = 10^{13} \tau_e \approx 10^{11} s \sim 6000 \ y \quad (8.36)$$

Hence, proton beams in a storage ring at multi-GeV total energy do not reach an equilibrium distribution in the sense of Eq. 8.4 because the emission of synchrotron radiation is suppressed by the large proton's rest mass.

The amount of energy radiated in the dipole magnets of LHC is calculated by means of Eq. 7.13, where $R_p = E_p[GeV]/(0.3 \cdot B_y[T]) = 1167$ m, and protons are in the ultra-relativistic limit ($\beta \approx 1$). We find $U_p \approx 16$ keV and $\tau_p \approx (27km/c) \cdot 7TeV/16keV = 39375$ s ≈ 11 h.

8.1.5.2 Discussion: Equilibrium Emittance of Multi-bend Lattices

How does the equilibrium emittance depend from the number of dipoles? What is the relationship between equilibrium emittance, momentum compaction, and betatron tune? Consider the equilibrium horizontal emittance in Eq. 8.22 in the approximation of small bending angle ($\theta_b = l_b/R \ll 1$), beam waist ($\alpha_x \approx 0$), and constant effective betatron function inside the dipoles of an isomagnettic lattice.

The dispersion function and its first derivative are both proportional to the bending angle (see Eq. 4.74, although their specific value depends on the magnetic focusing in between dipole magnets):

$$\frac{\langle H_x \rangle_R}{R} \approx \frac{1}{R} \left(\frac{1}{\beta_x} \langle D_x^2 \rangle + \beta_x \langle D_x'^2 \rangle \right) \approx \frac{\theta_b}{l_d} \left[\frac{l_d^2 \theta_b^2}{4\beta_x} + \beta_x \theta_b^2 \right] \approx \theta_b^3 \left(\frac{l_d}{\beta_x} + \frac{\beta_x}{l_d} \right) \approx \left(\frac{2\pi}{N_d} \right)^3$$

$$\Rightarrow \epsilon_{x,eq} = F \frac{C_e}{J_x} \frac{\gamma^2}{N_d^3}$$

(8.37)

where commonly the average betatron function in dipoles $\beta_x \approx l_d$, $\langle D_x \rangle \approx \theta_b l_b$, N_d is the number of dipoles in the ring, and F a constant of the order of unity which depends from the specific lattice design. In short, a large number of dipole magnets keeps the individual bending angle small, thus a small dispersion function is generated and, eventually, a small equilibrium emittance.

Equation 8.37 allows us to write:

$$\frac{\langle H_x \rangle}{R} \approx \theta_b^3 \approx \frac{\langle D_x \rangle^3}{l_d^3} \approx \frac{\langle D_x \rangle^3}{\beta_x^3} \approx \frac{\langle D_x^2 \rangle}{\beta_x R} \quad \Rightarrow \quad R \approx \frac{\beta_x^2}{\langle D_x \rangle}$$

(8.38)

The definition of betatron tune and momentum compaction is recalled, and the result of Eq. 8.38 is used to find:

$$Q_x = \frac{1}{2\pi} \oint \frac{ds}{\beta_x} \approx \frac{R}{\langle \beta_x \rangle} \approx \frac{\langle \beta_x \rangle}{\langle D_x \rangle}$$

$$\alpha_c = \frac{1}{C} \oint \frac{D_x}{R} ds \approx \frac{\langle D_x \rangle}{R} \approx \frac{\langle D_x \rangle^2}{\beta_x^2} \approx \frac{1}{Q_x^2}$$

(8.39)

Equation 8.39 is more and more accurate for $C \approx 2\pi R$ (like in multi-bend lattices with respect to double or triple-bend achromatic cells), $\langle D_x^2 \rangle \approx \langle D_x \rangle^2$ (i.e., the standard deviation of the dispersion function is small), and the average betatron function along the lattice is comparable to the average one in the dipoles, $\langle \beta_x \rangle \approx \beta_x$. The peculiarity of small $\langle D_x \rangle$ in multi-bend lattices implies α_c typically one order of magnitude smaller than in double-bend lattices, and consequently $\sim$ 3-times larger horizontal betatron tune.

8.1.6 Vlasov-Fokker-Planck Equation

Vlasov's equation, in Eq. 5.37, describes a Hamiltonian system. The Vlasov-Fokker-Planck's (VFP) equation is the extension of Vlasov's equation to the presence of dissipative and random perturbative forces [2].

In a synchrotron light source, radiation damping plays the role of a dissipative force that, in the absence of quantum excitation, would lead to the collapse of the phase space volume. According to it, the beam phase space density distribution function increases with time. For example, in the longitudinal plane $\frac{d\psi}{dt} = 2\alpha_\epsilon \psi$. Quantum excitation relies on the random emission of photons, and as such it leads to particles diffusion in the phase space, or $\frac{d\psi}{dt} = D \frac{d^2\psi}{dE^2}$, with D a diffusion coefficient.

The VFP equation for the longitudinal plane in the presence of these two perturbations to the single particle's Hamiltonian motion becomes:

$$\frac{d\psi}{dt} = \frac{\partial\psi}{\partial t} + \frac{\partial\psi}{\partial q}\dot{q} + \frac{\partial\psi}{\partial p}\dot{p} = \frac{\partial\psi}{\partial t} + \frac{\partial\psi}{\partial\phi}\dot{\phi} + \frac{\partial\psi}{\partial E}\dot{\epsilon} = 2\alpha_\epsilon\psi + D\frac{d^2\psi}{dE^2} \tag{8.40}$$

where we introduced the reduced variables (ϕ, ϵ).

The generic particle's energy deviation is the sum of the RF energy gain and the energy loss by synchrotron radiation emission. It is expanded at first order in the particle's energy, and its time-derivative taken on average in a turn:

$$\dot{\epsilon} \approx \frac{1}{T_0}\left(qV(\phi) - \frac{dU}{dE}\epsilon\right) \equiv \dot{\epsilon}_0 - 2\alpha_\epsilon\epsilon \tag{8.41}$$

By replacing Eq. 8.41 into Eq. 8.40:

$$\frac{\partial\psi}{\partial t} + \frac{\partial\psi}{\partial\phi}\dot{\phi} + \frac{\partial\psi}{\partial E}\dot{\epsilon}_0 = 2\alpha_\epsilon\psi + \frac{\partial\psi}{\partial E}2\alpha_\epsilon\epsilon + D\frac{d^2\psi}{dE^2} \tag{8.42}$$

The l.h.s. of Eq. 8.42 describes the Hamiltonian flow of the distribution function in the absence of radiation damping and quantum excitation, thus it must vanish by virtue of Eq. 5.37. The r.h.s. can be written as the first partial derivative with respect to the particle's energy, whose argument therefore has to be independent from energy:

$$2\alpha_\epsilon\frac{\partial}{\partial E}\left(\psi\epsilon + \frac{D}{2\alpha_\epsilon}\frac{\partial\psi}{\partial E}\right) = 0;$$

$$\psi\epsilon + \frac{D}{2\alpha_\epsilon}\frac{\partial\psi}{\partial E} = f(\phi); \tag{8.43}$$

$$\Rightarrow \psi(\phi, \epsilon) = F(\phi)e^{-\frac{1}{2}\frac{\epsilon^2}{(D/2\alpha_\epsilon)^2}} \equiv F(\phi)e^{-\frac{1}{2}\left(\frac{\epsilon}{\sigma_E}\right)^2}$$

We have found that the stationary solution of Eq. 8.40, the so-called "equilibrium" distribution, is Gaussian in the energy coordinate.

The explicit form of $F(\phi)$ is obtained below by recurring to the Hamiltonian for the longitudinal motion. At equilibrium, this has to correspond to oscillations in the longitudinal phase space (z, ϵ), see Eq. 8.2. We re-write the equations of motion for the relative energy deviation and the longitudinal coordinate internal to the bunch, for the independent variable s along the accelerator:

$$\begin{cases} z = -\frac{1}{\kappa_z}\sqrt{2J_z}\cos(\kappa_z s) \\ \delta = \sqrt{2J_z}\sin(\kappa_z s) \end{cases} , \quad \kappa_z := \frac{\Omega_s}{\alpha_c c} \tag{8.44}$$

It is straightforward to verify that these equations obey the Hamiltonian $H = J_z = \left(\frac{\kappa_z^2 z^2 + \delta^2}{2}\right)$. The longitudinal rms emittance results $\epsilon_z = \sigma_z\sigma_\delta = \frac{\langle J_z\rangle}{\kappa_z}$.

Since the system at equilibrium behaves as a Hamiltonian system, then according to the theorem in Eq. 5.39 the phase space distribution function has to be function of the Hamiltonian only, or $\psi(z, \delta) = \psi(J_z)$. Owing to the fact that it must be Gaussian in the energy coordinate as previously found in Eq. 8.43, it follows:

$$\psi(z, \delta) = \psi(J_z) = \psi(0, 0)e^{-\frac{1}{2}\left(\frac{\kappa_z^2 z^2}{\kappa_z^2 \sigma_z^2} + \frac{\delta^2}{\sigma_\delta^2}\right)}$$

$$\Rightarrow \psi(J_z) = \frac{1}{2\pi \sigma_z \sigma_\delta} e^{-\frac{1}{2}\left(\frac{z^2}{\sigma_z^2} + \frac{\delta^2}{\sigma_\delta^2}\right)}$$

$$\iint_{-\infty}^{+\infty} \psi(J_z) dz d\delta \equiv 1$$

$$(8.45)$$

Since radiation damping and quantum excitation behave similarly in the transverse planes, the same kind of stationary distribution function is expected in the transverse phase spaces:

$$\psi(u, u') = \frac{1}{2\pi \sigma_u \sigma_{u'}} e^{-\frac{1}{2}\left(\frac{u^2}{\sigma_u^2} + \frac{u'^2}{\sigma_{u'}^2}\right)}, \quad u = x, y \tag{8.46}$$

Alternatively, a Gaussian distribution function for the stationary state can be predicted by recurring to the Central Limit theorem, because emission of synchrotron radiation can be intended as an incoherent perturbation to the large population of particles in a bunch, as long as collective effects like particles interaction via self-induced fields are ignored (see later), and the linear approximation $U = U(\epsilon)$ allows one to neglect deformations of the Gaussian tails by nonlinear diffusive processes.

8.2 Lifetime

8.2.1 Quantum Lifetime

The 2-D phase space distribution function in a synchrotron tends to a Gaussian over a time scale of few damping times. In the most general case of non-zero correlation between particles' position and angle (u, u'), it results [4]:

$$\rho(u, u') = \frac{1}{2\pi \sigma_u \sigma_{u'}} e^{-\left(\frac{u^2}{2\sigma_u^2} + \frac{uu'}{2\sigma_u \sigma_{u'}} + \frac{u'^2}{2\sigma_{u'}^2}\right)} = \frac{1}{2\pi\epsilon} e^{-\frac{1}{2\epsilon}\left(\gamma u^2 + 2\alpha uu' + \beta u'^2\right)} = \frac{1}{\langle W \rangle} e^{-\frac{W}{\langle W \rangle}} \tag{8.47}$$

All variables above are intended in the u-phase space, and the suffix is suppressed for brevity of notation. $W = 2J$ is the single particle invariant defined by $2J = \pi\left(\gamma u^2 + 2\alpha uu' + \beta u'^2\right)$, $\langle J \rangle = \pi\epsilon$, and $\epsilon = \sigma_u \sigma_{u'}$ is the beam rms emittance (see also Eq. 4.132). The distribution function is normalized to unity, $\int_0^\infty \rho(W) dW = 1$.

A Gaussian distribution has infinitely long tails. If no limitation is imposed, a stationary situation exists where the number of particles crossing an arbitrary boundary due to quantum excitation (rate Q) equals the number of particles entering due to radiation damping (rate D). If a limitation is present in correspondence of the oscillation amplitude W_c, instead, the rate of lost particles crossing the border determines the beam "lifetime" (rate L). Because of the cut to the Gaussian tails, the distribution will be modified to some extent. But, if the restriction is far enough from the beam

core ($W_c \gg \epsilon$), the rate can be calculated by assuming that the number of lost particles per turn is small, and therefore the distribution is still approximately Gaussian.
Consequently, we can assume that the number of particles crossing W_c and being
lost, is very nearly the same as if there were no limitations. In particular, the particle
loss rate can be estimated as that due to radiation damping (rate L $\cong$ rate Q = rate
D).

To calculate the loss rate at $W = W_c$, we first consider the fraction of particles
in the differential amplitude range dW, i.e., $dN(W) = N\rho dW$. The variation of the
amplitude due to radiation damping in a characteristic damping time τ is $W(t) =$
$\hat{W}e^{-\frac{2t}{\tau}}$. By virtue of Eq. 8.47 we have:

$$\begin{cases} \frac{dN}{dW} = \frac{N}{\langle W \rangle}e^{-\frac{W}{\langle W \rangle}} \\ \\ \frac{dW}{dt} = -\frac{2}{\tau}W \end{cases} \tag{8.48}$$

$$\left(\frac{dN}{dt}\right)_{W_c} = \left(\frac{dN}{dW}\frac{dW}{dt}\right)_{W_c} = -\frac{2N}{\tau}\frac{W_c}{\langle W \rangle}e^{-\frac{W_c}{\langle W \rangle}} \Rightarrow \begin{cases} N(t) = N_0 e^{-\frac{t}{\tau_q}} \\ \\ \tau_q = \frac{\tau}{2}\frac{\langle W \rangle}{W_c}e^{\frac{W_c}{\langle W \rangle}} = \frac{\tau}{2}\frac{e^{\xi}}{\xi} \end{cases}$$
$$\tag{8.49}$$

The stored current decays exponentially with time. The characteristic constant of
decay, τ_q, is called *quantum lifetime* and it is equal to the damping time multiplied by a
large factor, in proportion to $\xi_u := \frac{W_c}{\langle W \rangle} = \frac{\pi u_{max}^2/\beta_u}{2\pi \epsilon_u} = \frac{u_{max}^2}{2\sigma_u^2}$. This brings to the golden
rule for the ratio of accelerator acceptance and beam size $\frac{u_{max}}{2\sigma_u} \geq 6.5$ to guarantee
$\tau_q \geq 100$ h. By virtue of the description of particle's longitudinal motion through
C-S parameters (see Eq. 4.141), a quantum lifetime can be similarly defined for the
motion in the longitudinal phase space in the presence of a dynamical boundary.

In summary, quantum lifetime origins in quantum fluctuations of the particles'
energy due to photon emission. Owing to the extension of the Gaussian tails of
the charge distribution in phase space, particles can exceed the transverse and/or
the longitudinal acceptance of the accelerator, thus reducing the stored current with
time. The two distinct cases of transverse and longitudinal limitation to the particles'
motion are treated below.

8.2.2 Dynamic Aperture

In the transverse planes, let u_{max} be the physical aperture of the vacuum chamber
and $\xi_{x,y}$ the smallest ratio of aperture and beam size along the accelerator. In common situations $\xi_{x,y} > 5$, hence τ_q is very large for any practical purpose. However,
nonlinearities in the particles' motion can substantially reduce the beam's lifetime.

This can be understood by recalling the definition of *dynamic aperture* (DA) as
the region in the configuration space (x, y) within which particle's motion remains
stable for a sufficiently long time. In other words, the DA (in each plane of motion,

Fig. 8.2 Correspondence of initial (A_i) and final oscillation amplitude (A_f) in the presence of linear (blue) and nonlinear motion (red)

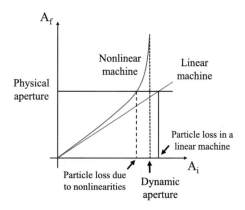

respectively) is the initial amplitude (A_i) in correspondence of which the particle's motion is unbounded ($A_f \to \infty$) after a large number of turns, as shown in Fig. 8.2.

The DA can either exceed or stay within the vacuum chamber. Particle dynamics is said to be linear when, despite the presence of nonlinear magnetic elements and machine errors, the correlation of final and initial amplitude of oscillation is approximately linear, or $A_f \propto A_i$. In this case τ_q is determined by physical restrictions. If the motion is nonlinear, instead, $A_f \propto (A_i)^n$, $n > 1$. Particles will always be lost on the chamber, but losses will now concern particles at smaller initial amplitudes (roughly speaking, particles closer to the bunch core). In this case τ_q is dominated by the DA. The smallest limitation among physical restriction and DA constitutes the *transverse acceptance* of the accelerator.

8.2.3   Overvoltage

In the longitudinal plane, $\xi_\epsilon = \frac{\delta_{acc}^2}{2\sigma_\delta^2}$. Since σ_δ is basically determined by the beam's energy and the dipole's bending radius (see Eq. 8.11), τ_q can be made larger by a larger RF energy acceptance [3]. According to Eq. 4.37, δ_{acc} can be made larger in turn by a larger peak RF voltage $e\hat{V}$, well in excess of the energy loss per turn U_0. The ratio $q := \frac{e\hat{V}}{U_0} > 1$ is denominated *overvoltage* factor.

At first, we express $\delta_{acc}(\psi)$ as function of q for the synchronous phase, where the energy gain per turn provided by the RF to balance the energy loss is $|\Delta E| = |e\hat{V}\cos\psi_s| \equiv |U_0|$. We also recall the peak accelerating gradient as the time derivative of the peak accelerating voltage, see Eq. 8.13. Then we have:

$$
\begin{cases}
\psi_s = \arccos(\frac{U_0}{e\hat{V}}) = \arccos(1/q) \\[2mm]
e\dot{V} = -\omega_{RF}e\hat{V}\sin\psi_s
\end{cases}
\tag{8.50}
$$

From the second equation:

$$\frac{(e\dot{V})^2}{\omega_{RF}^2} = (e\hat{V})^2(1 - \cos^2\psi_s) = (e\hat{V})^2 - \Delta E^2 = U_0^2\left(q^2 - 1\right) \qquad (8.51)$$

The RF energy acceptance (squared) defined in Eq. 4.37 is re-written:

$$\delta_{acc}^2 \approx 2\frac{e\hat{V}}{\pi h\alpha_c E_0}\left[(\psi_s - \pi)\cos\psi_s - \sin\psi_s\right] = \frac{2}{\pi h\alpha_c E_0}\left(-\psi_s\Delta E + e\dot{V}/\omega_{RF}\right) =$$

$$= \frac{U_0}{\pi h\alpha_c E_0}2\left[\sqrt{q^2 - 1} - \arccos(\tfrac{1}{q})\right] \equiv \frac{U_0}{\pi h\alpha_c E_0}F(q)$$

$$(8.52)$$

If we substitute U_0 from dipole magnets only (see Eq. 7.13) into Eq. 8.52, we obtain the following expression for electrons in an isomagnetic lattice:

$$\xi_\epsilon = \frac{\delta_{acc}^2}{2\sigma_\delta^2} = \left|\frac{J_\epsilon R}{2C_e\gamma^2}\frac{U_0}{\pi h\alpha_c E_0}F(q)\right| = \frac{64}{55\sqrt{3}}\frac{r_e}{\hbar c}\left|\frac{J_\epsilon E_0}{h\alpha_c}F(q)\right| \approx \left|\frac{J_\epsilon}{h\alpha_c}\right|F(q)E_0[GeV]$$

$$(8.53)$$

In most practical cases, $\left|\frac{J_\epsilon}{h\alpha_c}\right| \approx 1 - 10$.

Figure 8.3 shows $F(q)$ and δ_{acc} as function of q (Eq. 8.52) for typical parameters of a medium energy low-emittance electron storage ring. The presence of nonlinearities in the longitudinal dynamics, e.g. due to higher order momentum compaction, can easily reduce the RF energy acceptance predicted at first order by a factor 2–3. For this reason, the total RF peak voltage is commonly sized to have $q > 4$, which still guarantees $\delta_{acc}/\sigma_\delta > 10$ for usual values of the energy spread $\sigma_\delta \approx 0.1\%$.

Because of linear and nonlinear chromaticity, the betatron motion of off-energy particles can be distorted so that, at the end, the off-energy DA is smaller than the on-energy DA. If the off-energy DA contributes to particles loss more than the RF bucket height does, the quantum lifetime is determined by an effective "DA energy acceptance", usually named *momentum acceptance*.

According to Eq. 8.52, the contribution to U_0 by magnetic elements other than dipoles, such as wigglers and undulators (discussed in the following section), enlarges the RF energy acceptance, thus the quantum lifetime. The same result is obtained

Fig. 8.3 Overvoltage function $F(q)$ and RF energy acceptance of a 2.5 GeV, low emittance electron storage ring. The acceptance is calculated for a momentum compaction at first order

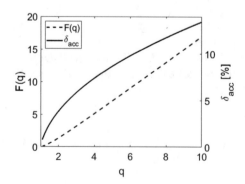

with a small $|\alpha_c|$. This, however, tends to shorten the acceptance in phase, i.e., the bunch length at equilibrium. From Eq. 8.14:

$$\sigma_{t,eq}^2 = 2\pi \frac{55}{32\sqrt{3}} \frac{\hbar}{m_e c} \gamma^3 \frac{\alpha_c m_e c^2}{J_\epsilon e V_0} = 2\pi \frac{55}{32\sqrt{3}} \frac{\hbar}{m_e c} \gamma^3 \frac{\alpha_c m_e c^2}{J_\epsilon \sqrt{q^2-1}} \frac{1}{\omega_{RF} U_0} =$$

$$= 2\pi \frac{55\sqrt{3}}{64} \frac{\hbar}{r_e m_e c^2} \frac{2\pi R}{2\pi c \gamma} \frac{\alpha_c}{J_\epsilon \omega_{RF} \sqrt{q^2-1}} = \left(\frac{55\sqrt{3}}{64}\right) \frac{\hbar c}{2\pi r_e} \frac{T_0}{E_0} \frac{\alpha_c}{J_\epsilon \omega_{RF} \sqrt{q^2-1}}$$

(8.54)

Since $\sigma_{t,eq} \propto \sqrt{\frac{\alpha_c}{q}}$, a trade-off for the value of q is usually found to maximize the RF bucket *area*.

8.2.3.1 Discussion: Transverse and Longitudinal Acceptance of a Light Source

Which of the three planes of motion first limits the quantum lifetime, if the on-energy dynamic aperture is larger than the vacuum chamber, and we neglect the contribution from the off-energy dynamic aperture? For a quantitative discussion, let us consider typical parameters of a medium-energy storage ring light source, such as 2.5 GeV beam energy, 0.2 nm rad horizontal emittance. The ellipsoidal vacuum chamber has size 12×4 mm, $\langle \beta_x \rangle \approx \langle \beta_y \rangle \approx 10$ m, the coupling factor is $\sim 1\%$, $h\alpha_c \approx 0.05$, and $\sigma_\delta = 0.1\%$. We also assume a dipole bending radius of 10 m, and $J_\epsilon \approx 2$. The total peak RF voltage is 2 MV.

In the transverse planes:

$$\xi_y = \frac{1}{2}\left(\frac{y_{max}}{\sigma_y}\right)^2 \approx \frac{1}{2}\left(\frac{x_{max}/3}{\sigma_x/10}\right)^2 \approx 9\xi_x,$$

(8.55)

$$\xi_x = \frac{1}{2}\left(\frac{12\,mm}{\sqrt{\epsilon_x \beta_x}}\right)^2 \approx 3.6 \cdot 10^4$$

In the longitudinal plane, $\xi_\epsilon = \frac{2 \cdot 2.5}{0.05} F(q)$. The overvoltage function is evaluated for $U_0 \approx 88.45 \frac{E^4[GeV]}{R[m]} = 346$ keV, therefore $q = 5.8$. According to Fig. 8.3, $F(q) \approx 8$, thus $\xi_\epsilon \approx 800$. The quantum lifetime results dominated by the longitudinal dynamics.

Let us now consider the off-energy DA and assume that, for example, it is internal to the vacuum chamber for relative energy deviation $|\Delta| \geq 2\%$. Since it results $\delta_{acc} = \sqrt{\left|\frac{U_0 F(q)}{\pi h \alpha_c E}\right|} \approx 8\%$, we expect the quantum lifetime to be dominated by momentum acceptance (still true in case of 3-fold reduction of δ_{acc} by nonlinearities in the longitudinal phase space).

8.2.4 Residual Gas Interactions

The beam lifetime can be degraded by scattering of stored particles on residual gas in the vacuum chamber. Such interaction is comprehensive of the following effects [4].

- *Large angle elastic (Coulomb) scattering*, which causes particles loss if the scattered particles hit a transverse physical aperture, or they are pushed outside the dynamic aperture. The interaction is described by the Rutherford's cross section:

$$\sigma \approx \frac{r_e^2 Z^2}{(\gamma^2 \theta^2)},$$
(8.56)

with Z the atomic number of the gas ion and θ the scattering angle. The latter can be expressed, with analogy to Eq. 6.2, as function of the maximum lateral displacement set by the vacuum chamber (A), the betatron function at the position of the aperture limitation, and the average betatron function along the ring (assuming a distributed interaction):

$$\sigma_{el} = \frac{2\pi r_e^2 Z^2}{\gamma^2} \frac{\langle \beta \rangle \beta_A}{A^2}$$
(8.57)

- *Inelastic scattering*, for electrons only, comprehensive of Bremsstrahlung (the electron is scattered by the atomic nucleus and emits a photon, but the atom is left unexcited) and inelastic atomic scattering (scattering by an atomic electron, the atom is left excited). Both these processes generate large particle's energy loss, possibly exceeding the RF or the momentum acceptance. The cross section is $\sigma \propto 4r_e^2 Z^2 \alpha$ ($\alpha = 1/137$ the fine structure constant), weakly dependent from the particle's energy.
- *Ion trapping*, i.e., the production of ions from scattering of electrons on residual gas. Ions are accumulated in specific regions of the vacuum chamber by focusing imposed by the circulating beam, until the repulsive space charge force of the ions starts limiting the ions concentration. The focusing strength imposed to the ions can be calculated with a formalism analogue to that used in colliders for evaluating the beam-beam tune shift (see later). We anticipate that, if $\Delta u' = a_{x,y} u$ is the angular divergence acquired by the ion as function of its (small) lateral distance u from the stored beam's axis, and a_k is the (linearized) focusing strength, we have:

$$a_{x,y} = \frac{Z}{A} \frac{2r_p N_e}{\sigma_{x,y}(\sigma_x + \sigma_y)}$$
(8.58)

where r_p, Z, A, N_e are, respectively, the classical proton radius, the ion atomic number, the ion mass number, and the number of electrons in a bunch of transverse sizes $\sigma_{x,y}$ at the interaction point. The ions are accumulated if the succession of focusing kicks due to the periodic spacing s_b of the stored electron bunches is such to guarantee a periodic motion. That is, the transfer matrix of the ions' motion has to have $|Tr(M)| < 2$. This implies a critical ion atomic mass, so that only heavier species are trapped:

$$M = \begin{pmatrix} 1 & s_b \\ 0 & 1 \end{pmatrix} \begin{pmatrix} 1 & 0 \\ -a_{x,y} & 1 \end{pmatrix} = \begin{pmatrix} 1 - a_{x,y} s_b & s_b \\ -a_{x,y} & 1 \end{pmatrix}$$

$$\Rightarrow \left(\frac{A}{Z} \right)_{trap} > \left(\frac{A}{Z} \right)_c = \frac{2r_p N_e s_b}{\sigma_y(\sigma_x + \sigma_y)}$$
(8.59)

where we have assumed $\sigma_x \gg \sigma_y$. Electron storage rings often implement "dark gaps" (several consecutive empty RF buckets) in the filling pattern to avoid stable resonances of the ions' motion.

- *Inelastic nuclear scattering*, for protons only, generates beam loss through nuclear reactions.

Gas scattering is counteracted with pumping systems aimed at obtaining low vacuum pressures (these can be as low as 10^{-10} bar in a storage ring). This is accompanied by a suitable preparation of the vacuum chamber to minimize desorption of gas molecules from the surface as possibly induced by synchrotron radiation.

If σ is the cross section of the interaction of ultra-relativistic beam particles with residual gas, and n_g is the number of gas atoms per unit volume, the number of particles traversing the unit volume is:

$$dN = -N\sigma n_g c dt \quad \Rightarrow \quad N(t) = N_0 e^{-\frac{t}{c\sigma n_g}} \equiv N_0 e^{-\frac{t}{\tau_g}} \tag{8.60}$$

If the gas concentration contains n_i molecules of type i, each molecule of type i made of $k_{i,j}$ atoms, then the total beam-gas scattering lifetime associated to the interaction σ becomes:

$$\frac{1}{\tau_g} = c\sigma n_g = c\sigma \sum_{i,j} k_{i,j} n_i = c\sigma \sum_{i,j} k_{i,j} \frac{p_i}{KT} = \frac{c\sigma}{KT} \sum_{i,j} k_{i,j} p_i \tag{8.61}$$

8.2.5 Touschek Lifetime

Touschek scattering, first explained by B. Touschek after the observation of current loss in the ADA storage ring in Frascati (Italy), describes the scattering of charged particles in the same bunch [4].

Collisions internal to the bunch happen all the time in all directions. However, if observed in the reference frame of the beam center of mass (c.m.), the particles' motion appears almost exclusively in the transverse planes (the longitudinal velocity relative to the c.m. being almost zero). In other words, the transverse momenta are much larger than the longitudinal one. *Touschek scattering* is momentum transfer from the transverse to the longitudinal plane in occasional large angle elastic scattering, which can therefore push particles off the RF or the momentum acceptance. Momentum transfer from the longitudinal to the transverse planes is also present (intrabeam scattering), but it is usually not harmful for the beam lifetime, as discussed in the following Section.

If two particles collide in the c.m. frame transferring their transverse momentum $\vec{p}_i' = (p_x', 0)$ to longitudinal momentum $\vec{p}_f' = (0, p_z') = (0, p_x')$, then the variation of longitudinal momentum in the laboratory frame is (see Fig. 8.4):

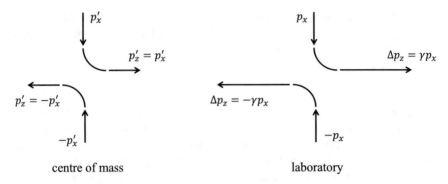

Fig. 8.4 Scattering of two particles transferring all their transverse momentum to longitudinal momentum, in the center of mass reference frame (left) and in the laboratory frame (right). The particle beam moves with a Lorentz factor γ in the laboratory

$$\Delta p_z = p_{z,f} - p_{z,i} = \gamma(p'_{z,f} + \frac{\beta}{c}E'_f) - \gamma(p'_{z,i} + \frac{\beta}{c}E'_i) = \gamma\Delta p'_z + \gamma\frac{\beta}{c}\Delta E' =$$

$$= \gamma(p'_{z,f} - p'_{z,i}) = \gamma p'_x = \gamma p_x = \gamma p_z \sigma_{u'}$$

$$\Rightarrow \frac{\Delta p_z}{p_z} \approx \gamma\sqrt{\frac{\epsilon_u}{\beta_u}}, \quad u = x, y$$

$$(8.62)$$

where we used Eqs. 1.17, 1.18 for the momentum transformation, $\Delta E' = 0$ for elastic scattering, and the definition of angular divergence. The momentum transfer is boosted by a factor γ in the laboratory frame. If the normalized divergence $\gamma\sigma_{u'}$ exceeds the RF or the momentum acceptance, then the particle gets lost.

The cross section of Touschek scattering calculated in the c.m. frame is the Moller cross section in the non-relativistic approximation, $\sigma_T \propto 2\pi r_0^2/\beta^4$. This suggests that the process is far more harmful for electron beams (lighter rest mass) with respect to proton beams, and that it is counteracted by a higher beam rigidity ($\beta \to 1$).

The Touschek lifetime is introduced by calculating the fraction dN_{loss} of particles lost with respect to the total number of particles dN_V contained in an element of volume $\sigma_T dl$, whose charge density is $\rho = dN_V/dV$:

$$\frac{dN_{loss}}{dN_V} = \int \rho\sigma dl = \int \rho\sigma v_z dt$$

$$(8.63)$$

$$\frac{dN_{loss}}{dt} = \frac{dN_{loss}}{dN_V}\frac{dN_V}{dt} = \int \rho\sigma v_z dN_V \approx \langle\sigma v_z\rangle \int \rho^2 dV = \langle\sigma v_z\rangle\frac{N^2}{8\pi^{3/2}\sigma_x\sigma_y\sigma_z}$$

The integration is done assuming a 3-D Gaussian distribution normalized to N. The average $\langle\sigma v_z\rangle$, which is assumed to be independent from the particles' coordinates, has to contain the information on the relative momentum deviation induced by the scattering or, equivalently, the normalized angular divergence as in Eq. 8.62. The additional dependence of the cross section from the beam's parameters can be inferred from the qualitative similarity with the scattering described by the Rutherford's cross section in Eq. 8.56. Here, $Z = 1$ and the scattering angle is replaced by the longitudinal acceptance δ_{acc}. We infer for the *Touschek lifetime* (assuming scattering in one plane only):

$$\frac{1}{\tau} = \frac{1}{N}\frac{dN}{dt} \sim \frac{1}{\gamma\sigma'_x}\frac{cr_0^2}{\gamma^2\delta_{acc}^2}\frac{N}{V} \tag{8.64}$$

An exact derivation for flat and round beam, respectively, provides:

$$\begin{cases} \dfrac{1}{\tau_{fl}} \approx \dfrac{r_0^2 c}{8\pi\gamma^3\sigma_{x'}\delta_{acc}^2}\dfrac{N}{\sigma_x\sigma_y\sigma_z} \\[3ex] \dfrac{1}{\tau_{ro}} \approx \dfrac{r_0^2 c}{4\sqrt{\pi}\gamma^4\sigma_{x'}\sigma_{y'}\delta_{acc}}\dfrac{N}{\sigma_x\sigma_y\sigma_z} \end{cases} \tag{8.65}$$

By virtue of Eq. 8.60, the *total lifetime* due to diverse beam-gas interactions and any other interaction of constant cross section, satisfies:

$$\frac{1}{\tau_{tot}} = \sum_m \frac{1}{\tau_m} \tag{8.66}$$

References

1. R.P. Walker, Radiation damping, Quantum excitation and equilibrium beam properties, in *Proceedings of CERN Accelerator School: 5th General Accelerator Physics Course*, CERN 94-01, vol. I, ed. by S. Turner (Geneva, Switzerland, 1994), pp. 461–498
2. J.M. Jowett, Introductory statistical mechanics for electron storage rings, in *Lectures Given at the U.S. Summer School on Physics of High Energy Particle Accelerators* (Stanford, CA, USA, 1985). Also SLAC-PUB-4033 (1986)
3. D.J. Thompson, D.M. Dykes, R.F. Systems, in *Synchrotron Radiation Sources—A Primer*, ed. by H. Winick (Published by World Scientific, Singapore, 1995), pp. 87–97. ISBN: 9810218567
4. C. Bocchetta, Lifetime and beam quality, in *Proceedings of CERN Accelerator School: Synchrotron Radiation and Free Electron Lasers*, CERN 98-04, ed. by S. Turner (Geneva, Switzerland, 1998), pp. 221–253

Perturbed Distribution

9

Perturbations to the regular 6-D particle distribution in linacs and in storage rings are discussed in this Chapter. "Regular" refers here to the charge density distribution function at the injection point in a linac, defined in terms of normalized emittances, and to the equilibrium distribution in a storage ring. Perturbations include (i) longitudinal-transverse coupling induced by variation of the particle's longitudinal momentum in a dispersive region (*synchro-betatron excitation*), (ii) particles diffusion by Coulomb scattering internal to the bunch (*intrabeam scattering*), and (iii) a variety of single-bunch and multi-bunch instabilities excited by the interaction of the beam particles with e.m. fields either internal to the beam, or associated to image currents on the vacuum chamber's surface (*collective effects*). A rigorous quantitative treatment of all these effects is beyond the aim of this book. Yet, a qualitative treatment and a classification of collective effects is provided. The main physical quantities are introduced in a single treatment for both linear and circular accelerators.

9.1   Synchro-Betatron Excitation

Owing to the linear superposition of betatron and dispersive motion in the complete solution of Hill's equation, the abrupt change of the particle's momentum in a dispersive region leads to a variation of the betatron amplitude: $\Delta x_\beta = -\Delta x_D = -D_x \Delta p/p$. This implies a local change of the C-S invariant which, extended to the beam particle distribution, can result into emittance growth.

This synchro-betatron excitation was observed to be at the origin of the equilibrium horizontal emittance in a synchrotron (see Eq. 8.17). Not surprisingly, $\epsilon_{x,eq}$ is proportional to the average dispersion function through H_x (see Eq. 8.22). In that case, an equilibrium can be reached because the shift of the radiating particle from

© The Author(s), under exclusive license to Springer Nature Switzerland AG 2022
S. Di Mitri, *Fundamentals of Particle Accelerator Physics*, Graduate Texts in Physics,
https://doi.org/10.1007/978-3-031-07662-6_9

an off-energy orbit, due to the change of its *total* momentum, is compensated (on average over a turn) by the increase of the only *longitudinal* momentum in RF cavities, thus by a reduction of the particle's angular divergence ($x' = \Delta p_x/p_z$), hence of the beam emittance.

These considerations suggest that, in general, RF cavities and magnetic insertion devices with high field should not be installed in dispersive regions because, similarly to emission of synchrotron radiation, the change in longitudinal momentum could induce emittance growth, and therefore to a modification of the equilibrium emittance defined in Eq. 8.22.

As previously discussed, the emission of radiation due to longitudinal acceleration in linacs is negligible. Nevertheless, coherent synchrotron radiation (CSR) can be emitted by relatively short electron bunches in dipole magnets of magnetic compressors and switchyard lines. In spite of the single-pass dynamics, the CSR intensity and the consequent change of particle's momentum can be large enough to induce emittance growth in the plane of non-zero dispersion.

The emittance growth induced by synchrotron radiation is an incoherent effect, i.e., the total radiated field is the linear superposition of the field radiated by each particle, and the change of particle's momentum is not correlated with the particle's longitudinal position internal to the bunch. On the contrary, the change of momentum by CSR is correlated with z. This implies a change of the particles' distribution in the transverse phase space, which is the resultant of a mismatch of individual bunch slices, each slice being associated to a different z-coordinate.

9.2   Intrabeam Scattering

Intrabeam scattering is the multiple small-angle Coulomb scattering of charged particles [1]. Unlike Touschek scattering, which is a single-scattering effect from the horizontal to the longitudinal direction of motion, leading to particle loss, intrabeam scattering is a diffusion process in all three dimensions. As such, it drives a growth with time of the transverse and the longitudinal beam emittance. In the following, the emittance growth rate will be reported without derivation, but a physical interpretation of its functional dependence from beam parameters is given, to discriminate intrabeam scatterng in linacs and in storage rings.

A common approach to derive intrabeam scattering growth rates starts from the assumption that the effect in an accelerated beam can be modelled as scattering of gas molecules in a closed box, where the box plays the role of magnetic and RF focusing in the accelerator, to keep the particles together. If we assume that the particles' three velocity components are independent, the scattering of particles leads to a Gaussian distribution in the 3-D momentum space.

Unlike in a closed box, however, the orbit curvature in a storage ring produces dispersion. Because of it, a sudden energy change translates into a change of the betatron amplitudes, thus to coupling of betatron and synchrotron motion, as discussed above. Moreover, the curvature leads to the so-called negative mass behaviour, which implies that an equilibrium condition above transition cannot exist. Finally, the derivation assumes that the particle's velocities are non-relativistic in the center-of-mass frame, which is strictly true for linacs only in specific cases. As of today, intrabeam scattering plays a major role in enlarging, for example, the bunch duration of proton beams in circular colliders, the transverse emittance of electron beams in multi-bend storage ring light sources, and the energy spread in highly dense electron beams in linacs for free-electron lasers.

9.2.1 Storage Rings

The intrabeam scattering growth rates are defined as the relative time-variation of beam's rms emittances [1]. Since in each plane the bunch dimensions are proportional to the square root of the emittance, a relative change of the bunch dimensions is half of the relative change of the emittance. In particular, we assume a round beam, no transverse coupling, and horizontal dispersion only. For the longitudinal plane we assume a bunched beam in the presence of synchrotron oscillations and constant mean energy. In case of small-amplitude oscillations (see Eq. 4.31), and with notation as in Eq. 8.44, the longitudinal emittance is proportional to the Hamiltonian, $\epsilon_z = \frac{\langle J_z \rangle}{\kappa_z} = \frac{\langle H \rangle}{\kappa_z}$. Then, according to Piwinski:

$$
\begin{cases}
\frac{1}{\tau_x^{ibs}} = \frac{1}{\epsilon_x} \frac{d\epsilon_x}{dt} = \frac{1}{2\langle x^2 \rangle} \frac{d\langle x^2 \rangle}{dt} \propto A_S f_x \left(\sigma_{x,y}^\beta, \sigma_{x',y'}^\beta, D_x, \sigma_\delta \right) \\
\frac{1}{\tau_y^{ibs}} = \frac{1}{\epsilon_y} \frac{d\epsilon_y}{dt} = \frac{1}{2\langle y^2 \rangle} \frac{d\langle y^2 \rangle}{dt} \propto A_S f_y \left(\sigma_{x,y}^\beta, \sigma_{x',y'}^\beta, D_x, \sigma_\delta \right) \\
\frac{1}{\tau_z^{ibs}} = \frac{1}{\epsilon_z} \frac{d\epsilon_z}{dt} = \frac{1}{2\langle H \rangle} \frac{d\langle H \rangle}{dt} \propto A_S f_z \left(\sigma_{x,y}^\beta, \sigma_{x',y'}^\beta, D_x, \sigma_\delta \right)
\end{cases}
\tag{9.1}
$$

$$
A_S = \frac{r_0^2 N_b [Clog]}{64\pi^2 \sigma_z \sigma_\delta \epsilon_{x,\beta} \epsilon_{y,\beta} \beta^3 \gamma^4}
\tag{9.2}
$$

The physics of intrabeam scattering is in the coefficient A_S. There, r_0 is the classical particle's radius, N_b the number of particles in a bunch, and all other symbols are self-explanatory. The term $[Clog]$ is said "Coulomb logarithm", and its argument is the ratio of maximum and minimum scattering angle relevant to the process, according to $\tan \left(\frac{\theta_{min,max}}{2} \right) = \frac{2r_0}{b_{max,min}\gamma^2 \sigma_{x'}^2}$, with b the scattering impact parameter.

In spite of some arbitrariness in the definition of θ_{min}, it is common to calculate it for $b_{max} \approx \sigma_x$. By definition $\theta_{max} < \pi$, but its upper limit is commonly restricted to $\sim 10\theta_{min}$. This approach discards single scattering events in the tails of the charge distribution, which may heavily bias the calculation of intrabeam scattering in the bunch core. The logarithmic dependence makes intrabeam scattering weakly dependent from the argument of $[Clog]$. For few-GeV energy electron storage rings, $[Clog] \sim 10$. Equation 9.2 suggests that intrabeam scattering is more effective for high charge density beams.

A manipulation of the growth rates in the three planes of motion leads to the following condition (see also Eq. 8.39):

$$\frac{d}{dt}\left[\langle\delta H\rangle\left(\frac{1}{\gamma^2}-\frac{D_x^2}{\beta_x^2}\right)+\frac{\langle\delta\epsilon_x\rangle}{\beta_x}+\frac{\langle\delta\epsilon_y\rangle}{\beta_y}\right]=0$$

$$\Rightarrow\langle H\rangle\left(\frac{1}{\gamma^2}-\alpha_c\right)+\frac{\langle\epsilon_x\rangle}{\beta_x}+\frac{\langle\epsilon_y\rangle}{\beta_y}\approx const.$$

(9.3)

and we used $\langle\delta q\rangle\approx\langle q\rangle dt/T_0$. Below transition energy, the sum of the three invariants is limited, namely, particles can exchange their oscillation energy among the three planes of motion. The beam behaves as gas molecules in a closed box or, in other words, an equilibrium distribution can exist, in which intrabeam scattering does not change the beam's dimensions any longer. Above transition energy, instead, the negative coefficient of $\langle H\rangle$ allows the oscillation energy to increase potentially in an indefinite way. In this case, an equilibrium distribution cannot exist.

9.2.2   Linacs

The advent of high brightness electron linacs for short wavelength free-electron lasers has only recently led to the detection of a noticeable effect of intrabeam scattering on the beam energy distribution [2]. The reason for this is that, if the beam is not stored in the accelerator for an extremely long time, a very high charge density would be required to make intrabeam scattering apparent. The relatively large transverse size of electron beams (from tens to few hundreds of microns) and the low growth rate of transverse emittance has so far allowed intrabeam scattering to be neglected in the transverse planes for any practical purpose. For this reason, the impact of intrabeam scattering on the beam energy spread only is treated in the following.

Equation 9.1 is re-written for the energy spread growth rate in a linac, in the approximation of ultra-relativistic, round beam ($\beta_x\approx\beta_y$, $\epsilon_x\approx\epsilon_y$) passing through a straight non-dispersive section, but in the presence of acceleration, $\gamma=\gamma_0+G\Delta s$:

$$\frac{1}{\sigma_\delta}\frac{d\sigma_\delta}{ds}=\frac{G}{\sigma_\delta}\frac{d\sigma_\delta}{d\gamma}=\frac{G}{\tau_{acc}}+\frac{1}{\tau_{ibs}}=\frac{G}{\tau_{acc}}+\frac{A_L}{\gamma^{3/2}\sigma_\delta^2},$$

$$A_L=\frac{r_0^2N_b[Clog]}{8\epsilon_n^{3/2}\beta_x^{1/2}\sigma_z}$$

(9.4)

G is the accelerating gradient, and in the following we neglect the weak dependence of $[Clog]$ from the beam energy. τ_{acc} is found by solving the equation for no intrabeam scattering ($\tau_{ibs}\to\infty$), and by imposing that the *absolute* energy spread does not change during acceleration (see e.g. Eq. 4.7 and discussion there):

$$\frac{d\sigma_\gamma}{ds} = \frac{d}{ds}(\gamma\sigma_\delta) = \frac{d\gamma}{ds}\sigma_\delta + \gamma\frac{d\sigma_\delta}{ds} \equiv 0;$$

$$\frac{1}{\sigma_\delta}\frac{d\sigma_\delta}{ds} = -\frac{1}{\gamma}\frac{d\gamma}{ds};$$

$$\frac{1}{\sigma_\delta}\frac{d\sigma_\delta}{d\gamma} = -\frac{1}{\gamma} = \frac{1}{\tau_{acc}}$$

$$\Rightarrow \sigma_\delta = \sigma_{\delta 0}\frac{\gamma_0}{\gamma}$$

(9.5)

With the prescription of Eq. 9.5 for τ_{acc}, Eq. 9.4 can be solved for $\sigma_\delta(\gamma)$:

$$\frac{d\sigma_\delta^2}{d\gamma} + \frac{2\sigma_\delta^2}{\gamma} - \frac{2A_L}{G}\frac{1}{\gamma^{3/2}} = 0$$

$$\Rightarrow \sigma_\delta^2 = \frac{c_0}{\gamma^2} + \frac{4}{3}\frac{A_L}{G}\frac{1}{\sqrt{\gamma}}$$

(9.6)

The coefficient c_0 is found by imposing $\sigma_\delta(0) \equiv \sigma_{\delta 0}$, and it results $c_0 = \gamma_0^2\sigma_{\delta,0}^2 - \frac{4A_L}{3G}\gamma_0^{3/2}$. This is substituted into Eq. 9.6 to get the solution:

$$\sigma_\delta^2(\gamma) = \sigma_{\delta 0}^2\frac{\gamma_0^2}{\gamma^2} + \frac{4}{3}\frac{A_L}{G}\frac{1}{\sqrt{\gamma}}\left[1 - \left(\frac{\gamma_0}{\gamma}\right)^{3/2}\right] \equiv \sigma_{\delta 0}^2\frac{\gamma_0^2}{\gamma^2} + \sigma_{\delta,IBS}^2(\gamma)$$

(9.7)

The simpler scenario of no acceleration, such as intrabeam scattering in a drift at constant energy $\gamma = \gamma_0$, is described by Eq. 9.4 for $\tau_{acc} \to \infty$:

$$\frac{d\sigma_\delta^2}{ds} = \frac{2A}{\gamma_0^{3/2}} = \frac{r_0^2 N_b[Clog]}{4\gamma_0^{3/2}\epsilon_n^{3/2}\beta_x(s)^{1/2}\sigma_z}$$

$$\Rightarrow \sigma_\delta^2 = \sigma_{\delta,0}^2 + \frac{r_0^2 N_b[Clog]}{4\gamma_0^{3/2}\epsilon_n^{3/2}\sigma_z}\int_0^L\frac{ds}{\sqrt{\beta_x(s)}} \approx \sigma_{\delta,0}^2 + \frac{r_0^2 N_b[Clog]}{4\gamma_0^2\epsilon_n\langle\sigma_x\rangle\sigma_z}L$$

(9.8)

The Coulomb logarithm in single-pass accelerators can be estimated in analogy to storage rings, with the prescription that its argument be proportional to the time τ the particles take to travel along the beamline:

$$\begin{cases} [Clog] = \ln\left(\frac{q_{max}\epsilon_n}{2\sqrt{2}r_0}\right) \\ q_{max} \approx \frac{c\tau N_b r_0^2}{2\gamma^{3/2}\epsilon_n^{3/2}\langle\beta_x\rangle^{1/2}\sigma_z} + o(\xi) \end{cases}$$

(9.9)

The approximated expression for q_{max} holds as long as $\xi = \sigma_\delta\sqrt{\frac{\langle\beta_x\rangle}{\gamma\epsilon_n}} \ll 1$, which in fact makes q_{max} independent from σ_δ.

Equation 9.7 points out that when $\gamma \gg \gamma_0$, and for any given G, (i) $\sigma_{\delta,IBS} \sim \gamma^{-1/4}$, i.e., the effect of intrabeam scattering on the *relative* energy spread depends weakly from the beam mean energy (the IBS-induced absolute energy spread goes like $\gamma^{5/4}$ instead), and (ii) the growth of *relative* energy spread evaluated at the end of acceleration is largely independent from the initial beam energy.

9.3   Collective Effects

Collective effects refer to all those phenomena in which the beam, being a collection of charges, acts back on itself via either direct particle-particle interaction, or the environment in which it travels [3]. They are current-dependent effects, and can either establish an instability of the particles' motion, or determine a new equilibrium distribution.

Collective effects due to direct particle-particle interaction comprise *direct space charge force* and *coherent synchrotron radiation instability*. In storage rings, they can be mediated by the interaction of the beam with the surrounding vacuum chamber, which makes this classification not that rigid.

The second class of interactions originates in the production of image charges on the wall of the vacuum chamber and of RF cavities. The finite conductivity of the metallic surroundings, as well as the abrupt change of the vacuum chamber profile, determines a causality principle according to which leading particles act back on trailing particles through the e.m. field associated to the image currents. The head-tail interaction is named *wake field*, if expressed in the time domain. Its counterpart in the frequency domain is called *impedance*. The classification remains ambiguous since an impedance can also be used to model a direct tail-head interaction such as CSR in single-pass accelerators.

All collective effects are potential sources of instability if the interaction of the beam with the e.m. fields establishes a positive feedback. This can lead to the deformation of the transverse, longitudinal and energy distribution of the accelerated beam. Damping mechanisms include Landau damping and negative feedback loops.

Landau damping originates in the spread of betatron tune, synchrotron tune and energy of the beam particles, in turn excited by nonlinearities in the transverse (higher order multipoles, field errors) and longitudinal motion (RF curvature, large synchrotron oscillation amplitudes). Landau damping is therefore intrinsic to the beam dynamics. It can be made even more effective via octupole magnets and higher harmonic RF cavities in storage rings, via off-crest acceleration and interaction with an external laser (to enlarge the beam energy spread) in linacs. The stochastic and possibly chaotic motion associated to Landau damping can lead to beam lifetime reduction in storage rings, which has therefore to be traded off with beam stability. Negative feedback loops rely on pick-ups (sensors and actuators), which sample the beam's motion and impose an e.m. kick, either transverse or longitudinal, to damp beam's coherent oscillations.

9.3.1   Wakefields

Let us consider an ultra-relativistic bunch of total charge Q, at longitudinal position s along the accelerator. For simplicity, the surrounding is assumed to have cylindrical symmetry. A "test" particle of charge q travels behind it at relative position (z, r) in cylindrical coordinates. In the presence of finite conductivity or discontinuities of the vacuum chamber (including e.g. RF cavities), the bunch charge Q acts as a source of e.m. field, which catches up with the trailing test charge q [3].

The longitudinal and transverse component of the Lorentz's force of interaction of the two charges, normalized to the source and to the test charge, provides a purely geometric function, called *wake function* or wake field, in that it only depends from the conductivity or the geometry of the elements surrounding the particles:

$$\begin{cases} w_\parallel(s, z) = -\frac{1}{qQ} \int_0^s ds' F_\parallel(s', z) = -\frac{1}{Q} \int_0^s ds' E_z \quad \left[\frac{V}{C}\right] \\[2mm] w_\perp(s, z) = -\frac{1}{qQ} \frac{d}{dr} \int_0^s ds' F_\perp(s', z, r) = -\frac{1}{Q} \frac{d}{dr} \int_0^s ds'(E_{x,y} \pm v_z \times B_{y,x}) \quad \left[\frac{V}{C \cdot m}\right] \end{cases}$$
$$(9.10)$$

The introduction of the e.m. field gives the alternative definition of wake function as the space-integrated field per unit of source charge. The integration is intended over the trajectory of the *test* particle, i.e., at the location where the e.m. field is sampled. The sign convention is such that a test particle at $z < 0$ is behind the beam. Sometimes the wake function is given per unit length of travel of the test particle, so that $w_\parallel'(s, z) = dw_\parallel/ds$ is in units of $[V/(C \cdot m)]$ and $w_\perp'(s, z, r) = dw_\perp/ds$ is in units of $[V/(C \cdot m^2)]$.

The Wake functions in Eq. 9.10 are truncated at first order in the lateral coordinate r; for this reason they are said to be "monopole" ($w_\parallel$) and "dipole" approximation ($w_\perp$) of the actual wakefield. It then emerges that $w_\parallel$ is non-zero independently from the radial position of the source. On the contrary, $w_\perp$ is non-zero only if the source is off the symmetry axis of the vacuum chamber or of the RF cavity ($r \neq 0$). Higher order terms of the wake field become important, for example, when the beam size is comparable to the vacuum chamber radius.

The wake function is a Green's function because it describes the response of the system (the vacuum chamber or the RF cavity) to a point-like and unitary stimulus. The convolution of the wake function with the source charge distribution gives the path-integrated field, namely, the e.m. potential generated by Q and sampled at the location occupied by q, during the time q has traveled a distance s. Such quantity is named *wake potential*:

$$\begin{cases} V_\parallel(s, z) = \iint dx'dy' \int_{-\infty}^z dz' \rho(x', y', z') w_\parallel(s, z - z') = \\[2mm] \qquad = \int_{-\infty}^z dz' \lambda(z') w_\parallel(s, z - z') \quad [V] \\[4mm] V_\perp(s, z) = \iint dx'dy' \int_{-\infty}^z dz' \rho(x', y', z') w_\perp(s, z - z') = \\[2mm] \qquad = \int_{-\infty}^z dz' \lambda(z') w_\perp(s, z - z') \quad \left[\frac{V}{m}\right] \end{cases}$$
$$(9.11)$$

where we defined $\iiint dr^3 \rho(\vec{r}) = \int dz \lambda(z) = Q$, with $\lambda(z)$ the longitudinal charge distribution function, or current profile.

The two-particle mathematical treatment adopted so far is identically valid if, for example, the test charge belongs to the source bunch. Then, s is the bunch coordinate along the accelerator, z the coordinate internal to the bunch. In reality, all bunch particles can play both roles of "source" and "test" particle, because they produce wakefield and are, in turn, affected by the wakefield generated by all other leading particles. By energy conservation, the e.m. energy stored in the cavity (or cavity-like insertion) in the form of wake field the total energy loss accumulated by the bunch through the element. For an element long L, the energy loss is the convolution of the wake potential with the charge distribution function. For the longitudinal wakefield in monopole approximation:

$$\Delta E(L) = \int_{-\infty}^{+\infty} dz \lambda(z) V_\|(L, z) \quad [J]$$

$$\Rightarrow k_\| = -\frac{\Delta E}{Q^2} \quad \left[\frac{V}{C}\right]$$

(9.12)

$k_\|$ is called *loss factor*, it is always positive and commonly of the order of few to several $\sim V/pC$ in single or multi-cell RF cavities. By definition, the loss factor depends only from the geometry of the RF cavity.

The analogous quantity for the transverse plane is called *transverse kick factor*. It measures the total change of transverse momentum accumulated by the bunch passing through, e.g., an RF cavity long L, per unit of lateral distance from the cavity electric axis:

$$c\frac{d}{dr}\Delta p_\perp(L) = \int_{-\infty}^{+\infty} dz \lambda(z) V_\perp(L, z) \quad \left[\frac{J}{m}\right]$$

$$\Rightarrow k_\perp = -\frac{c}{Q^2}\frac{dp_\perp}{dr} \quad \left[\frac{V}{C \cdot m}\right]$$

(9.13)

$k_\perp$ is typically of the order of few to several $\sim V/(pC \cdot mm)$ in single or multi-cell RF cavities.

9.3.2 Impedances

The Fourier transform (FT) of the wake function is called *impedance* [3]:

$$\begin{cases} Z_\|(\omega) = \frac{1}{c}\int_{-\infty}^{+\infty} dz e^{-i\omega z/c} w_\|(s, z) \quad [\Omega] \\ Z_\perp(\omega) = \frac{i}{c}\int_{-\infty}^{+\infty} dz e^{-i\omega z/c} w_\perp(s, z) \quad \left[\frac{\Omega}{m}\right] \end{cases}$$

(9.14)

The impedance is still a geometric property of the environment of the accelerated beam. Its use largely simplifies the calculation of collective effects in recirculating accelerators, (though being used also in single pass systems) by virtue of the following relationship:

$$V(\omega) = -I(\omega)Z(\omega)$$

(9.15)

with $V(\omega)$ and $I(\omega)$ the Fourier transform of the wakefield-induced potential and of the bunch current, respectively. Impedances show the following characteristics.

1. Eq. 9.10 states that wakefields are real functions. This implies $Z_\|(\omega)^* = Z_\|(-\omega)$ and $Z_\perp(\omega)^* = Z_\perp(-\omega)$.
2. The causality principle $w(z) = 0$ for $z > 0$ implies $Z(\omega) \to 0$ for $\omega \to \pm\infty$.
3. The Panofsky-Wenzel theorem establishes a one-to-one correspondence between the longitudinal and the transverse component of an impedance. The relation is exact for a smooth, infinitely long, cylindrical vacuum chamber of radius a, it is approximately valid for any chamber discontinuity provided its characteristic size is small compared to a, and valid only on average for large objects like RF cavities:

$$Z_\perp(\omega) \approx \frac{2c}{a^2}\frac{Z_\|(\omega)}{\omega} \approx \frac{2R}{a^2}\frac{Z_\|(\omega)}{n}, \quad n = \frac{\omega}{\omega_0} \tag{9.16}$$

The second equality of Eq. 9.16 applies to storage rings only, with R the equivalent storage ring radius, and ω_0 the revolution frequency hereafter.

In a cavity-like structure, the total wakefield can be described as the superposition of quasi-monochromatic waves, or "modes". Modes of characteristic wavelength smaller than the cavity apertures cannot be trapped. In other words, there exists a natural *cutoff frequency* $\omega_c \approx c/a$ above which there are no resonant modes in the cavity (modes beyond cutoff exist but they are associated to emission of synchrotron radiation, which is not considered here). Translating this picture into the frequency domain, we infer that the total longitudinal impedance of an accelerator can be modeled as the superposition of *single-resonator impedances*:

$$\begin{cases} Z_\|(\omega) = \sum_k Z_{\|,k}(\omega), \quad Z_{\|,k}(\omega) = \dfrac{R_{s,k}}{1+iQ_k\left(\frac{\omega_k}{\omega} - \frac{\omega}{\omega_k}\right)} \\[2ex] Z_\perp(\omega) = \sum_k Z_{\perp,k}(\omega), \quad Z_{\perp,k}(\omega) \approx \frac{2}{a}\frac{\omega_c}{\omega}Z_{\|,k}(\omega) \end{cases} \tag{9.17}$$

Each resonator or mode is characterized by its own quality factor (Q_k), shunt impedance ($R_{s,k}$, in units of Ω) and resonant angular frequency ω_k. Depending on the characteristic decay time scale of the wakefield, i.e., on the quality factor of the cavity-like element which allows the e.m. field to be stored, two classes of interactions can be discriminated.

Wakefields acting on the same source bunch are called *short range*. They correspond in the frequency domain to a *broad-band impedance* (Z_{bb}) or the superposition of low-Q resonators. For example, they are generated by the finite conductivity of the vacuum chamber (*resistive wall* wakefield), by discontinuities of the vacuum chamber (bellows, joints, etc.) and low-Q parasitic modes in RF cavities (*geometric or diffraction* wakefield). The total broad-band impedance is conventionally modeled through a broad-band resonator with $Q = 1$, $\omega_k = \omega_c$. The broad-band shunt impedance $R_{s,bb}$ is consequently retrieved from fit to experimental data or simulations:

$$\left.\frac{Z_\|}{n}\right|_{bb} \equiv \lim_{\omega\to 0}\left|\frac{Z_\|}{n}\right| = \frac{R_s}{Q}\frac{\omega_0}{\omega_k} \equiv R_{s,bb}\frac{\omega_0}{\omega_c} \tag{9.18}$$

The Longitudinal broad-band impedance of modern storage rings is typically in the range 0.1–1 Ω.

Wakefields lasting enough time to affect trailing bunches are called *long range*. They correspond in the frequency domain to a narrow-band impedance (Z_{nb}), or sum of high-Q resonators. They are for example stored in RF cavities or cavity-like insertions of the vacuum chamber. For this reason, they are commonly depicted as *Same* (of the fundamental) and *High Order Modes* (SOM and HOMs), or "parasitic RF modes" in general.

Since measurable effects of the longitudinal broad-band impedance, such as current-dependent bunch lengthening and energy spread growth, depend from the convolution of the accelerator impedance ($Z_{bb,\parallel}(\omega)$) and the bunch power spectrum ($|I(\omega)|^2$), it is convenient to introduce the *effective impedance*:

$$\left.\left|\frac{Z_\parallel}{n}\right|\right|_{eff} = \frac{\int_{-\infty}^{+\infty} \frac{Z_\parallel(\omega)}{n}|I(\omega)|^2 d\omega}{\int_{-\infty}^{+\infty}|I(\omega)|^2 d\omega} \tag{9.19}$$

9.3.3 Classification

Table 9.1 summarizes some common sources of collective effects in linacs and storage rings. The associated instability, either single bunch (SB) or multi bunch (MB), is depicted via established nomenclature in the literature A short description of each effect is given below.

Space Charge Force (SC)

The interaction can be intended as inelastic Coulomb scattering internal to the bunch. The strength of the average force depends from the 3-D spatial charge distribution. The electric and magnetic component of the space charge force cancel at very high energy. In fact, the transverse force evaluated in the laboratory reference frame, F_{sc}, is the Lorentz's force, with the electric field proportional to the total

Table 9.1 Classification of collective effects of single bunch (SB, or short range wakefield) and multi-bunch (MB, or long range wakefield)

Source	Linac		Storage ring	
	SB	MB	SB	MB
SC	ϵ_{6D}-dilution at injection		Low I_b at injection	
CSR	ϵ_x-growth, MBI		MWI, Power loss	
RW	Energy chirp, energy loss			
$Z_{bb,\parallel}$				
$Z_{bb,\perp}$	BBU		TMCI	
$Z_{nb,\parallel}$		Transient beam loading	Power loss	Transient beam loading, LCBI
$Z_{nb,\perp}$		BBU		TCBI

charge, $E_{sc} \sim N_b$, and the magnetic field generated by the beam current $B_{sc} \sim I_b \sim v_z N_b \sim v_z E_{sc}$. Hence, $F_{sc,\perp} = q(E_{sc} - v_z \times B_{sc}) \sim (1 - v_z^2)N_b \sim N_b/\gamma^2$. An exact derivation starting from the beam's rest frame is given in Eqs. 1.63 and 1.64.

In high charge density linacs, the direct space charge force dilutes the beam 6-D emittance. In electron linacs, its effect is counteracted by very high RF gradients in the first stage of acceleration (RF "Gun" cavities), to boost the beam to ultra-relativistic energies in relatively short distances.

In storage rings, the direct space charge force generates an incoherent tune shift which, coupled to synchrotron oscillations of the bunched beams, leads to a tune spread [4]. The effect goes like $\sim \gamma^{-3}$, and it can be harmful for the beam life-time if strong resonances are crossed. An indirect space charge force is produced by the interaction of the beam self-field with the vacuum chamber, in the presence of coherent betatron oscillations (i.e., of the bunch as a whole). In this case, a coherent-so-called *Laslet*—tune shift is observed, whose energy dependence is $\sim \gamma^{-1}$. The space charge force is particularly harmful in low energy proton storage rings. Being it proportional to the bunch charge, it limits the maximum current injected at low energy.

Coherent Synchrotron Radiation (CSR)

Long wavelength synchrotron radiation is emitted by trailing particles in a bunch on a curved path, and it catches up with leading particles of the same bunch. The intensity of radiation is enhanced in electron accelerators because of the lighter rest mass with respect to hadron beams, at the same total energy. Also, the effect is amplified by high charge, short bunches, because the intensity of the emitted radiation scales as $\sim N_b^2$, and the spectrum covers the range of wavelengths comparable to, or shorter than, the bunch length, i.e., commonly IR-THz frequencies.

In electron linacs, such tail-head interaction causes a net total energy loss, variation of longitudinal momentum along the bunch, and emittance growth in the bending plane, because the change of longitudinal momentum happens in a dispersive region. It also participates to the so-called *microbunching instability* (MBI), in synergy to the space charge force, i.e., the development of broad-band micron-scale longitudinal modulations.

In storage rings, such instability is mediated by the vacuum chamber and it is named *microwave instability* (MWI, also "turbulent bunch lengthening"). The inter-mittent emission of intense bursts of synchrotron radiation reflect the chaotic parti-cles' motion in the longitudinal phase space, where energy and density modulations develop at wavelengths in the IR-THz region. The bursts alternate to relaxation of the charge distribution.

Resistive Wall Wakefield (RW)

The resistive wall wake field is associated to the "skin effect" of the vacuum chamber. The corresponding longitudinal broad-band impedance is proportional to the skin depth $\delta(\omega) \sim c/\sqrt{\sigma|\omega|}$ of the chamber (whose conductivity is σ) and inversely proportional to the chamber radius.

In linacs adopting small gap chambers (e.g., long collimators or low gap magnetic devices), the wakefield generates a nonlinear correlation of the particle's energy along the bunch, and it can contribute substantially to additional mean energy loss. In storage rings, it contributes to the overall broad-band impedance of the accelerator, thus supporting the MWI.

Short Range Longitudinal Geometric Wakefield ($Z_{bb,\parallel}$)

The corresponding impedance collects the broad-band contribution of the vacuum chamber discontinuities (bellows, joints, etc.) and low-Q parasitic RF modes.

In linacs, it adds to the RW to deform the beam energy distribution, i.e., to lower the beam mean energy, and to induce a linear and, for high charge bunches in small iris cavities, nonlinear energy-chirp. The beam total energy spread is increased accordingly. The linear chirp can partly be compensated by off-crest acceleration, although at the expense of a lower total accelerating voltage.

In storage rings, the longitudinal wakefield superimposes to the RF potential. The *potential well distortion* can lead to either bunch lengthening or shortening, depending from the characteristic wavelength of the wakefield and the natural bunch length (i.e., in the limit of zero-current). The longitudinal charge distribution is deformed as well, and it can substantially deviate from a Gaussian.

The wakefield also adds to the RW in driving the MWI. The so-called *Keil-Schnell criterion*, extended by Boussard to short wavelength modulations in bunched beams, estimates the upper limit of the single bunch *peak* current to avoid that the instability build up:

$$\hat{I}_{b,th} \approx 2\pi \frac{|\eta|(E/e)\sigma_\delta^2}{|Z_\parallel/n|_{bb}} \tag{9.20}$$

The *average* current threshold for a Gaussian bunch is $\langle I_{b,th} \rangle = \hat{I}_{b,th} \frac{\omega_0 \sigma_t}{\sqrt{2\pi}}$, with σ_δ and $\sigma_t = \frac{|\alpha_c|}{\Omega_s}\sigma_\delta$ (see Eq. 8.12) the unperturbed rms relative energy spread and bunch duration.

Short Range Transverse Geometric Wakefield ($Z_{bb,\perp}$)

This is the transverse counterpart of $Z_{bb,\parallel}$ and it is proportional to the relative misalignment of the beam and the symmetry axis of the chamber or RF cavity. Leading particles in a bunch drive betatron oscillations of trailing particles. In other words, the wakefield determines a misalignment of bunch slices in the transverse phase space, or deformation of the transverse charge distribution.

In linacs, the misalignment of the bunch tail with respect to the head translates into an effective growth of the transverse emittance. If the wake potential is large enough, it can lead to beam loss. Since the single bunch *beam break-up instability* (BBU) is the result of a deterministic sum of transverse kicks collected along the accelerator (on top of trajectory jitter), it can be counteracted by trajectory bumps to make the kicks to cancel each other.

In storage rings, it is customary to analyse the transverse oscillations in terms of normal modes, referred to as "head-tail" modes. The instability is called *Transverse Mode-Coupling Instability* (TMCI) or *fast head-tail instability*. The m-th mode has m nodes along the bunch. For example, for mode $m = 0$ all particles have the same betatron phase (rigid dipole motion), whereas for $m = \pm 1$ the head and tail have opposite phases, etc. The angular frequency of each mode is modulated by the constant exchange of head and tail particles via synchrotron oscillations, i.e., $\omega_m = \omega_\beta + m\Omega_s$.

The threshold for the single bunch *average* current in order to avoid that the TMCI builds up is usually higher than for the MWI:

$$\langle I_{b,th} \rangle \approx \sqrt{2\pi} \frac{|\eta|(E/e)\sigma_\delta}{\langle \beta_{x,y}|Z_\perp|_{bb}\rangle_R} F(\sigma_z) \tag{9.21}$$

The form factor $F(\sigma_z)$ is basically the ratio of the total machine impedance to the effective impedance for the transverse plane (see Eq. 9.19). It is $F \approx 1$ for short bunches, and it grows linearly with the bunch length for longer bunches.

If the chromaticity is non-zero, the beam energy spread leads to the modulation of the betatron tune at the synchrotron tune, thus a phase shift between head and tail of the bunch and, eventually, to damping of the head-tail modes. Above transition energy, positive chromaticity damps the mode $m = 0$, but it excites the modes $m = \pm 1$. However, the characteristic time of these modes is usually longer than the damping time, which allows most electron storage rings to operate stably with slightly positive chromaticity.

Long Range Geometric Wakefields $(Z_{nb,\parallel}, Z_{nb,\perp})$

Among multi-bunch instabilities, *dipole coupled-bunch oscillations* usually dominate the beam dynamics. In linacs, $Z_{nb,\parallel}$ translates into the so-called *transient beam loading*, or differential (i.e., bunch-to-bunch) mean energy loss and energy chirp along the bunch train. In the transverse plane, $Z_{nb,\perp}$ generates the *beam break up* instability (BBU), i.e., trailing bunches in a train are pushed off-axis from the wakefield generated by leading bunches. Both instabilities are common in high-Q RF cavities traversed by many high charge bunches, such as in superconducting linacs.

In storage rings, the transient beam loading generates additionally a bunch-to-bunch synchrotron tune spread, whose range depends also from the kind of bunch fill pattern in the accelerator.

Dipole coupled-bunch oscillations translate into the motion of the bunches about their nominal centers as if they were rigid macroparticles. Such oscillations are called *Longitudinal Coupled-Bunch Instability* (LCBI) and *Transverse Coupled-Bunch Instability* (TCBI). For M bunches equally spaced, each multi-bunch mode, either longitudinal or transverse, is characterized by a bunch-to-bunch phase difference $\Delta\phi = 2\pi l/M$, where the "mode number" l can only take the values $l = 0, 1, 2, ..., M - 1$. The net phase advance of each mode is constrained by the periodic motion to be 2π. Each multi-bunch mode has a characteristic set of angular frequencies: $\omega_p = [pM \pm (l + \nu)]\omega_0$, with p integer and ν either the synchrotron or the betatron tune, depending if the mode is longitudinal or transverse, respectively.

LCBI can affect the brightness of light sources because energy oscillations at dispersive locations determine an effective growth of the beam size, as well as a spectral broadening of undulator emission. Similar effect arises from TCBI because of the enlargement of the effective size and angular divergence of the radiation source. Reduction of CBI is usually accomplished by means of a combination of Landau damping, negative multi-bunch feedback, and suppression of parasitic RF modes. The latter action can be taken through a suitable design of the RF cavity (e.g., including multiple ports or antennas to absorb undesired modes), and/or a mechanical deformation of the cavity geometry (via mechanical stress, positioning of tuning rods or temperature control, to induce de-tuning of the HOMs). Landau damping is favoured in the longitudinal plane by a lower voltage from the main RF and the adoption of higher harmonic cavities (leading to longer bunches, thus smaller synchrotron frequency), and by octupole magnets in the transverse plane (inducing a larger spread of the betatron tune).

Parasitic Power Loss

Although not directly related to beam instabilities, parasitic power loss due to the dissipation of energy via image currents can limit the stored beam current, because of excessive heating of the vacuum chamber. This is a common challenge in electron storage rings. For M identical bunches in a train and single bunch average current $\langle I_b \rangle$, the parasitic power loss is:

$$P = M \langle I_b \rangle^2 Z_{loss} \tag{9.22}$$

where Z_{loss} is the real part of the effective impedance causing the energy loss. It is usually contributed by $Z_{bb,\parallel}$ for $\frac{\omega_r}{Q_r} \gg M\omega_0$, and by $Z_{nb,\parallel}$ for $\omega_r \approx nM\omega_0, n \in \dot{\mathbb{N}}$ (namely, for the resonant frequency of the HOM close to a harmonic of the bunch frequency). The contribution of RW to Ohmic losses is usually negligible with respect to the one by $Z_{bb,\parallel}$.

9.3.4 Robinson's Instability

To minimize the effect of geometric wakefields, the vacuum chamber internal geometry has to be carefully designed to avoid trapping of resonant modes. RF cavities constitute one important exception to this, since they are built exactly to resonate at the fundamental mode of the accelerating field [5]. This means they are also perfect narrow-band resonators for the parasitic mode excited by the beam at the resonant fundamental frequency of the cavity, $\omega \approx \omega_{RF} = h\omega_0$. Even in case of a beam perfectly aligned onto the cavity electric axis, the LCBI is expected to be driven. How could one get rid of, or at least alleviate, the instability?

We have seen that the presence of $Z_{nb,\parallel}(\omega)$ drives longitudinal oscillations of the bunch as a whole. They are called coherent synchrotron oscillations because they

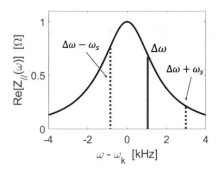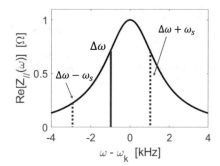

Fig. 9.1 Real part of the geometric impedance $Z_{nb,\parallel}(\omega)$ resonant at the fundamental frequency ω_k of the RF cavity, where $\Delta\omega = h\omega_0 - \omega_k$, ω_0 is the revolution frequency of the synchronous particle, and ω_s the synchrotron frequency. On the left, the RF cavity is tuned (the impedance is peaked) at a frequency slightly smaller than $h\omega_0$, which implies Robinson damping above transition energy. The opposite is on the right, which implies damping (excitation) below (above) transition energy

happen both in phase (arrival time at the RF cavity) and energy (mean value over the bunch particles). As a practical case, let us consider an electron storage ring above transition energy, i.e., $\eta < 0$ (see Eq. 4.19). The bunch revolution frequency is $\omega_{riv} = \omega_0(1 + \eta\delta)$, namely, $\omega_{riv} < \omega_0$ for a relative energy deviation (with respect to the synchronous bunch) $\delta > 0$.

The resistive part of $Z_{nb,\parallel}(\omega)$ or $\mathbb{R}e(Z_{nb,\parallel}(\omega))$ implies mean energy loss. In order to damp coherent synchrotron oscillations, i.e., to reduce the energy deviation turn-by-turn, we would like to have the impedance larger for those bunches which have $\delta > 0$ or $\omega_{riv} < \omega_0$, and smaller for those with $\delta < 0$ or $\omega_{riv} > \omega_0$.

In other words, $\mathbb{R}e(Z_{nb,\parallel}(\omega))$ should be centered—i.e., the cavity tuned—at a frequency slightly smaller than the nominal RF frequency, $\omega_k < h\omega_0$, as shown in Fig. 9.1-left plot. In particular, since the mode of oscillation induced by the impedance is modulated by the synchrotron frequency ω_s, we require $\mathbb{R}e(Z_{nb,\parallel}(\omega))$ to be smaller at $h\omega_0 + \omega_s$, and larger at $h\omega_0 - \omega_s$. The opposite happens for beams below transition. Such procedure is called *Robinson damping* or "Robinson's stability criterion".

References

1. A. Piwinski, Intra-Beam scattering, in *Proceedings of Joint US-CERN School on Particle Accelerators at South Padre Island*, Texas, USA, ed. by M. Month, S. Turner (Published by Springer-Verlag LNP 296, 1986), p. 297
2. S. Di Mitri et al., Experimental evidence of intrabeam scattering in a free-electron laser driver. New J. Phys. **22**, 083053 (2020); G. Perosa, S. Di Mitri, Matrix model for collective phenomena in electron beam's longitudinal phase space. Sci. Rep. **11**, 7895 (2021)
3. M. Furman, J. Byrd, S. Chattopadhyay, Beam instabilities, in *Synchrotron Radiation Sources—A Primer*, ed. by H. Winick (Published by World Scientific, Singapore, 1995), pp. 306–342. ISBN: 9810218567

4. A. Hofmann, Tune shifts from self-field images, in *Proceedings of CERN Accelerator School: 5th General Accelerators Physical Course*, Geneva, Switzerland. CERN 94-01, vol. I, ed. by S. Turner (1994), pp. 329–348
5. A. Chao, Beam dynamics of collective instabilities in high-energy accelerators, in *Proceedings of CERN Accelerator School: Intensity Limitations in Particle Beams*, Geneva, Switzerland, vol. 3/2017, ed. by W. Herr (2017)

Light Sources

10

Particle accelerators devoted to emission of radiation are named "light sources". Nowadays, the most advanced and powerful light sources driven by RF accelerators are electron storage rings and linac-driven free-electron lasers (FELs). Short electron linacs also drive Inverse Compton Scattering light sources, which extend the photon energy to γ-rays (up to $\sim$MeV photon energy). In the following, these three types of light sources are discussed.

From the middle of the XX century until 1980s, two generations of synchrotrons were developed for particle physics experiments. In spite of the attention of radiology industry already from the 1920s for x-ray emission, synchrotron radiation from dipole magnets of those machines was an undesired effect, reducing the beam energy and radio-activating vacuum and RF components of the accelerator. But, some pioneering scientists realized soon that x-rays could be used for optical experiments of matter physics. Since 1990s, 3rd generation synchrotron light sources started to be specifically designed and built for providing light pulses at high repetition rate, to several photon beamlines simultaneously. New magnetic elements, denominated *insertion devices* (IDs), were installed in the straight sections to enhance and shape the radiation emission, in addition to dipole radiation.

More recently, a 4th generation of storage ring light sources is being built worldwide. These machines are also called *diffraction limited storage rings*: a tight and strong focusing magnetic lattice based on multi-bend cells minimizes the electron beam emittance to unprecedented levels. Light pulses emitted by such collimated electron beams show a higher degree of transverse coherence in x-rays than in any preceding circular light source.

FELs started developing in their most powerful single-pass configuration since 1990s. Today, they are complementary to storage rings as for many aspects, and characterized by extremely high peak power, spectral brightness (6-D photon density), short pulse duration, and coherence. Next generation of high gain FELs is targeting

© The Author(s), under exclusive license to Springer Nature Switzerland AG 2022
S. Di Mitri, *Fundamentals of Particle Accelerator Physics*, Graduate Texts in Physics,
https://doi.org/10.1007/978-3-031-07662-6_10

sub-femtosecond pulse durations, TW-level peak power, full coherence at multi-keV photon energies, and MHz repetition rate. Since a single FEL facility usually serves one to few beamlines only simultaneously, compactness and larger experimental fan-out are also features worth of some attention in the community.

Compton sources are incoherent light sources based on scattering of a ~10–100s MeV energy electron beam and an IR or UV external laser. By virtue of the relativistic Doppler effect, the scattered radiation easily reaches 10s of keV to MeV photon energies. Compton sources are typically far more compact than x-ray FELs, but at the expense of peak power. Next generation of Compton sources aims at lager peak and average power, sub-picosecond pulse duration, and tuneable repetition rate up to MHz. So-called gamma-gamma colliders based on multi-GeV electron linacs have been conceived as Compton sources in which the back-scattered photon energy approaches the electron beam energy.

10.1   Brilliance

10.1.1 Practical Meaning

One of the figures of merit of light sources is the 6-D photon density [1,2]. It is named *spectral brightness* or *brilliance*, and it has some analogy to the particle beam brightness introduced in Eq. 4.149. The Brilliance describes the effective radiation flux per unit spectral bandwidth ($d\omega/\omega$), conventionally at the location of emission. It is meant to be a peak or an average quantity depending on if the flux is intended to be a peak (i.e., instantaneous) or time-averaged quantity (e.g., over the beam train duration, or a turn in a synchrotron).

Whenever the source, i.e. a charged particle beam, has non-zero transverse emittance, the spectral flux (or spatio-angular spectral density, defined as the spectral intensity ϕ per unit area and unit angular cone of emission) is diminished by the convolution of the intrinsic photon beam sizes and the charged particle beam sizes:

$$B_r = \frac{\phi}{4\pi^2 \Sigma_x \Sigma_{x'} \Sigma_y \Sigma_{y'}}, \tag{10.1}$$

where we defined:

$$\phi = \frac{dN_{ph}}{dt\, d\omega/\omega},$$

$$\Sigma_u = \sqrt{\sigma_u^2 + \sigma_r^2} \qquad \Sigma_{u'} = \sqrt{\sigma_{u'}^2 + \sigma_{r'}^2}$$

$$\begin{aligned}
\sigma_u &= \sqrt{\varepsilon_u \beta_u + (D_x \sigma_\delta)^2} \rightarrow \sqrt{\varepsilon_u \beta_u} \\
\sigma_{u'} &= \sqrt{\varepsilon_u \gamma_u + (D'_x \sigma_\delta)^2} \rightarrow \sqrt{\varepsilon_u / \beta_u} \\
\sigma_r &= \sqrt{\varepsilon_r \beta_r} \\
\sigma_{r'} &= \sqrt{\varepsilon_r / \beta_r}
\end{aligned} \tag{10.2}$$

$\varepsilon_u, \varepsilon_r$ are, respectively, the geometric emittance of the charged beam ($u = x, y$) and the "intrinsic" geometric emittance of the light pulse (identical in both planes). Hereafter, "intrinsic" refers to the properties of an ideal monochromatic ($\Delta\lambda \ll \lambda$) light pulse emitted by a point-like source ($\varepsilon_u \approx 0$). We anticipate that $\varepsilon_r = \sigma_r \sigma_{r'} = \frac{\lambda}{4\pi}$ (the derivation is postponed). This is the case, for example, of an ideal laser beam, which features a symmetric Gaussian intensity distribution in the transverse planes (TEM_{00} mode), and λ is the central wavelength of the narrow band laser pulse.

In the above expressions, the C-S formalism is identically applied to the charged and the photon beam. In this case, the betatron function is simply defined as function of the light beam's emittance and spot size, and it actually depends from the type of magnetic insertion used for stimulating the emission of radiation. In most cases, and for maximizing the brilliance, the charged beam is assumed to emit radiation in correspondence of a waist ($\alpha_u = 0$), in a dispersion-free region ($D_x = 0$, $D_{x'} = 0$).

If we describe a light pulse through a 6-D photon distribution, we can apply Liouville's theorem and find that brilliance is a conserved quantity in the absence of light absorption. If the phase space area is defined in terms of statistical parameters of the photon distribution (i.e., second order momenta or rms emittance), then the additional constraint of linear optics is required to make the brilliance a conserved quantity.

A photon beamline is an optical system aiming at manipulating and transporting the photon pulse to the experimental end-station. It includes a large variety of optical elements, as sketched in Fig. 10.1-left plot, to direct and focus radiation (mirrors), to filter it in wavelength (monochromators, such as gratings and crystals) or angle (mask, pinhole, slits). Unavoidable optical aberrations, micro-roughness, slope errors and thermal deformation of the optical surfaces, diffraction effects, and transmission efficiency of 0.1–1% are common challenges of real x-ray beamlines. They all contribute to a reduction of the brilliance *at the sample*.

But, why a high brilliance is important for experiments at light sources, and why should it be preserved down to the sample? The answer is in the need of high intensity at (sub-)micron-scale spot sizes at the sample for high spatial resolution, as well as at the entrance of a monochromator for high energy resolution.

If the source is not brilliant enough, a small spot size can still be produced by physically cutting the light pulse, but at the expense of intensity. Alternatively, the original spot size can be imaged to a smaller spot size ("demagnified"), but at the

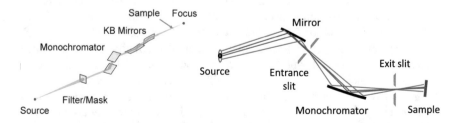

Fig. 10.1 Left: optical components of an x-ray beamline. Right: in red, a higher (in blue, smaller) brilliance pulse impinges on the optical elements with a smaller (larger) footprint

expense of large angular divergence. This would imply larger apertures, severe optical aberrations, larger size of the mirrors, hence more expensive specifications for the flatness of the mirrors' surface, and higher slope error.

We now understand that a higher brilliance at the source, i.e. smaller size and angular divergence, possibly accompanied by relatively narrow bandwidth, favors the preservation of the brilliance through the beamline. In short, a highly brilliant light source gives access to high energy resolution, small spot on the sample, accompanied by high intensity.

10.1.2 Optics Matching, Diffraction Limit

It is easy to show that for the realistic case of non-zero particle beam emittance, the brilliance in Eq. 10.1 is maximized by the electron beam transversely *matched* to the intrinsic photon beam sizes, or $\beta_{x,e} = \beta_{y,e} = \beta_r$. In this case:

$$\hat{B}_r = \frac{\phi}{4\pi^2(\varepsilon_{x,e}+\varepsilon_r)(\varepsilon_{y,e}+\varepsilon_r)} = \frac{2}{1+\kappa}\frac{\phi}{\lambda^2} \approx \begin{cases} \frac{2\phi}{\lambda^2}, \kappa \ll 1 \\[2mm] \frac{\phi}{\lambda^2}, \kappa \approx 1 \end{cases} \tag{10.3}$$

The r.h.s. of Eq. 10.3 is evaluated for a charged beam satisfying the so-called *diffraction limit* in the horizontal plane, $\varepsilon_{x,e} = \varepsilon_r = \frac{\lambda}{4\pi}$. The vertical emittance is expressed as function of the horizontal emittance through the coupling factor $\kappa \leq 1$ (see Eq. 5.53). Equation 10.3 highlights that, under proper matching, a flat beam ($\varepsilon_y \ll \varepsilon_x$) guarantees a brilliance as twice as that one of a round beam at full coupling ($\varepsilon_y \approx \varepsilon_x$). The absolute maximum of the brilliance is for an ideally zero-emittance— i.e., point-like—charged beam, and it amounts to $4\phi/\lambda^2$.

"Diffraction limit" refers to the capability of a particle beam of emitting radiation characterized by a high degree of transverse *coherence*, essentially by virtue of the source's small size and divergence. Coherence can be intended in turn as a high degree of flatness of the wavefront of the far field. This translates, for example, into the capability of radiation of producing interference fringes once diffracted through properly sized holes. Of course, a particle beam satisfying the diffraction limit at λ will produce even more coherent radiation at any longer wavelength.

Optics matching of a particle beam in a light source aims at forcing the transverse charge distribution to betatron functions which maximize the brilliance from IDs. We consider a specific class of IDs, called *undulators*. An undulator is basically made of two opposite arrays of magnetic poles of alternated polarity. Electrons wiggle through the undulator and emit synchrotron radiation stimulated by Lorentz's force. A detailed treatment of undulator radiation will be given later on. Here, we anticipate that the transverse intensity distribution of radiation emitted by a monochromatic Gaussian charged beam in a long undulator (L_u) is also approximately Gaussian, and its intrinsic transverse size and angular divergence can be approximated to:

$$\begin{cases} \sigma_r = \frac{\sqrt{\lambda L_u}}{2\pi} \\[2mm] \sigma_{r'} = \sqrt{\frac{\lambda}{4 L_u}} \end{cases} \Rightarrow \begin{cases} \sigma_r \sigma_{r'} = \varepsilon_r = \frac{\lambda}{4\pi} \\[2mm] \beta_r = \frac{\sigma_r}{\sigma_{r'}} = \frac{L_u}{\pi} \\[2mm] \gamma_r = \frac{\sigma_{r'}}{\sigma_r} = \frac{\pi}{L_u} \end{cases} \qquad (10.4)$$

Since ID segments in storage rings are long $\sim 1 - 5$ m for practical reasons, $\beta_r \approx 0.3 - 1.5$ m.

Unfortunately, optimal matching of the electron beam at the ID location is often prevented by other constraints to the optics design related to, e.g., minimum equilibrium emittance, chromaticity control, large dynamic aperture, etc. Consequently, the "geometric" component of B_r, i.e., the denominator of Eq. 10.1, can assume a variety of values depending from the actual value of the coupling factor and the ratio of the charged and photon beam betatron function. For the horizontal beam emittance at the diffraction limit, it results:

$$\chi := \frac{B_r}{\phi/(4\pi^2)} = \left[\sqrt{\varepsilon_x \beta_x + \varepsilon_r \beta_r}\sqrt{\frac{\varepsilon_x}{\beta_x} + \frac{\varepsilon_r}{\beta_r}}\sqrt{\varepsilon_y \beta_y + \varepsilon_r \beta_r}\sqrt{\frac{\varepsilon_y}{\beta_y} + \frac{\varepsilon_r}{\beta_r}}\right]^{-1} =$$

$$= \left[\sqrt{\varepsilon_r(\beta_x + \beta_r)}\sqrt{\varepsilon_r(\frac{1}{\beta_x} + \frac{1}{\beta_r})}\sqrt{\varepsilon_r(\kappa\beta_y + \beta_r)}\sqrt{\varepsilon_r(\frac{\kappa}{\beta_y} + \frac{1}{\beta_r})}\right]^{-1} = \qquad (10.5)$$

$$= \frac{1}{\varepsilon_r^2}\frac{\beta_r\sqrt{\beta_x \beta_y}}{(\beta_x + \beta_r)\sqrt{(\kappa\beta_y + \beta_r)(\beta_y + \kappa\beta_r)}}$$

We discriminate four cases:

1. $\kappa \approx 1$ (*full coupling*): $\chi = \frac{1}{\varepsilon_r^2}\frac{\beta_r\sqrt{\beta_x \beta_y}}{(\beta_x + \beta_r)(\beta_y + \beta_r)}$

   $\rightarrow$ max. for $\beta_x = \beta_y = \beta_r = L_u/\pi$, $\hat{\chi} = \frac{1}{4\varepsilon_r^2}$

2. $\kappa \ll 1$ & $\beta_y \approx \beta_r$ (*flat & matched beam*): $\chi = \frac{1}{\varepsilon_r^2}\frac{\sqrt{\beta_x \beta_r}}{(\beta_x + \beta_r)}$

   $\rightarrow$ max. for $\beta_x = \beta_y = \beta_r = L_u/\pi$, $\hat{\chi} = \frac{1}{2\varepsilon_r^2}$

3. $\kappa \ll 1$ & $\kappa\beta_y \approx \beta_r$ (*flat & mismatched beam*): $\chi = \frac{1}{\sqrt{2}\varepsilon_r^2}\frac{\sqrt{\beta_x \beta_r}}{(\beta_x + \beta_r)}$

   $\rightarrow$ max. for $\beta_x = \beta_r = L_u/\pi$, $\hat{\chi} = \frac{1}{2\sqrt{2}\varepsilon_r^2}$

4. $\kappa \ll 1$ & $\beta_y \approx \kappa\beta_r$ (*flat & over-matched beam*): $\chi = \frac{1}{\sqrt{2}\varepsilon_r^2}\frac{\sqrt{\beta_x \beta_r}}{(\beta_x + \beta_r)}$

   $\rightarrow$ max. for $\beta_x = \beta_r = L_u/\pi$, $\hat{\chi} = \frac{1}{2\sqrt{2}\varepsilon_r^2}$

As expected, the brilliance is maximized by a diffraction-limited, flat beam light source, with both horizontal and vertical betatron function matched to the intrinsic betatron function of the photon beam at the ID (case 2). Matching means here that the charge and the photon beam distributions can be represented in the transverse phase space by omothetic ellipses, see Fig. 10.2.

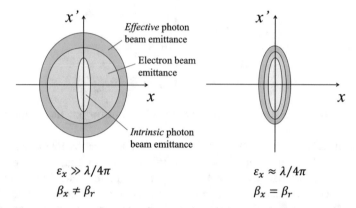

Fig. 10.2 Superposition of intrinsic photon beam (yellow) and electron beam (green) phase space area, and effective radiation envelope (blue). On the left, the electron beam is mismatched, and its emittance is larger than the intrinsic photon beam emittance. On the right, the electron beam is matched to the light pulse, and its emittance is close to the diffraction limit

10.1.3 Central Cone

Undulator radiation emitted on-axis is characterized by a higher brilliance and a larger degree of transverse coherence compared to far off-axis emission. For this reasons a beamline usually includes a "front-end" area, where optical elements limit the angular acceptance of the beamline in order to match the central angular cone of emission.

In case of undulator radiation, the *coherent flux* F_{coh} is estimated as the fraction of the total spectral flux $F = \phi/(\Sigma_x \Sigma_y)$ contained in the intrinsic central angular cone of emission:

$$
F_{coh} := F \frac{\delta\theta_x \delta\theta_y}{\Sigma_{x'} \Sigma_{y'}} = F \frac{\sigma_{r'}^2}{\sqrt{\frac{\varepsilon_x}{\beta_x} + \frac{\varepsilon_r}{\beta_r}} \sqrt{\frac{\varepsilon_y}{\beta_y} + \frac{\varepsilon_r}{\beta_r}}} = F \frac{\sigma_{r'}^2}{\frac{\varepsilon_r}{\beta_r} \sqrt{1 + \frac{\beta_r}{\beta_x} \frac{\varepsilon_x}{\varepsilon_r}} \sqrt{1 + \frac{\beta_r}{\beta_y} \frac{\varepsilon_y}{\varepsilon_r}}} =
$$

$$
= \frac{F}{\sqrt{\left(1 + \frac{\beta_r}{\beta_x} \frac{\varepsilon_x}{\varepsilon_r}\right)\left(1 + \frac{\beta_r}{\beta_y} \frac{\varepsilon_y}{\varepsilon_r}\right)}} \rightarrow \frac{F}{\sqrt{2}}
$$

(10.6)

The limit is for the coherent flux maximized by the same conditions of maximum brilliance, i.e., by a diffraction-limited, flat ($\varepsilon_y \ll \varepsilon_x$), and matched particle beam. $F_{coh} \rightarrow F$ when $\varepsilon_x, \varepsilon_y \rightarrow 0$, that is, radiation emitted by a point-like source is 100% transversely coherent.

10.2   Coherence

Coherence of light generally refers to the capability of predicting the e.m. field at any point P_2 of the radiation pattern from the knowledge of the field at a nearby point P_1. This implies a well-defined correlation of the e.m. field vectors, or phase relation, which is in turn at the origin of the interference pattern of electric fields of same polarization in the well-known Young's double slit experiment, see Fig. 10.3.

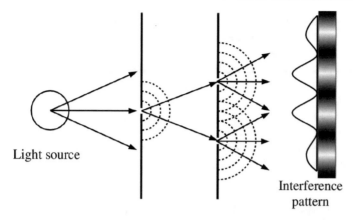

Fig. 10.3 An incoherent monochromatic light source is collimated through a pinhole. The size is assumed to be such that the emerging light is now (partially) coherent. As a result of diffraction at two downstream apertures, an interference pattern appears on the screen after a large number of pulses is collected

In such scheme, the total field intensity (normalized to $2Z_0$) recorded at the observation point P, and produced by the linear superposition $U(\vec{r}_1, \vec{r}_2, t_1, t_2)$ of two e.m. plane waves, is:

$$I_P = \langle UU^* \rangle = \left\langle \left[E_1 e^{i(\vec{k}\vec{r}_1 - \omega t_1)} + E_2 e^{i(\vec{k}\vec{r}_2 - \omega t_2)} \right] \left[E_1 e^{-i(\vec{k}\vec{r}_1 - \omega t_1)} + E_2 e^{-i(\vec{k}\vec{r}_2 - \omega t_2)} \right] \right\rangle =$$

$$= \langle |E_1|^2 \rangle + \langle |E_2|^2 \rangle + \langle E_1 E_2 \rangle e^{i(\vec{k}\Delta\vec{r} - \omega \Delta t)} + \langle E_2 E_1 \rangle e^{-i(\vec{k}\Delta\vec{r} - \omega \Delta t)} =$$

$$= \langle |E_1|^2 \rangle + \langle |E_2|^2 \rangle + 2\langle E_1 E_2 \rangle \cos\left(\vec{k}\Delta\vec{r} - \omega \Delta t\right) =$$

$$= I_1 + I_2 + 2\sqrt{I_1 I_2} \cos\left(\vec{k}\Delta\vec{r} - \omega \Delta t\right)$$

(10.7)

For simplicity, we assumed monochromatic waves of same wave number k. The two waves have amplitudes E_1, E_2. $\Delta\vec{r} = \vec{r}_1 - \vec{r}_2$, $\Delta t = t_1 - t_2$ are, respectively, the path length difference of the two waves to reach P, and the relative delay with which they were generated. This is usually zero in Young's experiment, and thereby $I_p = I_p(\Delta\vec{r})$. $\langle ... \rangle$ denotes an ensemble statistical average. For non-stationary states, such as light pulses, the average is calculated over many shots. In case of stationary states, such as plane parallel waves ($\vec{r} = z$), the ensemble average can be replaced with a time average, $\langle ... \rangle = \frac{1}{T} \int ... dt$.

Equation 10.7 shows that the maximum total intensity is obtained for relative phases multiple of 2π, i.e., the two waves are emitted "in phase". We also see that the relative phase—and therefore coherence—is, in general, a spatio-temporal quantity. By posing conditions to the smallness of either the pure spatial or the pure temporal relative phase, a distinction between longitudinal and transverse coherence can be made, although they remain different aspects of the same physical property.

Longitudinal coherence is often associated—but not limited—to a narrow spectral bandwidth. For a finite pulse duration, the narrowest bandwidth is determined by the Fourier limit, i.e. $\Delta t \Delta \nu \geq 1/2$. Transverse coherence is usually related to the capability of collimating the light pulse down to small spatial size and angular divergence. Hence, the pulse would ideally approach the transverse emittance of a laser beam.

Since an e.m. wave passing through a suitably small pinhole can produce a diffraction pattern analogue to that in Fig. 10.3, we infer that diffraction can be intended as a special case of interference of a wave with itself. Diffraction and interference are diagnostic tools to quantify the degree of coherence of a light source.

10.2.1 Correlation Functions

In the "classical" picture of far field e.m. radiation, the Glauber's normalized first order correlation function of the electric field is:

$$g_1(\vec{r}_1, t_1; \vec{r}_2, t_2) = \frac{\langle E^*(\vec{r}_1, t_1) E(\vec{r}_2, t_2) \rangle}{\sqrt{\langle |E(\vec{r}_1, t_1)|^2 \rangle \langle |E(\vec{r}_2, t_2)|^2 \rangle}} \tag{10.8}$$

In case of stationary states the result does not depend from t_1, but only from the delay $\tau = t_1 - t_2$:

$$g_1(\tau) = \frac{\langle E^*(t) E(t+\tau) \rangle}{\langle |E(t)|^2 \rangle} \leq \frac{\sqrt{\langle |E(t)|^2 \rangle \langle |E(t+\tau)|^2 \rangle}}{\langle |E(t)|^2 \rangle} = g_1(0) = 1 \tag{10.9}$$

The result $g_1(\tau) \leq g_1(0)$ is obtained by means of the Cauchy-Schwarz inequality $|\langle \vec{u}, \vec{v} \rangle|^2 \leq \langle \vec{u}, \vec{u} \rangle \cdot \langle \vec{v}, \vec{v} \rangle$. Then, by virtue of the oscillating behaviour of the electric field amplitude with time, $\langle |E(t+\tau)|^2 \rangle = \langle |E(t)|^2 \rangle \forall \tau$, and eventually $g_1(0) = 1$.

In a Michelson interferometer, the radiation pulse is split in two components. A time delay is introduced, and the two pulses are eventually recombined (see Fig. 1.1). The intensity of the resulting field at any position on the screen where the pulses impinge, can be measured as a function of the time delay. The *visibility* of the resulting interference pattern is:

$$v = \frac{I_{max} - I_{min}}{I_{max} + I_{min}} \tag{10.10}$$

where I_{max}, I_{min} are, respectively, the maximum and minimum intensity of the interference pattern. It can be shown that for monochromatic waves of same polarization:

$$v(\lambda) = \frac{2\sqrt{I_1 I_2}}{I_1 + I_2} |g_1(\vec{r}_1, t_1; \vec{r}_2, t_2)| \tag{10.11}$$

where I_1, I_2 are the total intensities of the plane waves. It is then apparent from Eqs. 10.11 and 10.9 that a fully coherent light pulse (e.g., a single frequency emission, such as an ideal laser or a monochromatic plane parallel wave) has $v = |g_1^{SF}(\tau)| = |g_1(0)| = 1$.

As a matter of fact, monochromatic light is characterized by $g_1^{SF}(\tau) = e^{-i\omega_0\tau}$, while for Gaussian chaotic light, $g_1^{GC}(\tau) = e^{-i\omega_0\tau - \frac{\pi}{2}\left(\frac{\tau}{\tau_c}\right)^2}$. The "coherence time" of light τ_c, treated later on, is generally inversely proportional to the spectral bandwidth. We have the following limits:

$$\begin{cases} \lim_{\tau\to\infty} g_1^{GC}(\tau) = 0, \\ \lim_{\tau_c\to\infty} g_1^{GC}(\tau) = g_1^{SF}(\tau)\forall\tau \end{cases} \quad \begin{cases} \lim_{\tau\to\infty} |g_1^{GC}(\tau)| = 0 \\ \lim_{\tau_c\to\infty} |g_1^{GC}(\tau)| = 1 \end{cases} \tag{10.12}$$

The visibility ranges from 0 for incoherent light pulses, to 1 for fully coherent pulses. Anything in between is described as "partially coherent".

Correlation functions of the electric field can be defined up to an arbitrarily high order. For example, the normalized second order correlation function is:

$$g_2(\vec{r}_1, t_1; \vec{r}_2, t_2) = \frac{\langle E^*(\vec{r}_1,t_1)E^*(\vec{r}_2,t_2)E(\vec{r}_1,t_1)E(\vec{r}_2,t_2)\rangle}{\langle |E(\vec{r}_1,t_1)|^2\rangle\langle |E(\vec{r}_2,t_2)|^2\rangle} \tag{10.13}$$

In the "classical" picture of electric fields, we can re-order them to express g_2 in terms of intensities. A plane parallel wave in a stationary state will have:

$$g_2(\tau) = \frac{\langle I(t)I(t+\tau)\rangle}{\langle I(t)\rangle^2} \Rightarrow g_2(0) = \frac{\langle I(t)^2\rangle}{\langle I(t)\rangle^2} \geq 1 \tag{10.14}$$

and the result on the r.h.s. is again by virtue of the Cauchy-Schwarz inequality (see Eq. 10.9 with $|v| = 1$). It then emerges that a fully coherent light pulse has $g_2(\tau) = g_2(0) = 1$.

It can be shown that chaotic light of all kinds has $g_2(\tau) = 1 + |g_1(\tau)|^2$. Hence, we have the following limits:

$$\begin{cases} \lim_{\tau\to\infty} g_2^{GC}(\tau) = 1, \\ \\ \lim_{\tau_c\to\infty} g_2^{GC}(\tau) = 2 \end{cases} \tag{10.15}$$

10.2.2 Transverse Coherence

At first order in the sense of Glauber's correlation functions, the *transverse coherence length* $L_{c,\perp}$ is the width (in each transverse plane) of the function g_1 in Eq. 10.8, evaluated as function of the lateral distance $\vec{r}_1 - \vec{r}_2$ for a constant τ, or $g_1(\vec{r}_1, \vec{r}_2)$. The "degree of transverse coherence" is [2]:

$$\xi_c = \frac{\iint |g_1(\vec{r}_1,\vec{r}_2)|^2\langle I(\vec{r}_1)\rangle\langle I(\vec{r}_2)\rangle d\vec{r}_1 d\vec{r}_2}{[\int\langle I(\vec{r}_1)\rangle d\vec{r}_1]^2} \leq 1 \tag{10.16}$$

In a more naive picture, we can think of $\xi_c \to 1$ like if there exist a maximum angular divergence of the light pulse which preserves a phase relation of the e.m.

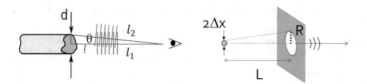

Fig. 10.4 Left: two rays emitted by a particle beam travel different path lengths up to the observer. If the path is long enough, they reach the out-of-phase condition. Right: a non-zero source size generates light diffraction at a downstream pinhole, as function of the angular acceptance determined by the pinhole size and the source-pinhole distance

field evaluated at two transverse locations of the wavefront. Such an angle is said *coherence angle*, and it is illustrated in Fig. 10.4-left plot.

In order to estimate the coherence angle, let us consider the path length difference of two rays emitted, respectively, on-axis and at the edge of the source. The two rays are emitted at the same time $t = 0$. If the observer is on-axis and at a distance $l_1 \gg d$ from the source, with d the source's full transverse size and such that $\theta \ll 1$, we have:

$$\Delta l = l_2 - l_1 = l_2(1 - \cos\theta) \approx \frac{l_2\theta^2}{2} \approx \frac{d\theta}{4} \tag{10.17}$$

The distance over which the two rays, initially in phase ($\Delta\phi = 0$), become out of phase ($\Delta\phi = \pi$), defines the coherence angle for the specified wavelength of emission:

$$\Delta\phi = \omega\Delta t = 2\pi\nu\frac{\Delta l}{c} \equiv \pi \Rightarrow \Delta l \approx \frac{\lambda}{2} \Rightarrow \theta_c = \frac{2\lambda}{d} \tag{10.18}$$

Such an estimate is consistent with, and made more accurate by, the Uncertainty Theorem of Fourier theory [3]. This states that the product of the effective widths of a function and of its Fourier transform (FT)—which thereby constitute a Fourier pair—must be equal or larger than 1/2. We apply the Theorem to the lateral spatial coordinate x of a photon. Since FT$(x) = 1/\lambda_x$, and by recalling $\frac{1}{\lambda_x} = \frac{k_x}{2\pi} = \frac{p_x}{h}$, we find:

$$\Delta x\Delta(\lambda_x^{-1}) = \Delta x\frac{\Delta\lambda_x}{\lambda_x^2} = \frac{\Delta x\Delta k_x}{2\pi} \geq \frac{1}{2} \Rightarrow \Delta x\Delta p_x \geq \frac{h}{2} \tag{10.19}$$

Owing to the fact that the angular divergence of radiation relates to the transverse momentum of the e.m. wave, which is assumed here to be monochromatic at the wavelength λ, we find the lower limit of the product of the spatial and angular divergence of the light pulse:

$$\Delta\theta = \frac{\Delta p_x}{p_z} \approx \frac{\Delta p_x}{p} = \frac{\Delta p_x}{(h\nu/c)} = \frac{\Delta p_x}{\hbar k} \Rightarrow \Delta x\Delta\theta\hbar k \geq \frac{h}{2} \Rightarrow \begin{cases} \sigma_{\theta,c} = \frac{\lambda}{4\pi}\frac{1}{\sigma_x} \\ \\ \theta_c = 0.44\frac{\lambda}{\Delta x} \end{cases} \tag{10.20}$$

We used $\Delta x = \sqrt{2\pi}\sigma_x$ and $\Delta\theta = \sqrt{2\pi}\sigma_{\theta,c}$ for the standard deviations, and $\Delta\theta = 2\sqrt{2\ln 2}\theta_c$ for fwhm quantities. One can notice that the same result is obtained by evaluating the minimum transverse rms phase space area occupied by a photon according to Heisenberg's Uncertainty Principle, i.e., $\sigma_x \sigma_{p_x} \geq \hbar/2$. The transverse coherence length is just $L_{c,\perp} = l_1 \theta_c$ (either in rms or fwhm sense).

The estimate of the coherence angle in Eq. 10.18 can be retrieved equivalently from the consideration that, according to geometric optics, the first minimum of the diffraction intensity pattern generated by a point-like source when the emitted light passes through a circular pinhole of diameter $2R$, is in correspondence of the angle $\theta = \lambda/(2R)$, see Fig. 10.4-right plot. This condition is smeared if the source has a non-zero transverse half-size Δx, because of the angular divergence determined by the source-pinhole distance L, $\Delta \theta = \Delta x / L$. In other words, the smaller the source size is, the more apparent the diffraction effect becomes at any given λ, and ideally at any λ for a zero-emittance source. We will observe diffraction at λ as long as $\Delta \theta = \frac{\Delta x}{L} \leq \frac{\lambda}{R}$. That is, as long as the angular acceptance determined by the pinhole is smaller than the coherence angle:

$$\frac{R}{L} \leq \frac{\lambda}{\Delta x} \equiv \theta_c \qquad (10.21)$$

In conclusion, two definitions of $L_{c,\perp}$ were given. The former one, related to Eq. 10.16, is often used to characterize a collected light beam by measuring the spatial visibility of interfering split pulses. The latter definition of $L_{c,\perp}$, related to the coherence angle in Eqs. 10.20 or 10.21, can be used to estimate the angular acceptance of a beamline in order to either exploit coherence properties of the central cone of radiation or, on the contrary, to avoid diffraction effects due to emerging transverse coherence.

As a matter of fact, Eq. 10.21 shows that transversely incoherent radiation can be made coherent (in a specific wavelength range) by physically selecting a small transverse portion of the incoming light beam, e.g., with a pinhole. Doing so, the downstream beamline will receive light from an "effective" source of smaller size than the actual one, thus closer to or below the diffraction limit, though at the expense of a reduced intensity.

10.2.3 Longitudinal Coherence

In analogy to first order transverse coherence, the *coherence time* τ_c is the width of the function g_1 in Eq. 10.8, evaluated as function of the relative delay τ for fixed points $\vec{r}_1, \vec{r}_2$ of the collected intensity pattern, or $g_1(\tau)$. The rms value of τ_c for an arbitrary intensity distribution is [2]:

$$\tau_{c,rms} = \int_{-\infty}^{\infty} |g_1(\tau)|^2 d\tau \qquad (10.22)$$

where for a Gaussian function $\tau_{c,rms} \cong 0.85 \tau_{c,hwhm}$. The longitudinal coherence length is $L_{c,\parallel} = c\tau_c$. Equation 10.22 is commonly adopted to retrieve τ_c from a set of measurements of the radiation spectral intensity pattern imaged on a screen. In analogy to Eq. 10.16, it can be used to characterize a collected light beam.

A more intuitive connection of τ_c to the spectral bandwidth of a light pulse can be found by looking the geometry in Fig. 10.4-right plot. Since the first minimum of the interference fringes is at $\theta = \lambda/(2R)$ for fully coherent radiation, we require that the spectral bandwidth $\Delta\lambda$ of the actual pulses be small enough not to perturb the pattern:

$$\Delta\theta = \tfrac{\Delta\lambda}{2R} \ll \theta \Rightarrow \tfrac{\Delta\lambda}{\lambda} = \tfrac{\Delta\nu}{\nu} \ll 1 \qquad (10.23)$$

Given the narrow bandwidth of partially coherent radiation, we wonder what is the corresponding fraction of the pulse duration over which the phase relation of electric fields of slightly different frequencies is still preserved. The two frequencies at spectral distance $\Delta\omega$ travelling for a time interval Δt accumulate a phase difference $\Delta\phi = \Delta\omega\Delta t$. The time the fields associated to the two frequencies take to become opposite in phase is, by definition, the coherence time:

$$\Delta\phi = \Delta\omega\Delta t = \Delta\omega\tau_c \equiv \pi \Rightarrow L_{c,\parallel} = c\tau_c = \tfrac{c}{2\Delta\nu} = \tfrac{\lambda^2}{2\Delta\lambda} \qquad (10.24)$$

As already for the transverse coherence length in Eq. 10.20, the longitudinal coherence time is equivalently retrieved from the Fourier's Uncertainty Theorem, $\Delta\nu\Delta t \geq 1/2$, with rms quantities $\Delta\nu = \sqrt{2\pi}\sigma_\nu$, $\Delta t = \sqrt{2\pi}\sigma_t$, or from Heisenberg's Uncertainty Principle applied to the longitudinal rms phase space area, $\sigma_E\sigma_t \geq \hbar/2$ [3].

Equation 10.24 implies that a single frequency has infinite coherence time. Viceversa, if τ_c is smaller than the finite duration of a particle beam source, and if the beam source has an internal structure that allows coherent (i.e., in phase) emission, we may imagine the beam emitting "spikes" of radiation, each spike being emitted independently from (i.e., not correlated to) the others, but individually coherent. This is the case of a free-electron laser in regime of so-called Self-Amplified Spontaneous Emission (SASE), as discussed later on.

So as a pinhole is used at beamlines to obtain some higher degree of transverse coherence, a monochromator is often adopted to further select a small spectral portion of the incoming broad-band radiation. Since the selection is physical in the dispersive plane of the monochromator, a larger coherence time, i.e. a narrower bandwidth, is obtained at the expense of reduced intensity.

10.2.3.1 Discussion: Degeneracy Parameter

Equations 10.18 and 10.24 for the coherence angle and the coherence time suggest that, for any given transverse and longitudinal size of the source, and therefore of the emitted light pulse, it is more and more difficult to obtain a high degree of coherence at shorter and shorter wavelengths. We corroborate this observation by exploiting the definition of brilliance and its dependence from the central wavelength of emission.

Let us consider a certain number of photons N_{ph} in a light pulse of central wavelength λ and spectral bandwidth $\Delta\lambda$. For simplicity, we assume the photons uniformly distributed in a pulse duration Δt. The number of photons contained in one longitudinal coherence length is:

$$N_{c,\parallel} = N_{ph} \frac{L_{c,\parallel}}{c\Delta t} = \frac{I_r}{c} \frac{\lambda^2}{2\Delta\lambda} \tag{10.25}$$

and I_r is the total intensity.

By substituting I_r from Eq. 10.25 in the definition of brilliance for a diffraction limited, matched, flat particle beam (Eq. 10.3), we find:

$$\hat{B}_r = \frac{2I_r}{\lambda^2(\Delta\lambda/\lambda)} = \frac{4}{\lambda^2(\Delta\lambda/\lambda)} \frac{cN_{c,\parallel}(\Delta\lambda/\lambda)}{\lambda} = \frac{4cN_c}{\lambda^3} \tag{10.26}$$

We introduced the notation $N_{c,\parallel} = N_c$ in the r.h.s. of Eq. 10.26 to stress out that, since the brilliance is estimated for a source at the diffraction limit, the emitted photons are by definition transversely coherent. Hence, N_c is the number of photons transversely *and* longitudinally coherent, and $\hat{B}_r$ is the total 6-D photon density emitted by a diffraction-limited source, evaluated in a "coherence volume" λ^3.

Equation 10.26 shows that, for any given brilliance of radiation centered at λ_0, the number of fully coherent photons becomes favorably larger at any $\lambda \geq \lambda_0$. The number of fully coherent photons N_c or, more precisely, the number of photons per coherence phase space volume and coherence time—i.e., the number of photons per "coherent mode"—is called *degeneracy parameter*, δ_w. If $D \leq 1$ is the light source duty cycle ($D = 1$ for δ_w of a single pulse, and B_r is the peak brilliance), it follows from Eq. 10.26:

$$\delta_w(\lambda) \equiv N_c(\lambda) = \frac{DB_r\lambda^3}{4c} = 8.34 \cdot 10^{-25} D\lambda^3[\text{Å}]B_r[\frac{\#ph}{mm^2 mrad^2 sec 0.1\%bw}] \tag{10.27}$$

Since photons are bosons, we can have $\delta_w \geq 1$. However, it is only with the advent of undulators in most recent storage ring light sources ($B_r(5\text{Å}) \sim 10^{22}$ in conventional units) and, especially, of free-electron lasers ($B_r(1\text{Å}) \sim 10^{32}$ in conventional units), that values $\delta_w \approx 1$ and $\delta_w \gg 1$, respectively, have been achieved in the x-ray region [1].

10.2.4 Intensity Enhancement

Equations 10.20 and 10.24 suggest that a charge distribution compact enough in the 3-D space tends to radiate coherently. Namely, the instantaneous emission of a distribution which behaves as a point-like charge—say, a "macro-particle"—tends to be transversely and longitudinally coherent because all photons are emitted with approximately the same spatio-temporal phase.

To be more rigorous, however, the degree of coherence of radiation emitted at a certain wavelength is established by the ratio of beam size and coherence length, either transverse or longitudinal. Thus, compactness *per se'*, defined with respect to the wavelength scale, is not a sufficient condition for coherent emission. We show below that, instead, it is so for enhancement of the radiation intensity, in proportion to the number of radiating beam particles.

We consider a bunch of $N_b \gg 1$ particles spatially distributed in space with vectors $\vec{r}_j, j = 1, ..., N_b$, emitting radiation at the central frequency ω. To simplify the math, we assume pure monochromatic radiation, $\Delta \omega \ll \omega$. The emission is said *incoherent* if the charges assume a random phase distribution. In this case the total radiated field (intensity) is the linear superposition of the field (linear sum of intensity) of single particle emissions: $I_{tot}(\omega) = N_b I_r(\omega)$.

The amplitude of the electric field radiated by the j-th particle is E_{0j}. The electric field in the far field approximation is described as a travelling wave with wave-vector $\vec{k}_j = \hat{n}_j \frac{\omega}{c}$. The total field vector is:

$$\vec{E}_{tot}(\omega) = \sum_{j=1}^{N_b} \vec{E}_j e^{i(\omega t - \vec{k}_j \vec{r}_j)} \tag{10.28}$$

The radiation intensity, or number of photons per unit of time, is proportional to the total e.m. energy and it can therefore be calculated by means of the Poynting vector (see Eqs. 7.6 and 7.7):

$$I_{tot}(\omega) = \frac{dN_{ph}}{dt} = \frac{1}{\hbar\omega}\frac{dE(\omega)}{dt} = \frac{1}{\hbar\omega}\int_\Omega |\vec{S} \cdot \hat{n}|R^2 d\Omega = \left(\frac{c\varepsilon_0}{2\hbar\omega}\right)\int_\Omega |\vec{E}_{tot}(\omega)|^2 R^2 d\Omega \tag{10.29}$$

We now expand the absolute value of the total electric field. The frequency dependence is collapsed into a normalized amplitude $E_j(\omega) = E_{0j}e^{i\omega t}$. Next, we assume particles' spacing in all directions be much smaller than the (central) wavelength of emission, or $|\Delta \vec{r}| \ll \lambda$:

$$|\vec{E}_{tot}(\omega)|^2 = |\sum_{j=1}^{N_b} E_j(\omega)e^{-ik\hat{n}_j\vec{r}_j}|^2 = \sum_{j=1}^{N_b} E_j(\omega)e^{-ik\hat{n}\vec{r}_j} \sum_{m=1}^{N_b} E_m^*(\omega)e^{ik\hat{n}\vec{r}_m} =$$

$$= \sum_{j=1}^{N_b} |E_j(\omega)|^2 + \sum_{j\neq m}^{N_b} E_j(\omega)E_j^*(\omega)e^{i\pi\frac{\hat{n}(\vec{r}_m-\vec{r}_j)}{\lambda}} \approx$$

$$\approx \sum_{j=1}^{N_b} |E_j|^2 + \sum_{j\neq m}^{N_b} E_j E_j^* = N_b|E_j|^2 + N_b(N_b - 1)|E_j|^2 = N_b^2|E_j|^2 \tag{10.30}$$

A more accurate analysis should consider a 3-D charge distribution function, whose Fourier transform would contribute through "form factors" to the coherent intensity. In the simplified assumption of no correlation between longitudinal and transverse distribution functions, and by substituting Eq. 10.30 into Eq. 10.29, we end up with [7]:

$$I_{tot}(\omega) = \frac{|\vec{E}_{tot}(\omega)|^2}{2Z_0} \approx N_b I_r(\omega) + N_b(N_b - 1)I_r(\omega)|f_\perp(\omega)|^2|f_\parallel(\omega)|^2 \approx \tag{10.31}$$

$$\approx N_b I_r(\omega)\left[1 + N_b|f_\perp(\omega)|^2|f_\parallel(\omega)|^2\right]$$

For example, the form factor of a Gaussian distribution function is $|f(\omega_u)|^2 = e^{-k_u^2\sigma_u^2/2}$, where $\omega_u = ck_u$ and $u = x, y, z$. $|f(\omega_u)| \to 0$ when $\sigma_u \gg \lambda_u$, and

$|f(\omega_u)| \to 1$ when $\sigma_u \ll \lambda_u$. In the former case, the intensity of the emitted radiation is just the single particle emission times the number of emitters ("incoherent emission"). On the contrary, if the beam charge is distributed over spatial scales smaller than the wavelength of interest, and for large number of emitters, the total intensity of emitted radiation at that wavelength goes like the intensity of the single particle emission times the *square* of the number of emitters ("coherent emission" or, more precisely, "intensity enhancement").

The latter effect is exploited, for example, in electron linacs, where the bunch is time-compressed and squeezed to small spot sizes. It is then sent through a small aperture or a metallic foil to generate coherent diffraction or transition radiation, respectively. Bunch shortening is sometimes approached in storage rings, where a nonlinear dependence of the intensity of dipole synchrotron radiation from the bunch charge is observed. In both cases, coherent emission can be obtained at IR-THz frequencies. EUV and x-ray free-electron lasers exploit the same physics, but in order to have large intensity at shorter wavelengths, electron bunches are internally micro- or nano-bunched. The electron clusters are separated by the central wavelength of emission, each cluster being much shorter than that.

10.3   Undulator Spontaneous Radiation

In its simplest configuration, an undulator [4,5] is made of two opposite periodic arrays of magnetic poles of alternated polarity, through which the charged particles, electrons hereafter, wiggle by virtue of the Lorentz's force—see Fig. 10.5. The magnetic poles can be either permanent magnets (e.g., rare earths) or electro-magnets, or hybrid elements. Depending if the electrons travel in a vacuum chamber internal to the undulator, or if the electrons and the undulator are both inside a tank under vacuum, the ID is said "out-of-vacuum" or "in-vacuum", respectively. "Planar" undulators have planar geometry of the arrays. Their relative shift in the longitudinal direction determines the plane of oscillation of the electrons, thus control of the electric field polarization (linearly horizontal, vertical, inclined, circular, etc.).

One of the most common designs is the out-of-vacuum variable gap, planar, variably polarized undulator (APPLE-II). More complex designs have up to 6 degrees of freedom (APPLE-X, with four independent arrays and variable gap both in the horizontal and in the vertical plane). Cryogenic and superconducting undulators have been developed with the main purpose of increasing the magnetic field at shorter undulator periods. Typical dimensions of an undulator are 1–5 m total length, 1–10 cm magnetic period, 3–20 mm gap. A single segment can be allocated in a storage ring straight section. Tens' of such segments in a series constitute the undulator line of a free-electron laser.

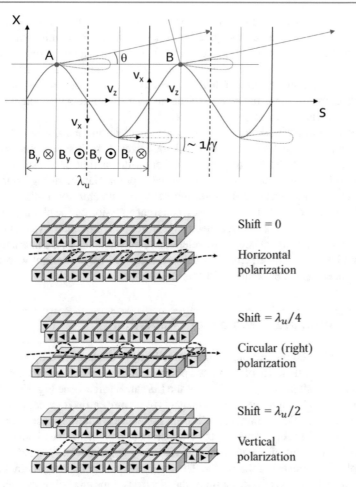

Fig. 10.5 Top view of electron trajectory in a planar linearly polarized insertion device (top), and APPLE-II type undulator configured for diverse polarization states

10.3.1 Central Wavelength

The central wavelength of undulator emission is determined by the constructive interference of field emitted at homologous points of consecutive periods. Being the *undulator spontaneous radiation* (spontaneous means here incoherent) a special case of synchrotron radiation, the radiated intensity is proportional to the magnetic field, and the central cone at each *local* source point has a characteristic aperture $\sim 1/\gamma$. However, this implies that the overlap of consecutive wavefronts, especially at high beam energies, is only possible if the electrons wiggle around the magnetic axis with a relatively small oscillation amplitude. This imposes in turn a moderate magnetic field.

Let us consider for simplicity a planar undulator for horizontally polarized light, see Fig. 10.5-top plot. The vertical magnetic field is ideally a sinusoidal function of the longitudinal coordinate $z = v_z t$ along the ID: $B_y = B_{0y} \sin\left(\frac{2\pi v_z t}{\lambda_u}\right)$, with $v_z \approx const$ the particle's longitudinal velocity, and λ_u the undulator period. The particle's horizontal velocity is calculated from the Lorentz's force:

$$F_x = \gamma m_0 a_x = m_0 \gamma \frac{dv_x}{dt} = -e v_z B_y = -e v_z B_{0y} \sin\left(\frac{2\pi v_z t}{\lambda_u}\right) ;$$

$$\int dv_x = -\left(\frac{e B_{0y}}{\gamma m_0}\right) \int dz' \sin\left(\frac{2\pi z'}{\lambda_u}\right) = \frac{1}{\gamma}\left(\frac{e B_{0y} \lambda_u}{2\pi m_0}\right) \cos\left(\frac{2\pi z}{\lambda_u}\right) ; \qquad (10.32)$$

$$\beta_x = \frac{v_x}{c} = \frac{1}{\gamma}\left(\frac{e B_{0y} \lambda_u}{2\pi m_0 c}\right) \cos\left(\frac{2\pi z}{\lambda_u}\right) \equiv \frac{K}{\gamma} \cos\left(\frac{2\pi z}{\lambda_u}\right)$$

where we introduced the *undulator parameter* $K = \left(\frac{e B_{0y} \lambda_u}{2\pi m_e c}\right) = 0.934 B_{0y}[T]\lambda_u[cm]$ in practical units, and for the electron rest mass.

The generic particle's longitudinal velocity is:

$$v_z = c\left(\beta^2 - \beta_x^2 - \beta_y^2\right)^{1/2} \approx c\left[1 - \frac{1}{\gamma^2} - \left(\frac{K}{\gamma}\right)^2 \cos^2(k_u z)\right]^{1/2} =$$

$$= c\left[1 - \frac{1}{\gamma^2} - \left(\frac{K}{\gamma}\right)^2 \frac{1+\cos(2k_u z)}{2}\right]^{1/2} = c\left[1 - \frac{1}{\gamma^2}\left(1 + \frac{K^2}{2}\right) - \frac{K^2}{2\gamma^2}\cos(2k_u z)\right]^{1/2} \approx$$

$$\approx c\left[1 - \frac{1}{2\gamma^2}\left(1 + \frac{K^2}{2}\right) - \frac{K^2}{4\gamma^2}\cos(2k_u z)\right] = \langle v_z \rangle - c\frac{K^2}{4\gamma^2}\cos(2k_u z)$$

$$(10.33)$$

and $\langle v_z \rangle = const.$ is the average velocity evaluated along one undulator period. It can be shown that for a helically polarized undulator of same peak field, $K \to \sqrt{2}K$ and $v_z = \langle v_z \rangle = const$ due to purely geometric considerations.

If $K < 1$, the radiation cones emitted along the trajectory overlap, so allowing constructive interference of the wavefronts to build up. The condition of constructive interference is illustrated in red in Fig. 10.5, where light is assumed to be emitted at a generic angle θ from the on-axis direction. During the time $\Delta t_1 = \frac{\lambda_u \cos\theta}{c}$ the wavefront emitted at A reaches B, the radiating electron, initially in A, reaches B in an interval $\Delta t_2 = \frac{\lambda_u}{v_z}$. Constructive interference happens if the path length difference $c(\Delta t_2 - \Delta t_1)$ is equal to one period of oscillation, λ, of the e.m. field, i.e., the two wavefronts result in phase:

$$c(\Delta t_2 - \Delta t_1) = c\frac{\lambda_u}{v_z} - \lambda_u \cos\theta \equiv \lambda \qquad (10.34)$$

In the ultra-relativistic limit $\gamma \gg 1$, small observation angles $\theta \ll 1$, and small oscillation amplitudes $K/\gamma \ll 1$, Eq. 10.34 is re-written as follows:

$$\frac{\lambda}{\lambda_u} + 1 - \frac{\theta^2}{2} = \frac{1}{\beta_z} \approx \left[\sqrt{\beta^2 - \beta_x^2 - \beta_y^2}\right]^{-1} \approx \left[\sqrt{1 - \frac{1}{\gamma^2} - \langle \beta_x^2 \rangle}\right]^{-1} =$$

$$\left[\sqrt{1 - \frac{1}{\gamma^2} - \frac{1}{2}\frac{K^2}{\gamma^2}}\right]^{-1} \approx 1 + \frac{1}{2\gamma^2}\left(1 + \frac{K^2}{2}\right) ; \qquad (10.35)$$

$$\Rightarrow \lambda = \frac{\lambda_u}{2\gamma^2}\left(1 + \frac{K^2}{2} + \gamma^2\theta^2\right)$$

Equation 10.35 establishes a one-to-one relationship between the central wavelength of undulator emission and the particle's energy. It is more and more accurate for a large number of periods, such that β_x in Eq. 10.32 can be replaced with its average value over z. By neglecting the vertical motion we also implied $\beta_y^2 \ll \frac{K^2}{\gamma^2}$.

The on-axis resonant wavelength and photon energy in practical units are:

$$\varepsilon_{ph}[eV] = h\nu_n = 9509\frac{E[GeV]^2}{\lambda_u[mm]\left(1+\frac{K^2}{2}\right)}$$

$$\lambda[nm] = \frac{c}{\nu} = \frac{1241.5}{\varepsilon_{ph}[eV]}$$

(10.36)

It is instructive to notice that Eq. 10.34 for on-axis emission can be written as $\frac{\lambda_u}{v_z} = \frac{\lambda}{c-v_z}$. This states that the time the electron takes to travel λ_u is equal to the time the light takes to slip ahead the electron by λ. The radiation *slippage* (relative to the source electron) after N_u periods is therefore $N_u\lambda$. This is the minimum duration of the radiation pulse, independently from the bunch duration.

For any given undulator period, the central wavelength can be continuously varied by tuning, for example, the beam energy. However, this usually implies a re-tuning of the entire accelerator, and might be not that practical for fine scans. Hence, a variable K is adopted through "variable gap" or "variable phase" undulators, in which the density of the magnetic field lines is varied with a remote control of, respectively, the undulator gap and the arrays relative phase, as shown in Fig. 10.5. It turns out that the higher intensity of radiation is at longer wavelengths, because these are obtained with a larger K, which corresponds to stronger centripetal acceleration and therefore larger radiated power (see Eq. 7.8). The shortest wavelength is emitted on-axis ($\theta = 0$, see Eq. 10.35).

10.3.1.1 Discussion: Doppler Effect in an Undulator

Equation 10.35 for the central wavelength of undulator emission can be retrieved starting from the particle's rest frame, where the undulator field is seen by the particle as a travelling wave. How long is the undulator period in the particle's rest frame? What is the wavelength of the scattered radiation in the laboratory frame, if elastic scattering is assumed and the particle's recoil is neglected?

In the particle's rest frame, the undulator period is shortened to $\lambda'_u = \lambda_u/\gamma$. The undulator field, static in the laboratory frame, is seen by the particle as an e.m. wave of period λ'_u. The wave scatters on the electron and, by virtue of the relativistic Doppler effect (see Eq. 1.48), its wavelength in the laboratory frame results :

$$\lambda = \gamma\lambda'_u(1 - \beta_z\cos\theta) = \lambda_u\left(1 - \sqrt{1 - \frac{1}{\gamma^2} - \beta_y^2 - \beta_x^2}\cos\theta\right) \approx$$

$$\approx \lambda_u\left[1 - \left(1 - \frac{1}{2\gamma^2} - \frac{K^2}{4\gamma^2}\right)\left(1 - \frac{\theta^2}{2}\right)\right] = \frac{\lambda_u}{2\gamma^2}\left(1 + \frac{K^2}{2} + \gamma^2\theta^2\right) + o\left(\frac{\theta^2}{\gamma^2}\right)$$

(10.37)

This derivation highlights the physical origin of the $\lambda \sim \gamma^{-2}$ dependence, i.e., relativistic length contraction of the undulator period (factor γ^{-1}), and the relativistic Doppler effect of the central frequency (factor $(2\gamma)^{-1}$).

10.3.2 Spectral Width, Angular Divergence

Equation 10.35 for the central wavelength is satisfied also by harmonics $\lambda_n = \lambda/n$, of the fundamental, $n \in \mathbb{N}$. Since the minimum duration of the spontaneous radiation pulse is the radiation slippage, and since this is determined by the fundamental wavelength of emission, the slippage of any hamonic is the same of the fundamental. Thus, we have [5,6]:

$$\Delta t_n = \frac{N_u n \lambda_n}{c} = \frac{N_u \lambda}{c} = \Delta t$$

$$\Rightarrow \Delta v_n = c \frac{\Delta \lambda_n}{\lambda_n^2} \approx \frac{1}{\Delta t_n} = \frac{c}{N_u \lambda} \tag{10.38}$$

$$\Rightarrow \frac{\Delta \lambda_n}{\lambda_n} = \frac{\Delta v_n}{v_n} \approx \frac{1}{n N_u}$$

In summary, higher harmonics ($n > 1$) are emitted in a pulse of same duration of the fundamental emission. Their absolute energy bandwidth is the same of the fundamental, while their *relative* bandwidth is $n-$times smaller.

The spectral intensity distribution—simply "spectrum" hereafter—is proportional to the Fourier transform of the electric field emitted along the N_u undulator periods. Let us assume for simplicity a monochromatic wave at the fundamental frequency $\omega_0 = 2\pi c/\lambda$. Since the number of periods is finite and the field amplitude is constant, the spectrum is the sinc function (see later Fig. 10.11 and inset):

$$E(t) = \begin{cases} E_0 e^{i\omega_0 t}, \, |t| \leq \frac{T}{2} = \frac{N_u \lambda}{c} \\ \\ 0, \, otherwise \end{cases} ;$$

$$\begin{cases} E(\omega) = \int_{-T/2}^{T/2} E_0 e^{-i(\omega-\omega_0)t} dt = -\frac{E_0}{i(\omega-\omega_0)} \left[e^{-i(\omega-\omega_0)T/2} - e^{i(\omega-\omega_0)T/2} \right] = \\ \\ = E_0 \frac{2\sin(\frac{\Delta\omega T}{2})}{\Delta\omega} = E_0 T \cdot sinc(\xi), \\ \\ \xi := \frac{(\omega-\omega_0)T}{2} = \pi N_u \frac{\Delta\omega}{\omega_0}; \end{cases}$$

$$I(\omega) = \frac{|E(\omega)|^2}{2Z_0} = \frac{E_0^2 T^2}{2Z_0} sinc^2(\xi).$$

$$\tag{10.39}$$

Since the fwhm bandwidth of the sinc function is $\Delta\omega \approx \frac{2\pi}{T} = \frac{2\pi c}{2N_u\lambda}$, the intrinsic relative bandwidth is $\frac{\Delta\omega}{\omega_0} \sim \frac{1}{N_u}$, as already in Eq. 10.38. That is, the larger the number of undulator periods is, the more effective the spectral filtering due to the constructive interference is, the narrower is the intrinsic relative bandwidth of the undulator. This, however, can be enlarged by non-zero dimensions of the electron beam, as discussed below.

Equation 10.35 shows that off-energy electrons will radiate at slightly different wavelengths. The effect can be neglected as long as the relative bandwidth enlargement due to the beam's energy spread is negligible with respect to the intrinsic relative bandwidth. This leads to an upper limit for the beam's energy spread:

$$\frac{\Delta\lambda_n(\gamma)}{\lambda_n} = 2\frac{\Delta\gamma}{\gamma} \approx \sigma_\delta \ll \frac{\Delta\lambda_n}{\lambda_n} \approx \frac{1}{nN_u} \tag{10.40}$$

This prescription is usually met at electron synchrotrons for fundamental emission, where $\sigma_\delta \leq 0.1\%$ and $\frac{1}{N_u} \approx 1\%$. Some bandwidth enlargement is expected at higher harmonics, e.g. $n \geq 10$.

Equation 10.35 also shows that electrons travelling with angular divergence $\Delta\theta$ will radiate at slightly different wavelengths. At such off-axis directions of observation, some red-shifted bandwidth enlargement is expected. By recalling the intrinsic bandwidth in Eq. 10.38, this effect is negligible as long as:

$$\frac{\Delta\lambda_n(\theta)}{\lambda_n} = \frac{1}{\lambda_n}\frac{\lambda_u}{2n}\Delta\theta^2 = \frac{\lambda_u}{2\lambda}\Delta\theta^2 \ll \frac{1}{nN_u}$$

$$\Rightarrow \sigma_\theta \ll \sqrt{\frac{2\lambda}{\lambda_u}\frac{1}{nN_u}} = \frac{1}{\gamma}\sqrt{\frac{1+K^2/2}{nN_u}} \approx \frac{1}{\gamma}\sqrt{\frac{1}{nN_u}} \tag{10.41}$$

The r.h.s. of Eq. 10.41, obtained for $K < 1$, describes the characteristic angular divergence of the n-th harmonic of undulator radiation, or central cone. It suggests that, although the local emission has angular divergence $1/\gamma$, the constructive interference imposes a tighter condition to the overall overlap of the wavefronts, so that the angular divergence of the radiation pulse exiting the undulator is smaller than that of synchrotron radiation, in proportion to the number of undulator periods. Higher harmonics, i.e. shorter wavelengths, show smaller angular divergence.

10.3.2.1 Discussion: Intrinsic Size of Undulator Radiation

Obtain the expressions for the *intrinsic* transverse size and angular divergence of the fundamental transverse mode [5,6] anticipated in Eq. 10.4, starting from the spectral and angular distribution of the undulator spontaneous radiation.

The rms angular divergence $\sigma_{r',n}$ is calculated from Eq. 10.41 for the n-th harmonic by assuming $K < 1$ and by replacing the relativistic γ-factor with the expression for the central wavelength (see Eq. 10.35). An undulator length $L_u = N_u\lambda_u$ is used:

$$\sigma_{r',n} \approx \frac{1}{\alpha}\sqrt{\frac{2\lambda}{\lambda_u}\frac{1}{nN_u}} \approx \frac{1}{2\sqrt{2}}\sqrt{\frac{2\lambda_n}{\lambda_u}\frac{\lambda_u}{L_u}} = \sqrt{\frac{\lambda_n}{4L_u}}, \tag{10.42}$$

where $\alpha = 2\sqrt{2} = 2.8$ is a numerical factor to pass from full width $\Delta\theta$ to rms value of the angular divergence, intermediate to 3.46 (uniform) and 2.36 (Gaussian). According to Eq. 10.20, the rms emittance of a transversely coherent light beam is $\varepsilon_r = \sigma_r\sigma_{r'} = \lambda/(4\pi)$. It follows $\sigma_{r,n} = \frac{\varepsilon_r}{\sigma_{r',n}} \approx \frac{\sqrt{\lambda_n L_u}}{2\pi}$.

10.3.3 Dipole, Wiggler, Undulator

Historically, IDs were introduced in synchrotrons to complement synchrotron radiation from dipole magnets, and in particular:

- to shift the critical photon energy to higher values by virtue of a larger magnetic field (see Eq. 7.24);
- to increase the radiated power, in proportion to a larger number of magnetic poles;
- to increase the spectral brightness, e.g. by producing narrower intrinsic spectral bandwidth.

In practice $B_{0y} \propto \lambda_u$, which typically leads to $K < 10$ for $\lambda_u < 5$ cm (still, higher K can be produced at shorter λ_u with specific magnetic designs and technologies). At such period lengths, $N_u \geq 100$ periods can be assembled in an ID to be accommodated in a synchrotron's straight section, which is commonly long ~ 5 m or so. The moderate deflection of the electrons' orbit for $K \leq 1$ classifies the ID as an *undulator*.

Longer periods allow $K > 10$, and the ID can be classified as a *wiggler*. Owing to the larger magnetic field, electrons in a wiggler have large oscillation amplitude, and the wavefront interference, otherwise present in undulators, gets lost. That is, a wiggler behaves as a series of (strong) dipole magnets.

Given the similarity in the physics of emission shared by dipole, undulator and wiggler magnets, we can estimate the total average (over one turn) radiated power in the three cases starting from Eq. 7.15, assuming N_b electrons in a bunch of average current $\langle I \rangle$ [1,4–6].

For the dipole magnet, it is simple to find the radiated power in a turn and, from that, along a magnet of length $L = R\theta_d$, assuming an isomagnetic lattice:

$$P_{d,turn}[kW] = \frac{N_b U_0}{T_0} = \frac{\langle I \rangle U_0}{e} = 26.5 E^3[GeV]B[T]\langle I \rangle[A]$$
(10.43)

$$P_{d,L}[kW] = P_{d,turn}[kW] \cdot \frac{\theta_d}{2\pi} = 14.06 E^4[GeV]\langle I \rangle[A]L[m]/R^2[m]$$

The wiggler total radiation intensity is the incoherent sum of synchrotron radiation emitted at N_u-poles, thus the total intensity is N_u-fold that from a single dipole of same magnetic field. The total power from a wiggler magnet long L_w (a dipole magnet is equivalent to 2 wiggler periods) is calculated as the fraction $\frac{L_w/2}{2\pi R}$ of the synchrotron radiation power emitted along the whole circumference:

$$P_w[kW] = \frac{N_b U_0}{T_0} \frac{L_w}{4\pi R} = \frac{\langle I \rangle[A]U_0[kV]}{4\pi} \frac{0.3 B[T]L_w[m]}{E[GeV]} = 0.633 E^2[GeV]\langle B^2[T] \rangle \langle I \rangle[A]L_w[m]$$
(10.44)

and $\langle B^2[T] \rangle$ is the rms wiggler field squared.

The average power from an undulator is estimated through Eq. 7.12 with the prescription of spectro-angular filtering provided by the ID. According to Eq. 10.35, the relative bandwidth observed off-axis at the generic angle θ is:

$$\frac{\Delta\lambda}{\lambda} = \frac{\lambda(\theta) - \lambda(0)}{\lambda(0)} = \frac{\gamma^2\theta^2}{1 + \frac{K^2}{2}}$$
(10.45)

To estimate the power emitted only in the central cone, we restrict the observation angle to the characteristic angular divergence of undulator spontaneous radiation $\theta \approx \frac{1}{\gamma}\sqrt{\frac{1}{N_u}}$ (see Eq. 10.41), to obtain $\frac{\Delta\lambda}{\lambda} \approx \frac{1}{N_u\left(1+\frac{K^2}{2}\right)}$. The dipole radius is replaced by the undulator magnetic field according to $R = \frac{p_z}{eB} \approx \frac{\lambda_u}{2\pi m_e c K}\frac{E}{c}$. Finally, if particles take the time interval Δt to pass the undulator length $L_u \approx c\Delta t$, the turn-averaged power has to be re-scaled by the fraction $\Delta t/T_0$:

$$P_u = N_b P_{sr}\frac{\Delta\lambda}{\lambda}\frac{\Delta t}{T_0} = \frac{1}{4\pi\varepsilon_0}\frac{2}{3}e^2 c\gamma^4\frac{4\pi^2 m_e^2 c^2}{\lambda_u^2 E^2}\frac{K^2}{N_u\left(1+\frac{K^2}{2}\right)}\frac{N_b\Delta t}{T_0}\frac{ec}{ec} = \frac{2\pi}{3}\frac{e}{\varepsilon_0}\frac{\langle I\rangle N_u\gamma^2}{\lambda_u}\frac{K^2}{\left(1+\frac{K^2}{2}\right)}$$
(10.46)

An ab initio calculation would provide the following expression, in practical units, for the spectral intensity of undulator spontaneous emission at the n-th harmonic of the fundamental (n odd), within the central angular cone $2\pi\sigma_{r'}^2$:

$$\phi_n = \frac{dN_r}{dtd\omega/\omega} = 2\pi\sigma_{r'}^2\left(\frac{d\phi_n}{d\Omega}\right)_{\theta=\psi=0} = \frac{1}{2}\pi\alpha N_u\frac{\langle I\rangle}{e}\left(\frac{K^2}{1+\frac{K^2}{2}}\right)n[JJ]_n^2 =$$
$$= \frac{1.43}{2}\cdot 10^{14}N_u\left(\frac{K^2}{1+\frac{K^2}{2}}\right)n[JJ]_n^2\langle I\rangle[A]\quad\left[\frac{\#ph}{sec\cdot 0.1\%bw}\right]$$
(10.47)

where $\alpha = 1/137$ is the fine structure constant. $[JJ]_n = 1$ for helically polarized undulators, while it is a combination of Bessel functions for linearly polarized light:

$$[JJ]_n = (-1)^{\frac{n-1}{2}}J_{\frac{n-1}{2}}\left(\frac{na_u^2}{2(1+a_u^2)}\right) - J_{\frac{n+1}{2}}\left(\frac{na_u^2}{2(1+a_u^2)}\right),\ a_u = K^2/2$$
(10.48)

When the lowest harmonics ($n \leq 3$) are included in the total intensity, the transverse mode of radiation cannot be any longer approximated by a Gaussian, and the factor 1/2 in Eq. 10.47 is removed.

By recalling Eq. 10.4, the central angular cone has amplitude:

$$2\pi\sigma_{r'}^2 = \frac{\pi\lambda}{2nL_u} = \frac{\pi}{N_u}\frac{\left(1+\frac{K^2}{2}\right)}{4n\gamma^2}$$
(10.49)

Inserted into Eq. 10.47, this shows that in the central cone $\frac{d\phi_n}{d\Omega} \propto N_u^2$. On the contrary, emission from a wiggler or a dipole shows $\phi \propto N_u$ ($N_u = 1$ in a dipole, see also Eqs. 7.29 and 10.44). The N_u^2-dependence of the central cone of undulator radiation is the result of constructive interference, and a signature of partial coherence. Still, the total intensity is linearly proportional to the number of particles (so as in a wiggler and a dipole, see Eqs. 7.31 and 10.44) and therefore it is *not* coherent emission in the sense of Eq. 10.31. This can be rephrased by stating that undulator spontaneous radiation is produced as a coherent sum of fields emitted by one electron (Gaussian laser-like mode with rms size and angular divergence in Eq. 10.4)

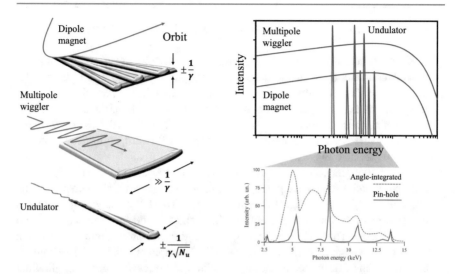

Fig. 10.6 Electron trajectory, characteristic angular divergence (left) and spectrum (right) of radiation from dipole magnet, multipole wiggler, undulator. Zoom: spectral intensity from a planar undulator optimized for the 3rd harmonic of 2.7 keV, in the presence of pin-hole for angular selection of on-axis radiation (red, solid), and without angular selection (blue, dashed). A large angular acceptance leads to red-shifted bandwidth enlargement of the odd harmonics, as well as to observation of even harmonics off-axis (see next Section)

over many consecutive periods ($I_r \propto N_u^2$). However, it remains an incoherent sum of wave trains emitted by the population of electrons ($I_r \propto N_e$).

In summary, dipole magnet synchrotron radiation, multipole wiggler emission and undulator spontaneous radiation have the same physical origin, but the output spectral and angular characteristics are eventually determined by the specificity of the electrons' trajectory in the magnetic device. A qualitative comparison of typical angular and spectral distribution of radiation from a dipole magnet and IDs is illustrated in Fig. 10.6.

10.3.4 Harmonic Emission

Despite Eq. 10.35 is satisfied by any harmonic of the fundamental wavelength, emission of *sub*-harmonics from wigglers or undulators is suppressed by the electron trajectory through the ID [5]. Indeed, the particle's trajectory with local versor $\hat{n} = (-\hat{x}\cos\phi\sin\theta, -\hat{y}\sin\phi\sin\theta, -\hat{z}\cos\theta)$ in the Frenet-Serret coordinates system, is embedded in the expression of the radiated electric field (see Eq. 7.5). The spectral distribution of the emitted radiation is proportional to the Fourier transform of the electric field (see Eq. 7.25), hence of the particle's trajectory:

$$\frac{dU}{d\Omega} = \int \frac{dP}{d\Omega} dt = c\varepsilon_0 \int |R\vec{E}(t)|^2 dt = 2c\varepsilon_0 \int |R\vec{E}(\omega)|^2 d\omega \Rightarrow \frac{d^2U}{d\Omega d\omega} = 2c\varepsilon_0 |R\vec{E}(\omega)|^2$$

$$(10.50)$$

For a *planar, linearly polarized* ID, we draw the following observations.

- If $K \ll 1$ (undulator), the particle's longitudinal velocity in Eq. 10.33 is approximately constant, $v_z \cong \langle v_z \rangle = const$. The trajectory is a pure sinusoidal oscillation, whose FT is a monochromatic emission. Since in the particle's reference frame the motion is a purely transverse harmonic oscillation, the angular distribution of radiation is that one of an electric dipole (see Fig. 10.7-top row). In the laboratory reference frame, radiation is boosted to a forward, collimated, centered emission (see also Fig. 7.2).

- If $K \gg 1$ (wiggler), Eq. 10.33 describes the superposition of a pure sinusoidal oscillation and a *higher* frequency perturbation to the particle's velocity, of the order of cK^2/γ^2 (see Fig. 10.7-bottom row) . The FT of such trajectory has higher harmonic contents, namely, sub-harmonics of the fundamental emission are suppressed. The particle's motion in the reference frame moving at $\langle v_z \rangle$ with respect to the laboratory has a net longitudinal component: the orbit is a "figure-8", which can be described as the superposition of a transverse and a longitudinal electric dipole. In the laboratory reference frame, they generate respectively forward off-axis and on-axis emission, at higher harmonics of the fundamental (see also Fig. 7.2). But, since the motion is anti-symmetric in proximity of the ID axis and symmetric off-axis, only odd harmonics are permitted on-axis, and only even harmonics off-axis.

- Any intermediate value of K determines a situation intermediate to the two extreme cases above. For example, the emission from a linearly polarized undulator with $K \geq 1$ shows odd harmonics on-axis and even harmonics off-axis. The higher the K is, the larger the contribution of higher harmonics to the whole radiated energy will be. At large K-values, the distinction between undulator and wiggler becomes more vague.

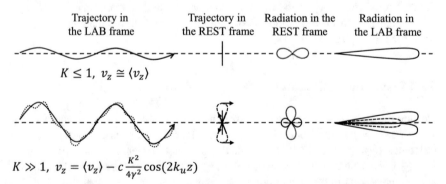

Fig. 10.7 Top view of particle's trajectory through an undulator (top) and a wiggler magnet (bottom), and angular distribution of radiation, in the particle's rest frame and in the laboratory frame, where the particle moves at ultra-relativistic velocity. The geometry of the particle's trajectory determines the emission of higher odd and even harmonics of the fundamental

Another way of looking to the spectral properties of emission from IDs is by considering the duration of the light flash seen by the observer. For an ideal undulator, the observer—assumed on-axis—is illuminated by radiation emitted along the device with no interruption. Since the spectral bandwidth is inversely proportional to the time duration of the pulse, it will be narrow in proportion of the number of undulator periods (see Eq. 10.38).

In a wiggler, the observer will receive most of the light when the particles travel parallel to the axis of the device, i.e., along portions of the trajectory far off-axis. This way, the radiation pulse at the exit of a wiggler can be seen as the incoherent superposition of light flashes transversely separated but periodically spaced. In the frequency domain, this is equivalent to a rich harmonic content, in proportion to the deflection parameter. At very high frequencies, the natural enlargement of each spectral line, as already discussed for a dipole magnet, smears the spectrum, which in fact results a *continuum* (see Fig. 10.6-right plot). The intensity is amplified with respect to a dipole magnet of same magnetic field, by just the number of wiggler periods.

10.3.4.1 Discussion: Transverse Coherence of Light Sources

We want to show that the intrinsic degree of transverse coherence of undulator radiation is larger than that of dipole radiation at the same wavelength. What is the size of a pinhole at distance $L = 15$ m from the ID, to make the undulator radiation fully transversely coherent? What electron beam emittance would be required to ensure fully transversely coherent radiation?

Let us consider an electron beam energy $E = 2.5$ GeV, and transverse beam size at the dipole magnet and at the ID $\sigma_{x,y} \approx 30\,\mu$m. The undulator fundamental emission is at $\lambda = 0.5$ nm, from $N_u = 100$ periods, each period long $\lambda_u = 2$ cm. For simplicity, synchrotron radiation from the dipole magnet is assumed to be at the critical frequency.

The relative degree of transverse coherence is evaluated in terms of the ratio of the typical angular divergence of radiation and the coherence angle $\theta_c \approx \frac{\lambda}{4\pi\sigma_{x,y}} = 1.3\,\mu$ rad (see Eq. 10.20). The ratio for the dipole and the undulator amounts to, respectively, $(\theta_c\gamma)^{-1} \approx 150$ and $\left(\theta_c\gamma\sqrt{N_u}\right)^{-1} \approx 15$ (see Eq. 10.41). It results that both sources are far from transverse coherence at 0.5 nm, but undulator radiation is $\sim$10-fold closer to that than synchrotron radiation.

The pinhole radius needed to push the undulator radiation to the diffraction limit, i.e., to provide full transverse coherence, is estimated through Eq. 10.21, giving $R \leq \frac{\lambda L}{4\pi\sigma_{x,y}} = 20\,\mu$m. Worth to notice, the intrinsic angular divergence of undulator radiation is approximately 6-fold larger than the coherence angle (see Eq. 10.4).

Transversely coherent radiation could be naturally produced if the electron beam transverse emittances satisfy the diffraction limit condition at 0.5 nm, i.e., $\varepsilon_{x,y} < \frac{\lambda}{4\pi} = 40$ pm rad, or $\varepsilon_n < \frac{\gamma\lambda}{4\pi} = 0.2\,\mu$m rad.

10.3.4.2 Discussion: Longitudinal Coherence in Storage Rings

What is the wavelength range in which an electron bunch in a storage ring could radiate in regime of intensity enhancement? Does intensity enhancement imply longitudinal coherence, or viceversa, in an undulator and in a dipole magnet, assuming no monochromatization of radiation?

For a quantitative answer, typical parameters at medium energy storage ring light sources are considered, such as 3 GeV-energy, 1.6 nC-bunch charge, and rms bunch length of 9 mm. For simplicity, we ignore the transverse beam dimensions ("pencil beam" approximation). Let us assume a dipole curvature radius $R = 10$ m, number of undulator periods $N_u = 100$, and undulator emission optimized at $\lambda = 0.1$ nm in $n = 5$th harmonic of the fundamental.

According to Eq. 10.31, intensity enhancement is present for $(N_b - 1)|f_\parallel(\omega)|^2 > 1$. For a Gaussian beam $|f_\parallel(\omega)|^2 = e^{-\omega^2 \sigma_t^2}$, hence:

$$N_b e^{-\omega^2 \sigma_t^2/2}$$
$$N_b > e^{\frac{(2\pi c\sigma_t)^2}{2\lambda^2}} \tag{10.51}$$
$$\sigma_z < \tfrac{\lambda}{2\pi}\sqrt{\ln N_b}$$

Coherent emission in the sense of Eq. 10.30 starts appearing at $\lambda > 1$ cm (microwaves). As a matter of fact, the formation of density modulations internal to the bunch due to the so-called "microwave instability" can amplify the emission at wavelengths shorter than σ_z, typically down to the IR-THz region.

A large degree of longitudinal coherence can be obtained if the bunch length is shorter than the longitudinal coherence length. In the limit of vanishing bunch length, a single mode would be emitted, i.e., a fully coherent pulse. By recalling Eq. 10.24 and the definition of critical frequency of synchrotron radiation in Eq. 7.24, we impose:

$$2\pi\sigma_z \le \tfrac{\lambda^2}{2\Delta\lambda} = \tfrac{c}{2\Delta\nu} \approx \tfrac{c}{4\nu_c} = \tfrac{c}{4}\tfrac{1}{\frac{3}{4\pi}\frac{c\gamma^3}{R}} \approx \tfrac{R}{\gamma^3} = 0.05 \text{ nm} \tag{10.52}$$

A similar consideration for the undulator leads to:

$$2\pi\sigma_z \le \tfrac{\lambda^2}{2\Delta\lambda} \approx n N_u \tfrac{\lambda}{2} = 25 \text{ nm} \tag{10.53}$$

It emerges that both synchrotron radiation and undulator spontaneous radiation are longitudinally incoherent radiation for realistic bunch durations. Since the longitudinal coherence of dipole radiation is constrained by the critical frequency (Eq. 10.52), there is no practical combination of parameters which guarantee some intrinsic degree of longitudinal coherence. Instead, an undulator can be configured and tuned to a central wavelength approaching the bunch length (Eq. 10.53). For example, partial longitudinal coherence could be generated by a far-IR undulator traversed by a ~mm long beam.

By comparing Eqs. 10.52, 10.53 with Eq. 10.51 we deduce that, for typical beam and ID parameters at storage ring light sources:

- for synchrotron radiation in proximity or *above* the critical frequency and for undulator spontaneous radiation, the condition of intensity enhancement automatically implies some degree of longitudinal coherence;
- for synchrotron radiation *below* the critical frequency, longitudinal coherence implies intensity enhancement; this is not necessarily true for undulator radiation.

10.3.4.3 Discussion: Longitudinal Coherence of Harmonic Emission

Let us demonstrate that the longitudinal coherence length of undulator fundamental and harmonic emission is the same, and that it can be lengthened by tuning the undulator to a sub-harmonic of the fundamental.

$L_{c,\parallel}$ is calculated from Eq. 10.24 for both the fundamental emission at λ and harmonic emission at $\lambda_n = \lambda/n$. The intrinsic relative bandwidth of undulator radiation satisfies $\frac{\Delta\lambda_n}{\lambda_n} = \frac{1}{n}\frac{\Delta\lambda}{\lambda}$ (see Eq. 10.38). Hence:

$$L_{c,\parallel}(\lambda_n) = \frac{\lambda_n^2}{2\Delta\lambda_n} = \frac{n\lambda}{2\Delta\lambda}\lambda_n = \frac{\lambda^2}{2\Delta\lambda} = L_{c,\parallel}(\lambda)$$

$$\Rightarrow L_{c,\parallel}(\lambda_n) = \frac{N_u\lambda}{2}$$

(10.54)

We suppose that the undulator can be tuned to a sub-harmonic $\lambda' = n\lambda$ of the original fundamental, which is now generated as a higher harmonic of order n. In this case, the radiation slippage along the undulator changes because the fundamental wavelength has changed, and it amounts to $\Delta t = N_u\lambda' = nN_u\lambda$. Equation 10.38 prescribes an n-fold smaller energy bandwidth, $\Delta\nu_n = \frac{c}{nN_u\lambda}$. The relative bandwidth at the wavelength $\lambda_n = \lambda$ is:

$$\frac{\Delta\lambda}{\lambda} \equiv \frac{\Delta\lambda_n}{\lambda_n} = \frac{\Delta\nu_n}{\nu_n} = \frac{c}{nN_u\lambda}\frac{\lambda}{c} = \frac{1}{nN_u}$$

$$\Rightarrow L_{c,\parallel}(\lambda) = \frac{\lambda^2}{2\Delta\lambda} = n\frac{N_u\lambda}{2}$$

(10.55)

As demonstrated in Eq. 10.54, it still holds $L_{c,\parallel}(\lambda') = L_{c,\parallel}(\lambda)$. However, the absolute value of the longitudinal coherence length evaluated at a *given* wavelength is n-fold longer when radiation is generated as a higher harmonic (compare Eqs. 10.54 and 10.55).

10.4 Inverse Compton Scattering

Electron linear accelerators find application in so-called Inverse Compton Scattering light sources (ICS). Accelerated electrons interact in (quasi) head-on collision with an external IR or UV laser. Back-scattered photons are boosted in frequency by virtue of the relativistic Doppler effect, which applies first to the incoming radiation in the electrons' rest frame, then to the scattered wavelength in the laboratory frame. Owing to the similarity of this interaction with that one of electrons and undulator field in the electron's rest frame, an expression for the central wavelength in ICS similar to that of spontaneous undulator emission is obtained.

10.4.1 Thomson Back-Scattering

We assume a laser of wavelength λ_L is impinging on a relativistic electron beam at a positive angle $\theta_i \ll 1$ in the laboratory frame. We apply Eq. 1.47 to calculate the laser wavelength in the electron's rest frame (where the laser source is moving), so that accented quantities in Eq. 1.47 refer here to the laboratory frame (where the laser source is at rest):

$$\lambda'_L = \frac{\lambda_L}{\gamma(1+\beta_z \cos\theta_i)} \tag{10.56}$$

Let us assume an elastic scattering on an ultra-relativistic electron ($\gamma \gg 1$) with no electron recoil, or *Thomson scattering* [7]. In other words, radiation instantaneously scattered at a small negative angle θ_s in the laboratory frame, is seen by the electron with a wavelength $\lambda'_s \approx \lambda'_L$ (the photon energy does not change). In the laboratory frame, the source of scattered radiation is moving at the electron's velocity and we therefore apply Eq. 1.48 to retrieve the wavelength of the wavefront propagating at the angle θ_s (accented quantities in the equation refer here to the electron's rest frame):

$$\lambda_s \approx \lambda'_L \gamma (1 - \beta_z \cos\theta_s) \approx \lambda_L \frac{1-\beta\cos\theta_s}{1+\beta\cos\theta_i} \approx \lambda_L \frac{1-\sqrt{1-\frac{1}{\gamma^2}\left(1-\frac{\theta_s^2}{2}\right)}}{1+\beta} \approx$$

$$\approx \frac{\lambda_L}{2}\left[1-\left(1-\frac{1}{2\gamma^2}\right)\left(1-\frac{\theta_s^2}{2}\right)\right] \approx \frac{\lambda_L}{4\gamma^2}\left(1+\gamma^2\theta_s^2\right) + o\left(\frac{\theta_s^2}{\gamma^2}\right) \tag{10.57}$$

The no recoil approximation implies that in the frame where the electron is initially at rest, the electron energy—i.e., its rest energy—is not changed. That is, the electron's rest energy is much larger than the photon energy. The invariant mass of the electron-photon system is approximately the electron's rest energy. The initial 4-momentum in the laboratory frame is (see Eq. 1.36) $p_t^\mu = \left(\frac{E_e+E_L}{c}, \vec{p}_e + \vec{p}_L\right)$. The invariant mass in the laboratory frame is:

$$\sqrt{p_t^\mu p_{t,\mu}c^2} = \sqrt{E_e^2 + E_L^2 + 2E_e E_L - p_e^2 c^2 - p_L^2 c^2 - 2\vec{p}_e \vec{p}_L c^2} =$$

$$= \sqrt{m_e^2 c^4 + 2E_e E_L + 2\frac{E_e}{\beta_z}E_L \cos\theta_i} \approx \sqrt{m_e^2 c^4 + 4E_e E_L} = m_e c^2 \sqrt{1 + \frac{4E_e E_L}{m_e^2 c^4}} =$$

$$= m_e c^2 \sqrt{1 + \frac{4\gamma^2 E_L}{E_e}} = m_e c^2 \sqrt{1 + \frac{\hat{E}_s}{E_e}} \rightarrow m_e c^2 \iff E_e \gg \hat{E}_s \tag{10.58}$$

A "recoil parameter" $X := \hat{E}_s/E_e \ll 1$ defines the transition from the exact "Compton regime" to the approximated "Thomson regime" (no recoil).

10.4.2 Angular and Spectral Distribution

Because of the analogy of ICS with the generation of undulator radiation, Eq. 10.57 has the same functional dependence of Eq. 10.35. But, by virtue of a laser wavelength much shorter ($\leq 1\mu$m) than any magnetic undulator period (≥ 1 cm), ICS easily boosts the on-axis scattered photon energy to hard x-rays and up to γ-rays ($\sim$MeV photon energy range), for only few 100s MeV electron energy. In spite of this, the Thomson regime is still a good approximation of the process in many practical cases ($X \leq 1\%$).

The total Thomson cross section is pretty small, $\sigma_T \simeq \frac{8\pi}{3}r_e^2 = 6.7 \cdot 10^{-25}\text{cm}^2 \approx$ 1 barn. Hence, high charge and large laser photon densities are required at the interaction point (IP) to produce substantial scattered radiation intensity. They are usually accompanied by small transverse sizes of the two beams at the IP, and by small incident scattering angle.

Since the emission is incoherent and no interference process intervenes, the characteristic angular divergence of the scattered radiation is $1/\gamma$ and, so as for synchrotron radiation, $\frac{(d\sigma/d\Omega)_{\theta_s=1/\gamma}}{(d\sigma/d\Omega)_{\theta_s=0}} = \frac{1}{8}$ (see Eq. 7.22). But, for ICS it holds:

$$\sigma_{T,cone} = \int_0^{2\pi} d\phi \int_0^{1/\gamma} \frac{d\sigma}{d\Omega} \sin\theta d\theta \simeq \frac{4\pi}{3}r_e^2 = \frac{1}{2}\sigma_T \qquad (10.59)$$

Namely, about half of the scattered photons are in the central cone of emission. By virtue of Eq. 10.57, these are also the most energetic photons.

In summary, photons are scattered at any angle and the spectral bandwidth is large. However, both intensity and photon energy distribution are peaked on-axis (so-called "Compton edge"). Here, the bandwidth is typically of the order of $\sim 1 - 10\%$, dominated by the angular acceptance $\sim \gamma\theta_s$ of the system, the electron beam relative energy spread, and the laser relative spectral bandwidth.

The correlation between photon energy and angle of emission is described by Eq. 10.57, which can be re-written as:

$$E_s = E_L \frac{1+\beta\cos\theta_i}{1-\beta\cos\theta_s} \simeq 2\gamma^2 E_L \frac{1+\cos\theta_i}{1+\gamma^2\theta_s^2} \leq 4\gamma^2 E_L \qquad (10.60)$$

For head-on collision, the correlation is illustrated in Fig. 10.8. We identify two relevant points (for $\beta \to 1$):

$$\begin{cases} E_{s,\theta_s=0} = 4\gamma^2 E_L \equiv \hat{E}_s \\[2ex] E_{s,|\theta_s|\approx 1/\gamma} = 2\gamma^2 E_L = \frac{1}{2}\hat{E}_s \end{cases} \qquad (10.61)$$

This allows a spectral selection of ICS radiation by means of a collimation system. Photons which are not scattered and propagate straight have energy $E_{s,\theta_s=\pi} = E_L$.

Fig. 10.8 Spectral distribution of Compton-scattered photons. The photon energy is correlated to the scattering angle in the laboratory frame, where $\theta = 0$ corresponds to back-scattering (or "inverse scattering", i.e., opposite to the direction of propagation of the laser)

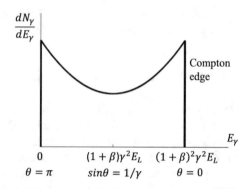

10.4.3 Compton Back-Scattering

The relevance of electron's recoil can be appreciated in the electron's reference frame, i.e., in the reference frame where the electron is *initially* at rest [7]. We assume scattering happening in the x-y plane. We use the accent for quantities after scattering.

A photon of initial energy $h\nu$ is scattered to an angle θ with respect to the original direction of motion; the final energy is $h\nu'$. The electron recoils with a velocity u at an angle ϕ, see Fig. 10.9. The conservation laws of energy and momentum state:

$$
\begin{cases}
h\nu + m_e c^2 = h\nu' + \gamma m_e c^2 \\
p_{x,p} = p'_{x,p} + p'_{x,e} \\
0 = p'_{y,p} + p'_{y,e}
\end{cases}
\quad
\begin{cases}
\gamma m_e c = \frac{h}{c}\left(\nu - \nu'\right) + m_e c \\
\frac{h\nu}{c} = \frac{h\nu'}{c}\cos\theta + \gamma m_e u \cos\phi \\
0 = \frac{h\nu'}{c}\sin\theta + \gamma m_e u \sin\phi
\end{cases}
\quad (10.62)
$$

By squaring and then adding the second and the third equation to remove ϕ, we find:

$$
\begin{cases}
\gamma^2 m_e^2 c^2 = \left[\frac{h}{c}\left(\nu - \nu'\right) + m_e c\right]^2 \\
\gamma^2 m_e^2 u^2 = \left(\frac{h}{c}\right)^2 \left(\nu^2 - 2\nu\nu'\cos\theta + \nu'^2\right)
\end{cases}
\quad (10.63)
$$

To further remove u and therefore $\gamma(u)$, the second equation is divided by the first one to get:

$$
\beta^2 = \frac{u^2}{c^2} = \frac{\left(\frac{h}{c}\right)^2\left(\nu^2 - 2\nu\nu'\cos\theta + \nu'^2\right)}{\left[\frac{h}{c}(\nu - \nu') + m_e c\right]^2},
$$

$$
(10.64)
$$

$$
\gamma^2 = \frac{1}{1 - \beta^2} = \frac{\left[\frac{h}{c}(\nu - \nu') + m_e c\right]^2}{\left[\frac{h}{c}(\nu - \nu') + m_e c\right]^2 - \left(\frac{h}{c}\right)^2\left(\nu^2 - 2\nu\nu'\cos\theta + \nu'^2\right)}
$$

Fig. 10.9 Compton scattering in the frame where the electron is initially at rest

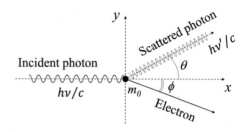

Finally, γ^2 from Eq. 10.64 is substituted back into the first line of Eq. 10.63:

$$\frac{m_e^2 c^2}{\left[\frac{h}{c}(\nu-\nu')+m_e c\right]^2 - \left(\frac{h}{c}\right)^2 (\nu^2 - 2\nu\nu'\cos\theta + \nu'^2)} = 1;$$

$$m_e^2 c^2 = \left(\frac{h}{c}\right)^2 (\nu-\nu')^2 + m_e^2 c^2 + 2m_e h(\nu-\nu') - \left(\frac{h\nu}{c}\right)^2 - \left(\frac{h\nu'}{c}\right)^2 + 2\left(\frac{h}{c}\right)^2 \nu\nu'\cos\theta;$$

$$m_e h\nu - m_e h\nu' - \frac{h}{c}\nu\nu'(1-\cos\theta) = 0;$$

$$\nu' = \nu\frac{m_e c}{m_e c + \frac{h\nu}{c}(1-\cos\theta)}.$$

$$\Rightarrow \lambda' = \lambda + \frac{h}{m_e c}(1-\cos\theta). \tag{10.65}$$

In conclusion, the scattered wavelength in the electron's rest frame is shifted by the Compton wavelength $\lambda_C = \frac{h}{m_e c}$ (see also Eq. 4.157). As expected, the wavelength of non-interacting photons ($\theta = 0$) is not changed, while the maximum shift is by $2\lambda_C$ for back-scattered light.

10.4.3.1 Discussion: Gamma-Gamma Collider

Gamma-gamma colliders have been proposed as Compton sources driven by multi-GeV-energy electron linacs, for high intensity-high energy photonic interactions for particle physics experiments. We demonstrate below that when the recoil parameter cannot be neglected ($X \geq 1$), the maximum energy of back-scattered photons in the laboratory frame approaches (from below) the ultra-relativistic electron energy.

At first, the back-scattered photon energy is calculated in the electrons' rest frame, as given by Eq. 10.65:

$$\lambda_f^e = \lambda_i^e + 2\lambda_C;$$

$$\frac{c}{\lambda_f^e} = \frac{c}{\lambda_i^e}\frac{\lambda_i^e}{\lambda_i^e + 2\lambda_C} = \frac{c}{\lambda_i^e}\frac{1}{1+\frac{2\lambda_C}{\lambda_i^e}}; \tag{10.66}$$

$$\nu_f^e = \nu_i^e\frac{1}{1+\frac{2\nu_i^e}{\nu_C}}$$

The relativistic Doppler effect in Eq. 1.48 is applied to Eq. 10.66 to express the photon frequencies in the laboratory frame, assuming head-on collision ($\theta_i = \pi$) and back-scattering ($\theta_s = 0$):

$$v_i^{lab} = v_i^e \frac{1}{\gamma(1+\beta)}, \qquad v_f^{lab} = v_f^e \frac{1}{\gamma(1-\beta)}$$

$$\Rightarrow v_f^{lab}\gamma(1-\beta) = v_i^{lab}\gamma(1+\beta)\frac{1}{1+\frac{v_i^{lab}}{v_C}2\gamma(1+\beta)} \tag{10.67}$$

We define $E_e = \gamma m_e c^2$ the electron's energy before scattering, $\hat{E}_{S,C} = h v_f^{lab}$ the maximum Compton back-scattered photon energy, $\hat{E}_L = h v_i^{lab}$ the incident laser photon energy, and $\hat{E}_{S,T} = h v_i^{lab} 2\gamma^2(1+\beta) \approx 4\gamma^2 E_L$ the maximum back-scattered photon energy in the Thomson regime (small recoil approximation, see Eq. 10.60). Then, Eq. 10.67 becomes:

$$\hat{E}_{S,C} = \hat{E}_L \frac{1+\beta}{1-\beta}\frac{1}{1+\frac{\hat{E}_L}{\gamma m_e c^2}2\gamma(1+\beta)} \approx \frac{4\gamma^2\hat{E}_L}{1+\frac{4\gamma^2\hat{E}_L}{E_e}} = \frac{\hat{E}_{S,T}}{1+\frac{\hat{E}_{S,T}}{E_e}} = \frac{X}{X+1}E_e \leq E_e \tag{10.68}$$

10.5   Free-Electron Laser

Free-electron lasers (FELs) are advanced light sources ranging from IR–THz to hard x-rays. They are characterized by high peak intensity, large degree of transverse and longitudinal coherence, accompanied by sub-picosecond to attosecond pulse duration. The low gain regime of FEL emission, also named "small signal", is introduced first. Although not explicitly discussed, it applies, for example, to optical-cavity FELs driven by electron storage rings. The formalism is then extended to the high gain regime, commonly driven by high brightness electron linacs. The FEL theory is limited to the longitudinal beam dynamics, i.e., for vanishing transverse emittances. 3-D effects are finally recalled for completeness.

10.5.1 Resonance Condition

The FEL process relies on the energy exchange between electrons and undulator radiation emitted by the electrons themselves. The exchange happens by virtue of the collinearity of the electron's transverse velocity (Eq. 10.32) and the transverse electric field of the co-propagating e.m. wave. In a planar undulator for horizontally polarized light, the electrons wiggle in the horizontal plane, and the amount of radiated power is [8]:

$$\frac{dE}{dt} = -ec[JJ]\vec{\beta}\cdot\vec{E} = -ec[JJ]\beta_x E_x = -ec\frac{\hat{K}E_{x,0}}{\gamma}\cos(k_u z)\cos(kz - \omega t + \phi_0) =$$

$$= -ec\frac{\hat{K}E_{x,0}}{2\gamma}\{\cos[(k_u+k)z - \omega t + \phi_0] + \cos[(k_u-k)z + \omega t - \phi_0]\} \equiv$$

$$\equiv -ec\frac{\hat{K}E_{x,0}}{2\gamma}[\cos(\psi+\phi_0) + \cos\chi], \tag{10.69}$$

where the minus sign is conventionally adopted to indicate energy loss by the electron when the scalar product is positive, $\hat{K} = [JJ]K$ and $[JJ] = J_0(\xi) - J_1(\xi) < 1$ is the electron-radiation "coupling factor", given by difference of Bessel functions with argument $\xi = K^2/(4 + 2K^2)$. The coupling factor takes into account the oscillation of the electron's longitudinal velocity in the horizontally polarized undulator (see Eq. 10.33). It applies to the electron-radiation interaction only, not to the definition of the central wavelength of emission. In a helically polarized undulator, the particle's longitudinal velocity is constant and $[JJ] = 1$.

Equation 10.69 can be intended as the superposition of the electrons with a forward $(\cos\psi)$ and a backward $(\cos\chi)$ travelling wave. The phase of the latter one oscillates faster than the forward wave, therefore it can be neglected over many undulator periods:

$$\chi = z\left(k_u - k + \frac{\omega}{v_z}\right) = z\left(k_u - k + \frac{k}{\beta_z}\right) = z\left[k_u - k\left(1 - \frac{1}{\beta_z}\right)\right] \approx$$
$$\approx z\left[k_u - k\left(1 - 1 - \frac{1}{2\gamma^2} - \frac{K^2}{4\gamma^2}\right)\right] = 2k_u z \tag{10.70}$$

The former is a *ponderomotive phase*, i.e., the phase of a wave co-propagating with the electrons, given in turn by the superposition of the travelling wave of the undulator spontaneous radiation (k) and the stationary wave describing the undulator magnetic field (k_u).

Since both the electron's velocity and the electric field oscillate with time, the average energy exchange over many undulator periods would be null. Thus, a net energy exchange is guaranteed only by a synchronization of the two vectors, i.e., a constant ponderomotive phase:

$$\frac{d\psi}{dt} = 2\pi \frac{d}{dt}\left(\frac{z}{\lambda_u} + \frac{z}{\lambda} - \frac{ct}{\lambda}\right) = 2\pi c\left(\frac{\beta_z}{\lambda_u} + \frac{\beta_z - 1}{\lambda}\right) \equiv 0;$$
$$\Rightarrow \lambda = \frac{\lambda_u}{\beta_z}(1 - \beta_z) \approx \lambda_u \left(\frac{1}{\sqrt{1 - \frac{1}{\gamma^2} - \frac{K^2}{2\gamma^2}}} - 1\right) \approx \frac{\lambda_u}{2\gamma^2}\left(1 + \frac{K^2}{2}\right) \tag{10.71}$$

We conclude that only radiation at the wavelength of the on-axis undulator spontaneous emission sustains the interaction with the electrons (see Eq. 10.35). In particular, Eq. 10.71 identifies the particle's "resonant energy", $\gamma \equiv \gamma_r$, which guarantees emission at a certain λ for any given undulator period and undulator field.

The synchronization implied by Eq. 10.71 can be re-written as:

$$k_u + k = \frac{\omega}{v_z} \Rightarrow v_z = \frac{\omega}{k_u + k} \equiv v_p \tag{10.72}$$

That is, the particle's longitudinal velocity is equal to the phase velocity of the ponderomotive wave. In general, the synchronization or the resonant emission implies that at the end of the undulator, some electrons will have been subject to net energy loss (emission of radiation), others to net energy gain (absorption of radiation). This, however, does not imply any amplification of the output radiation yet, which indeed needs the electron dynamics to be taken into account, as explained below.

10.5.2 Pendulum Equation

The dynamics of the j-th electron is described in the longitudinal phase space through the ponderomotive phase ψ_j and the relative energy deviation with respect to the resonant energy, $\eta_j = \frac{\gamma_j - \gamma_r}{\gamma_r}$. The amplitude of the radiated electric field is assumed to be approximately constant at this stage (regime of "small signal") [4].

The time-derivative of the phase is calculated by considering the practical case $\eta_j \ll 1$:

$$
\begin{aligned}
\frac{d\psi_j}{dt} &= c\beta_{z,j}(k_u + k) - k \approx ck_u + ck\left[1 - \frac{1}{2\gamma_j^2}\left(1 + \frac{K^2}{2}\right) - 1\right] = \\
&= ck_u - \frac{ck}{2\gamma_r^2}\left(1 + \frac{K^2}{2}\right)\frac{\gamma_r^2}{\gamma_j^2} = ck_u - \frac{ck}{2\gamma_r^2}\left(1 + \frac{K^2}{2}\right)\frac{1}{(1+\eta_j)^2} = \\
&\approx ck_u - \frac{ck}{2\gamma_r^2}\left(1 + \frac{K^2}{2}\right)(1 - 2\eta_j) = ck_u - ck_u + 2ck_u\eta_j = \\
&= 2ck_u\eta_j
\end{aligned}
\tag{10.73}
$$

The time-derivative of the relative energy deviation is calculated by using Eq. 10.69:

$$
\frac{d\eta_j}{dt} = \frac{d}{dt}\frac{\gamma_j}{\gamma_r} = \frac{1}{\gamma_r m_e c^2}\frac{dE_j}{dt} \approx -\left(\frac{eE_{x,0}\hat{K}}{2\gamma_r^2 m_e c}\right)\cos\psi_j = -a\sin\phi_j
\tag{10.74}
$$

where the amplitude a and the phase $\phi_j = \psi_j + \pi/2$ were defined. By replacing Eq. 10.74 into the time-derivative of Eq. 10.73, one obtains the "pendulum equation":

$$
\begin{cases}
\frac{d\psi_j}{dt} = 2ck_u\eta_j \\
\frac{d\eta_j}{dt} = -a\sin\phi_j
\end{cases}
\Rightarrow \frac{d^2\phi_j}{dt^2} = 2ck_u\frac{d\eta_j}{dt} = -\left(\frac{eE_{x,0}k_u\hat{K}}{\gamma_r^2 m_e}\right)\sin\phi_j \equiv -\Omega^2\sin\phi_j
\tag{10.75}
$$

In the approximation of small signal, the system behaves as a conservative one, and the Hamiltonian is a constant of motion. Hamilton's equations $\frac{d\eta}{dt} = -\frac{\partial H}{\partial\phi}, \frac{d\phi}{dt} = \frac{\partial H}{\partial\eta}$ are satisfied by the following Hamiltonian:

$$
H(\phi, \eta) = ck_u\eta^2 + \frac{\Omega^2}{2ck_u}(1 - \cos\phi)
\tag{10.76}
$$

The Hamiltonian along the separatrix and the separatrix equation are, respectively:

$$
H_{sep} = H(\pm\pi, 0) = \frac{\Omega^2}{ck_u}
$$

$$
H(\phi, \eta) \equiv H_{sep} \Rightarrow \eta_{sep} = \pm\frac{\Omega}{ck_u}\cos\left(\frac{\phi}{2}\right) \propto \sqrt{E_{x,0}}
\tag{10.77}
$$

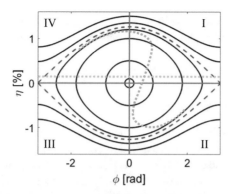

Fig. 10.10 Electron phase space trajectories (black solid) according to Eq. 10.76 in small signal regime, in the range $\phi \in [-\pi, \pi]$. The separatrix is in dashed red. An initial electron distribution slightly above the resonant energy (dotted light green) evolves towards positive phases (dotted dark green). The opposite happens for electrons initially below resonance (not shown)

Figure 10.10 illustrates the phase space trajectories $\eta(\phi)$ retrieved from Eq. 10.76, where the Hamiltonian is evaluated for different initial conditions ($\phi_0 = 0$, $\eta_0 \neq 0$). Particles in quadrant I have $\eta > 0$ and therefore they move forward in ϕ. Since their phase is $0 < \phi < \pi$, it turns out $\frac{d\eta}{dt} < 0$, i.e., particles are losing energy (see Eq. 10.75). For energy conservation, their energy is transferred to the radiated field. Similarly, particles in quadrant II are moving back in phase but they still yeld energy to the field. Particles in quadrants III and IV are absorbing energy from the field. The area delimited by the separatrix is called "FEL bucket".

10.5.2.1 Discussion: Undulator Momentum Compaction
What is the linear momentum compaction, or R_{56} transport matrix term, of an undulator made of N_u periods and resonant at the wavelength λ?

According to the definition of momentum compaction in Eq. 4.15, the longitudinal relative shift of off-momentum particles in a bunch is $\Delta z = R_{56}\delta$. The shift can be calculated explicitly from Eq. 10.73, in the ultra-relativistic limit and under on-resonance condition:

$$\frac{d\psi}{dt} = 2ck_u\eta \approx 2ck_u\delta;$$

$$\int d\psi = \int_0^t \omega dt' \approx -\int_0^z \frac{\omega}{c}dz' \approx 2k_u\delta \int_0^{L_u} ds;$$

$$\Delta z = -2\frac{k_u}{k}N_u\lambda_u\delta = -2N_u\lambda\delta$$

$$\Rightarrow R_{56} = -2N_u\lambda = -\frac{L_u}{\gamma^2}\left(1 + \frac{K^2}{2}\right)$$

(10.78)

This is basically the R_{56} of a drift section long $L_u = N_u\lambda_u$ (see Eq. 4.21), but modified by the wiggling trajectory of the particles in the presence of the undulator magnetic field.

10.5.3 Low Gain

The amplification of FEL radiation is calculated in terms of energy lost by the electrons. The small-signal approximation [6,9] implies a small variation of the radiated field amplitude during the interaction with the electrons. In other words, the electrons' synchrotron oscillation period ($\sim 1/\Omega$) is assumed to be much longer than the time taken to travel the undulator length $L_u = N_u\lambda_u$. This allows us to describe the variation with time of the generic particle's coordinates in the framework of a 2nd order perturbation theory, where the perturbation coefficient is $\varepsilon = (\Omega L_u/c)^2 \ll 1$:

$$\begin{cases} \phi(t) = \phi_0(t) + \varepsilon\frac{d\phi_0}{dt}dt + \varepsilon^2\frac{d^2\phi_0}{dt^2}dt^2 + o(\varepsilon^3) \equiv \phi_0(t) + \varepsilon\phi_1(t) + \varepsilon^2\phi_2(t) \\ \\ \eta(t) = \eta_0 + \varepsilon\eta_1(t) + \varepsilon^2\frac{d\eta_1}{dt}dt + o(\varepsilon^3) \equiv \eta_0 + \varepsilon\eta_1(t) + \varepsilon^2\eta_2(t) \end{cases}$$

(10.79)

The 0th order solution assumes constant particle energy, therefore from Eq. 10.75:

$$\eta(t) \approx \eta_0 = const. \Rightarrow \phi_0(t) = \int_0^t \frac{d\phi_0}{dt'}dt' = 2ck_u\eta_0 t + \theta_0 \equiv \xi + \theta_0 \qquad (10.80)$$

The 1st order solution is:

$$\begin{cases} \dot{\phi}_1 = 2ck_u\eta_1 \\ \\ \dot{\eta}_1 = -\frac{\Omega^2}{2ck_u}\sin(\phi_0) = -\frac{\Omega^2}{2ck_u}\sin(\xi + \theta_0) \end{cases}$$

(10.81)

$$\Rightarrow \begin{cases} \eta_1(t) = \int_0^t \frac{d\eta_1}{dt'}dt' = \frac{\Omega^2}{(2ck_u)^2\eta_0}[\cos(\xi + \theta_0) - \cos(\theta_0)] \\ \\ \phi_1(t) = \int_0^t \frac{d\phi_1}{dt'}dt' = \int_0^t 2ck_u\eta_1(t')dt' = \frac{\Omega^2}{2ck_u\eta_0}\left[\frac{\sin(\xi+\theta_0)-\sin(\theta_0)}{2ck_u\eta_0} - t\cos(\theta_0)\right] \end{cases}$$

(10.82)

The 2nd order solution is:

$$\dot{\eta}_2 = \frac{d}{dt}\left(\frac{d\eta_1}{dt}dt\right) = \frac{d\dot{\eta}_1}{dt}dt = -\eta_0\Omega^2\cos(\xi+\theta_0)dt =$$

$$= -\left(\frac{1}{2ck_u}\frac{d\phi_0}{dt}dt\right)\Omega^2\cos(\xi+\theta_0) = -\frac{\Omega^2}{2ck_u}\phi_1\cos(\xi+\theta_0);$$

$$\Rightarrow \eta_2(t) = -\frac{\Omega^2}{2ck_u}\int_0^t dt'\phi_1(t')\cos(\xi(t')+\theta_0) =$$

$$= -\frac{\Omega^4}{(2ck_u)^3\eta_0^2}\int dt'\left[\sin(\xi+\theta_0)-\sin(\theta_0)\right]\cos(\xi+\theta_0)$$

$$+\frac{\Omega^4}{(2ck_u)^2\eta_0}\int dt't'\cos(\theta_0)\cos(\xi+\theta_0) =$$

$$= -\frac{\Omega^4}{(2ck_u)^3\eta_0^2}\left\{\frac{1}{4ck_u\eta_0}\left[\sin(\xi)\sin(\xi+2\theta_0)\right]-\frac{\sin(\theta_0)}{2ck_u\eta_0}\left[\sin(\xi+\theta_0)-\sin(\theta_0)\right]\right\}+$$

$$-\frac{\Omega^4}{\eta_0}\frac{\cos(\theta_0)}{(2ck_u)^4\eta_0^2}\left[\xi\sin(\xi+\theta_0)+\cos(\xi+\theta_0)-\cos(\theta_0)\right] =$$

$$= -\frac{\Omega^4}{(2ck_u)^4\eta_0^3}\left[\tfrac{1}{2}\sin(\xi)\sin(\xi+\theta_0)-\sin(\theta_0)\sin(\xi+\theta_0)+\sin^2(\theta_0)+\right.$$

$$\left.-\xi\sin(\xi+\theta_0)\cos(\theta_0)-\cos(\theta_0)\cos(\xi+\theta_0)+\cos^2(\theta_0)\right]$$

$$(10.83)$$

Finally, we calculate the relative energy variation averaged over all electrons' initial phase θ_0:

$$\langle\eta\rangle_{\theta_0} = \frac{1}{2\pi}\int_0^{2\pi}\eta(t;\theta_0)d\theta_0 = \eta_0 + 0 + \langle\eta_2\rangle_{\theta_0} =$$

$$= \eta_0 + \frac{\Omega^4}{(2ck_u)^4\eta_0^3}\left[\tfrac{\xi}{2}\sin(\xi)+\cos(\xi)-1\right] \qquad (10.84)$$

The *small-signal gain* is the amount of energy transferred on average from all electrons to the radiated field, normalized to the resonance energy:

$$G_{ss}(t) = \eta_0 - \langle\eta\rangle_{\theta_0} = -\frac{\Omega^4}{(2ck_u)^4\eta_0^3}\left[\tfrac{\xi}{2}\sin(\xi)+\cos(\xi)-1\right] \qquad (10.85)$$

At the end of the undulator it results $\xi/2 = ck_u\eta_0 \cdot N_u\lambda_u/c = 2\pi N_u\frac{\Delta\gamma}{\gamma_r} = \pi N_u\frac{\Delta\omega}{\omega_r} \equiv x$ (see also Eq. 10.39). The paremeter x is called *detuning parameter*, and Eq. 10.85 becomes:

$$G_{ss}(L_u) = -\frac{\Omega^4 L_u^3}{16ck_u}\frac{1}{(ck_u\eta_0 L_u)^3}\left[\tfrac{\xi}{2}\sin(\xi)+\cos(\xi)-1\right] \propto$$

$$(10.86)$$

$$\propto -\frac{1}{x^3}\left[x\sin(2x)+\cos(2x)-1\right] = -\frac{d}{dx}sinc^2(x)$$

Equation 10.86 is called *Madey's theorem* and it says that, by recalling Eq. 10.39, the small-signal gain is proportional to the negative derivative of the undulator spontaneous emission. Figure 10.11 shows that the gain is positive, i.e. the field is amplified, only if electrons' initial energy is slightly above the resonant energy (see also Fig. 10.10). The gain is maximum for $\xi \approx 2$.

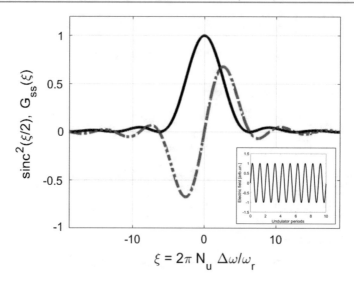

Fig. 10.11 In the inset, constant amplitude electric field along 10 undulator periods. The FT of the radiated pulse of finite duration is a $sinc^2$ function, shown in the main plot versus the detuning parameter (black solid). It represents the intensity of undulator spontaneous radiation normalized to the single particle's emission. The small-signal gain (red dashed) is proportional to its derivative

A monochromatic electron beam above resonance energy evolves as illustrated in Fig. 10.10. As the beam travels along the undulator, electrons shift towards positive phases (quadrants I and II), where they continue yielding energy to the radiated field. Electrons within the separatrix are trapped and perform closed orbits in the longitudinal phase space. As the radiated field amplitude increases, the separatrix height increases (see Eq. 10.77), and initially untrapped electrons can be trapped, so contributing to the gain. The radiative process starts being amplified with respect to the undulator spontaneous emission by the build-up of periodic λ-spacing of electron clusters into consecutive FEL buckets (intensity enhancement, see Eq. 10.31). Such a positive feedback instability justifies the naming of *stimulated emission*.

The abrupt growth of the field amplitude and the corresponding electrons' nonlinear dynamics is said "high gain" regime of the FEL. It normally requires a relatively long undulator traversed by a high brightness electron beam, as discussed in the following Section.

10.5.4 High Gain

The variation of the transverse electric field amplitude generated by the electron's transverse current density is described by the following 1-D wave equation [4,8,9]:

$$\left(\frac{\partial^2}{\partial z^2} - \frac{1}{c^2}\frac{\partial^2}{\partial t^2}\right) E_x(z,t) = \mu_0 \frac{\partial j_x}{\partial t} \tag{10.87}$$

A solution of the form $E_x(z,t) = E_x(z)e^{i(kz-\omega t)}$ is plugged into Eq. 10.87 to find:

$$\begin{cases} \frac{\partial^2}{\partial z^2} E_x(z)e^{i(kz-\omega t)} = \frac{\partial}{\partial z}\left[E_x' + ikE_x(z)\right]e^{i(kz-\omega t)} = \left(E_x'' + 2ikE_x' - E_x k^2\right)e^{i(kz-\omega t)} \\ \\ -\frac{1}{c^2}\frac{\partial^2}{\partial t^2}\left[E_x(z)e^{i(kz-\omega t)}\right] = \frac{\omega^2}{c^2}E_x(z)e^{i(kz-\omega t)} \end{cases}$$

(10.88)

Equation 10.88 can be simplified by virtue of the following assumptions.

- The field amplitude varies slowly, i.e., $E_x''(z) \ll kE_x'(z)$ or $\frac{dE_x(z)}{dz} \ll \frac{E_x(z)}{\lambda}$ ("slow variable field approximation" or SVEA).

- The radiation slippage $N_u\lambda$ is much shorter than the bunch length. Therefore, the wave is superimposed to the ultra-relativistic electrons along the whole undulator, through which the dispersion relation for the on-resonance wave is preserved, $k^2 - \frac{\omega^2}{c^2} = 0$.

- In the ultra-relativistic limit, the ratio of transverse and longitudinal current density results $j_x/j_z = v_x/v_z \cong v_x/c = \beta_x = \frac{\hat{K}}{\gamma}\cos(k_u z)$ (see Eq. 10.32, and the generalized undulator parameter for $v_z \neq const$ in Eq. 10.69). Since v_z has a DC and an AC component (see Eq. 10.33), we also have $j_z = j_0 + j_1(z)$. In particular, the electron-wave synchronization leading to the resonance condition implies that j_1 oscillates with the electron's ponderomotive phase introduced in Eq. 10.71. In conclusion, $j_x = j_z\frac{\hat{K}}{\gamma}\cos(k_u z) = \frac{\hat{K}}{\gamma}\left[j_0 + j_1(\psi)e^{i[(k+k_u)z-\omega t]}\right]e^{ik_u z}$ in Euler's notation.

With the aforementioned approximations applied to Eq. 10.88, Eq. 10.87 becomes:

$$\left[E_x''(z) + 2ikE_x'(z) + \left(\frac{\omega^2}{c^2} - k^2\right)E_x(z)\right]e^{i(kz-\omega t)} \approx 2ikE_x'(z)e^{i(kz-\omega t)} =$$

$$= \mu_0 j_1(\psi)\frac{\hat{K}}{\gamma}\frac{\partial}{\partial t}e^{i[(k+2k_u)z-\omega t]} = -i\omega\mu_0 j_1(\psi)\frac{\hat{K}}{\gamma}e^{i[(k+2k_u)z-\omega t]};$$

$$E_x'(z) = -\mu_0 j_1\frac{\omega}{k}\frac{\hat{K}}{2\gamma}e^{i[(k+2k_u)z-\omega t]}e^{-i(kz-\omega t)} = -\mu_0 c j_1\frac{\hat{K}}{2\gamma}e^{i2k_u z} \approx -\frac{\hat{K}}{4\varepsilon_0 c}\frac{\langle j_1\rangle}{\langle\gamma\rangle},$$

(10.89)

and the very last approximated equality is after averaging over many undulator periods, over all beam particles' phases.

The electron current density $\langle j_1\rangle$ in Eq. 10.89 is the bunch peak current averaged over the transverse beam sizes; it is proportional to the correlation of the phases of N electrons in a bunch:

$$\langle j_1\rangle = -\frac{I}{2\pi\sigma_\perp^2}\frac{1}{N}\sum_{n=1}^{N}e^{-i\psi_n}$$

(10.90)

and therefore proportional to the FT of the current distribution or *bunching factor* $b = |\langle e^{-i\psi}\rangle|$. This is 0 for randomly phased electrons, and 1 for phase separation between two electrons multiple of 2π.

It is convenient to re-write Eq. 10.89 in terms of an electric field amplitude normalized to the average current density. Doing so, the oscillation frequency in the FEL bucket for the pendulum equation in Eq. 10.75 results proportional to the "plasma frequency", $\Omega^2 \equiv (4ck_u)^2 A\rho^3 \propto \omega_p^2$, where we introduce:

$$
\begin{cases}
\omega_p^2 = \dfrac{n_e e^2}{m_e \varepsilon_0} = \dfrac{N_e e}{2\pi\sigma_\perp^2 c\sigma_t}\dfrac{e}{m_e \varepsilon_0} = \dfrac{I}{I_A}\dfrac{2c^2}{\sigma_\perp^2} \\[3mm]
A = \varepsilon_0 E_x(z)\dfrac{\langle\gamma\rangle}{\hat{K}}\dfrac{ck_u}{\frac{I}{2\pi\sigma_\perp^2}} \\[3mm]
\rho = \dfrac{1}{2\langle\gamma\rangle}\left(\dfrac{I}{I_A}\right)^{\frac{1}{3}}\left(\dfrac{\hat{K}}{k_u\sigma_\perp}\right)^{\frac{2}{3}} = \dfrac{1}{\langle\gamma\rangle}\left(\dfrac{\hat{K}\omega_p}{4ck_u}\right)^{\frac{2}{3}}
\end{cases}
\tag{10.91}
$$

and $I_A = 4\pi\varepsilon_0 m_e c^3/e = 17045$ A is the Alfven current.

The so-called *FEL parameter* or "Pierce's parameter" ρ is proportional, though weakly, to the electron beam brightness. For typical electron beam and undulator parameters for x-ray FELs like $\gamma \sim 1-10$ GeV, $I \sim kA$, $\sigma_\perp \sim 10-100$ µm, $\hat{K} \sim 1$, $\lambda_u \sim 1-5$ cm, we find $\rho \sim 0.01\% - 0.1\%$ (still, some FEL designs may target one order of magnitude smaller or larger ρ).

Finally, the 1-D coupled Newton-Lorentz's equations of motion for the electrons and Maxwell's equation for the electric field can be written in the following compact form:

$$
\begin{cases}
\dfrac{d\psi}{dz} = 2k_u\eta \\[3mm]
\dfrac{d\eta}{dz} = -\dfrac{\Omega^2}{2c^2 k_u}\cos\psi = -8k_u\rho^3 A\cos\psi \\[3mm]
\dfrac{dA}{dz} = k_u\langle e^{-i\psi}\rangle
\end{cases}
\tag{10.92}
$$

The normalized field amplitude is derived further to get a third order differential equation:

$$
\dfrac{d^3 A}{dz^3} = \dfrac{d}{dz}\left(-2ik_u^2\langle\eta e^{-i\psi}\rangle\right) = -2ik_u^2\left(\langle\tfrac{d\eta}{dz}e^{-i\psi}\rangle - i\langle\eta e^{-i\psi}\tfrac{d\psi}{dz}\rangle\right) =
\tag{10.93}
$$

$$
= i16k_u^3\rho^3 A\langle e^{-i\psi}\cos\psi\rangle - 4k_u^3\langle\eta^2 e^{-i\psi}\rangle
$$

For electrons' energy on resonance, the second term on the r.h.s. of Eq. 10.93 is null, $\langle\gamma\rangle = \gamma_r \Rightarrow \eta = 0$. The first term averaged over all phases reduces to $i8k_u^3\rho^3$. We then adopt the short notation $\Gamma = 2k_u\rho$, and search a solution of the form $E_x(z) = E_{x,0}e^{\alpha z}$:

$$E_x''' - i\Gamma^3 E_x = 0 \Rightarrow \alpha^3 = i\Gamma^3 \Rightarrow \alpha_j = \begin{cases} \frac{i+\sqrt{3}}{2}\Gamma \\ \frac{i-\sqrt{3}}{2}\Gamma \\ -i\Gamma \end{cases} \tag{10.94}$$

The three roots of the cubic equation describe, respectively, a growing, a decaying, and an oscillatory mode of the field amplitude. At the beginning of the undulator, the three modes compete with one another, i.e., the total electric field amplitude grows slowly with z ("lethargy"). For $z \geq 1/\Gamma$, the exponentially growing mode (α_1) dominates and the field intensity can be written as:

$$I = \frac{|E_x(z)|^2}{2Z_0} = \frac{1}{2Z_0}\left(\frac{E_{x,in}}{3}\right)^2 \cdot (e^{\alpha_1 z} + e^{\alpha_2 z} + e^{\alpha_3 z})^2 \approx \frac{1}{9}\frac{|E_{x,in}|^2}{2Z_0}e^{\sqrt{3}\Gamma z}$$

$$\tag{10.95}$$

$$\Rightarrow P(z) \approx \frac{P_0}{9}e^{\frac{z}{L_g}}, \, L_g = \frac{1}{\sqrt{3}\Gamma} = \frac{\lambda_u}{4\pi\sqrt{3}\rho}$$

Equation 10.94 points out that, unlike the small-signal regime, the high gain FEL is optimized by an ideally monochromatic electron beam at the resonant energy $\langle\gamma\rangle = \gamma_r$. It can be shown that this applies also in the presence of detuning as long as $\rho \ll 1$, which is the common regime of operation of short wavelength FELs. The characteristic length L_g is said *1-D gain length*, and the regime is said "high gain" because the radiated energy at the exit of a long undulator ($z \gg 1/\Gamma$) is orders of magnitude larger than that of the initial undulator spontaneous emission ($z \leq 1/\Gamma$).

10.5.5 Pierce's Parameter

The high gain FEL modelled by Eq. 10.95 is named *Self-Amplified Spontaneous Emission* (SASE) because the equation describes the amplification of undulator spontaneous radiation emitted in the first segments of the undulator line (an estimate for the initial power P_0 will follow). The model assumes that each new bunch of electrons accelerated in a single-pass linac reaches the undulator with a different, randomly distributed configuration of phases (see $\langle j_1 \rangle$ in Eq. 10.90). Hence, a statistical characterization of the FEL spectral intensity can be carried out, where many intense light pulses, each light pulse radiated by a large number of electrons, are considered.

The longitudinal coherence length (sometimes also *cooperation length*) of a SASE FEL is determined by the average radiation slippage over the characteristic scale of the FEL power growth, which is the gain length L_g. In this sense, the slippage identifies the bunch slice through which the field phase is approximately constant:

$$l_{coh} \approx \frac{N_u\lambda}{L_u/L_g} \approx \frac{\lambda}{4\pi\rho} \tag{10.96}$$

For example, if a bunch long $\Delta z_b = 30$ µm is lasing at $\lambda = 1$ nm with $\rho = 0.1\%$, we expect on average $\Delta z_b / l_{coh} \approx 400$ spikes in a pulse, each spike representing a longitudinally coherent mode. The 400 modes, however, are mutually incoherent. It follows that when $\Delta z_b \leq l_{coh}$, a single longitudinal mode (single narrow spectral line) can be produced.

By virtue of the definition of longitudinal coherence length in Eq. 10.24, we find:

$$l_{coh} \approx \frac{\lambda}{4\pi\rho} = \frac{1}{2\pi}\frac{\lambda^2}{2\Delta\lambda} \Rightarrow \rho \approx \frac{\Delta\lambda}{\lambda} \tag{10.97}$$

that is, the FEL parameter quantifies the FEL rms intrinsic relative bandwidth. Since the relative spectral width of each SASE spike is of the order of $\lambda/\Delta z$, the average number of spikes in the spectrum will be $\rho\Delta z/\lambda \approx 30$.

The resonance condition says that the FEL relative bandwidth is proportional to the electron beam's relative energy spread, since off-resonance electrons will tend to emit radiation at slightly different wavelengths. Hence, the intrinsic bandwidth is preserved as long as it is much larger than the relative energy spread evaluated on the scale of the cooperation length:

$$\frac{\Delta\lambda}{\lambda} = 2\frac{\Delta\gamma}{\gamma} \approx \sigma_\delta \Rightarrow \sigma_\delta < \frac{\Delta\lambda}{\lambda} \approx \rho, \tag{10.98}$$

This defines ρ as the upper limit to the beam rms *uncorrelated* relative energy spread for not degrading the FEL brilliance.

As a matter of fact, a small energy spread ensures efficient electrons trapping into the FEL bucket (see Fig. 10.10). But, since FEL emission implies energy modulation of the trapped electrons, the beam's energy spread grows along the undulator, till electrons escape from the bucket (de-bunching), or they are too far from the resonance condition to contribute substantially to the gain. At this point, the amplification of the radiated field stops. Since the amount of energy transferred from the electrons to the radiated field is of the order of the energy modulation amplitude, the *saturation power* level is:

$$P_{sat} \approx \frac{I\Delta E_{sat}}{e} \approx \frac{I\langle E \rangle}{e}\sigma_{\delta,sat} \approx \rho P_b \tag{10.99}$$

and P_b is just the electron beam power. In conclusion, ρ is also the efficiency of energy transfer of the FEL process.

The initial radiation power level P_0 introduced in Eq. 10.95 is estimated as follows. We observe that a fraction ρ of the electron beam power P_b is converted into radiation. Since the initial emission is governed by the initial bunching factor b_0, it has to be $P \sim |b_0|^2$. Let us define $|b_0|^2$ for a Poisson-like electron distribution within the FEL spectral bandwidth $\Delta\nu$. We get:

$$|b_0|^2 = \frac{2e}{I}\Delta\nu = \frac{2ec}{I}\frac{\Delta\lambda}{\lambda^2} \approx 2ec\frac{\rho}{\lambda}\frac{\rho E}{eP_{sat}} \Rightarrow P_0 \approx \rho P_b |b_0|^2 \approx 2c\rho^2 E/\lambda. \tag{10.100}$$

Short wavelength SASE FELs are commonly characterized by shot-noise and saturation power of the order of $P_0 \sim 10 - 100$ W and $P_{sat} \sim 1 - 10$ GW, respectively.

At the saturation point, electrons have completed one-half of the synchrotron period in the FEL bucket. After that, they start absorbing energy from radiation. But, since the motion becomes chaotic, the FEL power oscillates around the saturation power level (assuming an undulator longer than the saturation length). The saturation length can be estimated from Eqs. 10.95 and 10.100 as $L_{sat} \approx L_g \ln(9\frac{I}{2ec}\frac{\lambda}{\rho})$. In EUV and x-ray SASE FELs, the power saturation length is typically in the range 15–22 L_g.

In summary, FEL emission in high gain regime is so intense because it is coherent in the sense of Eq. 10.31. Electrons are first modulated in energy by the radiated field $(d\eta/dz)$. The undulator momentum compaction ($R_{56} = 2N_u\lambda$) shifts the electrons back and forth depending on the sign of their relative energy deviation, so that they distribute in clusters inside the FEL bucket with periodic spacing of λ ($d\psi/dz$). The sharper the electron distribution is in the FEL bucket, the larger the bunching is, the larger the field amplitude becomes (dA/dz). The FEL bucket grows in height along the undulator by virtue of the growing field amplitude, and it starts trapping more and more electrons. The process stops when the electron beam uncorrelated energy spread is so enlarged by the FEL instability to prevent any additional net energy transfer.

10.5.6   Electron Beam Quality

The high gain regime of FELs requires a persistent interaction of electrons and radi-ated field along the whole undulator line. In the transverse plane, this is guaranteed by (i) negligible light diffraction along the characteristic length of power growth, or $L_g < L_R = 4\pi\sigma_r^2/\lambda$, with L_R the Rayleigh length, and by (ii) matched spot sizes of the two beams. The relative phase slippage due to electrons' betatron motion and energy spread could lead to loss of synchronism, which is therefore minimized by (iii) small electron beam's angular divergence, and (iv) small relative energy spread.

Conditions (ii)–(iii) translate into an electron beam ideally below the diffraction limit, $4\pi\varepsilon_{x,y} \leq \lambda$. This determines in turn a high degree of transverse coherence (on-axis Gaussian mode), as suggested by Eq. 10.20. For an electron beam at the diffraction limit, condition (i) implies $L_g < \langle\beta_{x,y}\rangle$, the latter being the average beta-tron function along the undulator, assuming smooth optics.

In summary, light diffraction, electron beam's transverse emittance and energy spread (so-called "3-D effects"), all contribute to a reduction of the lasing efficiency. This can be quantified by means of an effective 3-D gain length $L_{g,3D} \approx L_{g,1D}(1 + \chi)$, where typically $\chi \sim 0.1 - 0.3$. At this point, one could note that the transverse momentum spread due to betatron motion scales as $\Delta p_\perp \sim \sqrt{\varepsilon/\beta}$, whereas the transverse beam size scales as $\sigma_\perp \sim \sqrt{\varepsilon\beta}$. These opposing scaling laws suggest there is some optimal value of β that, on top of light diffraction, will minimize $L_{g,3D}$ for a given set of radiation and electron beam parameters.

10.5.6.1 Discussion: Brilliance of FELs and Synchrotron Light Sources

The peak brilliance of high gain x-ray FELs is several orders of magnitude larger than that of undulator emission at synchrotron light sources. Roughly speaking, this can be attributed to coherent emission in FELs during the exponential power growth, with respect to the incoherent emission in a short undulator (compare Eqs. 10.46 and 10.95).

We would like to quantify the ratio of peak brilliance at the two light sources in a more rigorous way, considering the undulator spontaneous radiation at the source point and in the presence of monochromatization. Typical parameters for emission in soft x-rays are considered, such as bunch charges $Q_{fel}/Q_{sr} = 0.1nC/1nC$, $\rho = 10^{-3}$, $N_u = 100$ in the short undulator, resolving power $RP = 10^4$ and transmission efficiency $TR = 1\%$ through the monochromator for the undulator spontaneous radiation. We assume same central wavelength in soft x-rays, similar spot size and angular divergence of the light pulse at the source. This allows us to reduce the brilliance ratio to the ratio of the spectral power.

With the short notation $b = d\lambda/\lambda$, the ratio of peak brilliance *at the source* (e.g., undulator exit) and after *monochromatization* of the spontaneous emission, is:

$$\left.\frac{\hat{B}_{fel}}{\hat{B}_{sr}}\right|_{source} \approx \frac{(d\hat{P}/db)_{fel}}{(d\hat{P}/db)_{sr}} \approx \frac{(N_b^2/b)_{fel}}{(N_b/b)_{sr}} \approx \frac{N_{b,fel}/\rho}{10^2 N_u} \approx 10^8$$

$$(10.101)$$

$$\left.\frac{\hat{B}_{fel}}{\hat{B}_{sr}}\right|_{mono} \approx \frac{(d\hat{P}/db)_{fel}}{(d\hat{P}/db)_{mono}} \approx \frac{N_{b,fel}/\rho}{10^2 TR \cdot RP} \approx 10^8$$

where we used $b_{fel} \approx \rho$ and $b_{sr} \approx 1/N_u$. Equation 10.101 points out that some larger degree of longitudinal coherence of the spontaneous emission is obtained through monochromatization at the expense of a lower flux.

An explicit and more rigorous calculation can be done by means of Eq. 10.46 for undulator spontaneous emission (re-written by means of the resonance condition), and of Eq. 10.99 for FEL emission at saturation. Doing so, we assume a consistent set of parameters for the FEL such as $E = 2$ GeV and $K \approx 1$ to get $\lambda \approx 1$ nm. The ratio of linac-to-ring peak current is of the order of $I_{lin}/I_{sr} \approx 1000\,A / 10\,A$:

$$\frac{\hat{B}_{fel}}{\hat{B}_{sr}} \approx \frac{(d\hat{P}/db)_{fel}}{(d\hat{P}/db)_{sr}} \approx \frac{\rho I_{lin} E}{e\rho} \frac{3\varepsilon_0}{\pi e} \frac{\lambda}{I_{sr} N_u K^2} \approx 10^8 \qquad (10.102)$$

In conclusion, the large increase of peak brilliance in a high gain FEL compared to other x-ray light sources has to be attributed to coherent emission, intended as intensity enhancement, *and* to a high degree of longitudinal coherence. One can notice, however, that the FEL power *at saturation* shows a weaker dependence from the total number of particles than in fully coherent emission: $P_{sat} \propto \rho I \propto N_b^{4/3}$. This reflects the loss of longitudinal coherence, or "debunching", associated to synchrotron oscillations of electrons in the FEL bucket. Namely, the maximum power is reached at the expense of some reduced longitudinal coherence.

Equation 10.102 is now revised to estimate the ratio of the *average* brilliance. This is simply done by multiplying the above expression by the ratio of FEL-to-storage ring bunch duration, and FEL-to-storage ring repetition rate. Repetition frequencies of high brightness linacs span 10^2–10^6 Hz, while medium size, multi-GeV-energy storage ring light sources have harmonic number $h \sim 500$ and revolution frequency of the order of 1 MHz or so. We find:

$$\frac{\langle B_{fel} \rangle}{\langle B_{sr} \rangle} \approx \frac{\hat{B}_{fel}}{\hat{B}_{sr}} \cdot \frac{0.1 ps}{100 ps} \cdot \frac{(10^2 - 10^6) Hz}{10^9 Hz} \approx 10^{-2} - 10^2, \tag{10.103}$$

with the lower limit for NC linacs, the upper limit for SC ones.

10.5.6.2 Discussion: Is a Free-Electron Laser a Laser?

The FEL acronym points out that, on the one hand, the electrons are unbounded or "free" from atomic levels. On the other hand, it suggests a similarity with a conventional atomic laser by virtue of the large degree of transverse and longitudinal coherence accompanying lasing in the high gain regime. Nevertheless, the coherence of an atomic laser is properly described in terms of *quantistic* states of light, while the coherence of an FEL has so far been characterized only through Eqs. 10.9 and 10.14 in a *classical* formalism.

With the due aforementioned differences in mind, we may establish, with some imagination, a similitude between the process of electrons' manipulation and emission of radiation in an atomic laser, and in a high gain FEL, as sketched by the 4-step process in Table 10.1.

Table 10.1 Process leading to lasing in atomic lasers and in high gain FELs

Step	Atomic Laser	Free-Electron Laser
1	Stable configuration of atomic levels	Non-relativistic photo-electrons
2	Inversion of electron population between two atomic levels by means of external energy pumping	Energy increase to GeV-level by means of RF acceleration
3	Spontaneous emission of radiation by electrons migrating from upper to lower atomic level	Undulator spontaneous emission in the first undulator segments
4	Stimulated radiation emission, nonlinearly amplified in intensity, narrow-band, and highly collimated in the direction of the seeding. An optical cavity narrows the spectral bandwidth	Radiation emission stimulated by spontaneous emission (or external laser in seeded-FELs), exponentially amplified in intensity, highly collimated. The long resonant undulator selects the central frequency, the spectral bandwidth narrows

References

1. A. Balerna, S. Mobilio, Introduction to synchrotron radiation, in *Synchrotron Radiation: Basics, Methods and Applications*, ed. by S. Mobilio, F. Boscherini, C. Meneghini (Published by Springer, Berlin, Heidelberg, 2015), pp. 3–28. ISBN: 978-3-642-55314-1
2. D. Attwood, A. Sakdinawat, *X-rays and Extreme Ultraviolet Radiation* (Published by Cambridge University Press, 2016), pp. 110–147, 227–278. ISBN: 9781107477629
3. P.A. Millette, The Heisenberg Uncertainty Principle and the Nyquist-Shannon Sampling Theorem. Progress Phys. **3**, 9–14 (2013)
4. M.R. Howells, B.M. Kincaid, The properties of undulator radiation, LBL-34751, in *Proceedings of the NATO Advanced Study Institute*, Maratea, Italy, ed. by A.S. Schlacher, F.J. Wuilleumier (1992)
5. R.P. Walker, Insertion devices: undulators and wigglers, in *Proceedings of CERN Accelerator School: Synchrotron Radiation and Free Electron Lasers*, Geneva, Switzerland. CERN 98-04, ed. by S. Turner (1998), pp. 129–190
6. K.-J. Kim, Z. Huang, R. Lindberg, *Synchrotron Radiation and Free-Electron Lasers: Principles of Coherent X-ray Generation* (Published by Cambridge University Press, 2017), pp. 74–138. ISBN: 9781316677377
7. V. Berestetskii, E. Lifshitz, L. Pitaevskii, Quantum electrodynamics, in *Landau and Lifshitz, Course of Theoretical Physics*, vol. 4 (Published by Pergamon Press, New York, 1982)
8. C. Pellegrini, The history of X-ray free-electron lasers. Eur. Phys. J. H **37**, 659–708 (2012)
9. G. Dattoli, E. Di Palma, S. Pagnutti, E. Sabia, Free electron coherent sources: from microwave to X-rays. Phys. Rep. **739**, 1–51 (2018)

Colliders

11

In high energy colliders, two counter-propagating beams collide to produce new particles ("events"), aiming at investigate the nuclear and sub-nuclear structure of matter, verify theoretical particle models, increase the experimental precision of particles' parameters, etc. This Chapter treats the physics of beam-beam collision by introducing the concept of luminosity in a circular collider, strategies to maximize it, and limitations related to beam stability. It is complemented by considerations on linear colliders.

11.1 Luminosity

If n_1, n_2 are the number of particles involved in a two-beam collision, the number of events produced in the unit of time is expected to be proportional to the beams' population, $\dot{n}_3(t) \propto n_1 \cdot n_2$. Moreover, the closer the two beams are one another at the interaction point (IP)—namely, the smaller the effective interaction area A is—, the higher the probability of their interaction will be. Hence, the event rate shall be proportional to the convolution of the beams' charge densities, $\dot{n}_3(t) \propto \rho_1 \rho_2 \sim \frac{n_1 n_2}{A}$.

In general, a specific class of events could only be produced with a certain probability among several "production channels" permitted by the beams interaction. In particle physics, such a probability is expressed via the *cross-section* of the interaction, commonly in units of cm^2, and thereby $\dot{n}_3(t) \propto \sigma$. While n_1, n_2, A are all features of the accelerator system, σ is independent from the way the interaction is put in place, i.e., it depends only from the physics of the particles' interaction.

In summary, the production rate of specific events can be expressed as the product of quantities related to the accelerator and the event cross-section [1]:

$$\frac{dn_3}{dt} = L(t)\sigma \quad \Rightarrow \quad n_3 = \int_{t_1}^{t_2} \frac{dn_3}{dt} dt = \int_{t_1}^{t_2} \sigma L(t) dt \equiv \sigma \mathbb{L}(\Delta t) \tag{11.1}$$

© The Author(s), under exclusive license to Springer Nature Switzerland AG 2022
S. Di Mitri, *Fundamentals of Particle Accelerator Physics*, Graduate Texts in Physics,
https://doi.org/10.1007/978-3-031-07662-6_11

$L(t) \propto \frac{n_1 n_2}{A}$ is the *instantaneous luminosity* of the collider, in units of events/sec/ cm^2. State-of-the-art values at circular colliders (storage rings) are $L \sim 10^{32} - 10^{34}$ cm^{-2}s^{-1}. In general, we might expect the beam parameters, and therefore L, to vary with time during the beam storage of several hours, according to the beams' lifetime. The time-integral of $L(t)$ defines the *integrated luminosity* $\mathbb{L}$. This is commonly expressed in units of "inverse barn" (symbol b^{-1}), where $1b = 10^{-24}$ cm^2. For example, two of CERN's LHC experiments, ATLAS and CMS, reached over 5 inverse femtobarn of proton-proton data in 2011 alone. This means an expectation of 5 events per femtobarn of cross-section (10^{-39} cm^2) within those data.

As anticipated, $L(t)$ is determined by the convolution of the two colliding beams' distributions, which is reasonable to approximate in most cases with a Gaussian centered in $x = 0$, $y = 0$ and $s = \pm s_0$, s_0 being the distance of the two bunch centroids from the IP. For the moment, let us consider *head-on collision* (null crossing angle), with ultra-relativistic particles' velocities $\vec{v}_{z,1} = -\vec{v}_{z,2} \approx \vec{c}$. Although the two beams may have different transverse sizes at the IP, we assume for simplicity same bunch duration. Finally, the total frequency of interaction is the product of the single bunch revolution frequency and the number of IPs along the accelerator, $f_{ip} = f_0 \cdot N_{ip}$. For a multi-bunch fill pattern, f_{ip} is multiplied by the number of bunches in each of the two beam trains, n_b. The instantaneous luminosity results:

$$L = \frac{|\vec{v}_{z,1} - \vec{v}_{z,2}|}{c} n_1 n_2 n_b f_{ip} \iiiint_{-\infty}^{+\infty} \rho_1(x)\rho_1(y)\rho_1(s - s_0)\rho_2(x)\rho_2(y)\rho_2(s + s_0) d^4\vec{u} =$$

$$= \frac{2n_1 n_2 n_b f_{ip}}{(\sqrt{2\pi})^6 \sigma_{x,1}\sigma_{x,2}\sigma_{y,1}\sigma_{y,2}2\sigma_s^2} \iiiint e^{-\frac{x^2}{2\sigma_{x,1}^2}} e^{-\frac{x^2}{2\sigma_{x,2}^2}} e^{-\frac{y^2}{2\sigma_{y,1}^2}} e^{-\frac{y^2}{2\sigma_{y,2}^2}} e^{-\frac{(s-s_0)^2}{2\sigma_s^2}} e^{-\frac{(s+s_0)^2}{2\sigma_s^2}} d^4\vec{u} =$$

$$= \frac{2n_1 n_2 n_b f_{ip}}{(\sqrt{2\pi})^6 \sigma_{x,1}\sigma_{x,2}\sigma_{y,1}\sigma_{y,2}2\sigma_s^2} \iiiint e^{-x^2\frac{\sigma_{x,1}^2+\sigma_{x,2}^2}{2\sigma_{x,1}^2\sigma_{x,2}^2}} e^{-y^2\frac{\sigma_{y,1}^2+\sigma_{y,2}^2}{2\sigma_{y,1}^2\sigma_{y,2}^2}} e^{-\frac{s^2}{\sigma_s^2}} e^{-\frac{s_0^2}{\sigma_s^2}} ds_0 ds dx dy =$$

$$= \frac{2n_1 n_2 n_b f_{ip}}{(\sqrt{2\pi})^6 \sigma_{x,1}\sigma_{x,2}\sigma_{y,1}\sigma_{y,2}2\sigma_s^2} \frac{(\sqrt{2\pi})^2(\sqrt{\pi})^2 \sigma_{x,1}\sigma_{x,2}\sigma_{y,1}\sigma_{y,2}2\sigma_s^2}{\sqrt{\sigma_{x,1}^2+\sigma_{x,2}^2}\sqrt{\sigma_{y,1}^2+\sigma_{y,2}^2}} =$$

$$= \frac{n_1 n_2}{2\pi\sqrt{\sigma_{x,1}^2+\sigma_{x,2}^2}\sqrt{\sigma_{y,1}^2+\sigma_{y,2}^2}} n_b f_{ip} \rightarrow \frac{n_1 n_2}{4\pi\sigma_x\sigma_y} n_b f_{ip} \tag{11.2}$$

The derivation used the well-known equivalence $\int_{-\infty}^{+\infty} e^{-au^2} du = \sqrt{\pi/a}$. The last r.h.s. expression is for comparable beam sizes of the two beams at the IP.

The denominator in Eq. 11.2 quantifies the effective interaction area of the two beams. When the two beams obey to the same optics, have similar emittances, and the interaction is assumed to be in a dispersion-free region, they both have $\sigma_u = \sqrt{\beta_u \epsilon_u}$ ($u = x, y$). By introducing the coupling factor $\kappa = \epsilon_y/\epsilon_x$ (see Eqs. 5.51), Eq. 11.2 becomes:

$$L = \frac{n_1 n_2 n_b f_{ip}}{4\pi\sqrt{\epsilon_x \epsilon_y \beta_x^* \beta_y^*}} = \frac{n_1 n_2 n_b f_{ip}}{4\pi\epsilon_x\sqrt{\kappa\beta_x^*\beta_y^*}} \tag{11.3}$$

and β_u^* are betatron functions at the IP.

Equation 11.3 demonstrates the importance of a small beam geometric emittance to maximize the luminosity. Specially strong quadrupole magnets are commonly arranged in proximity of the IP to squeeze the betatron functions. Their high gradients become responsible of a large part of the storage ring chromaticity (see Eq. 6.29).

11.1.1 Discussion: Lifetime, Run Time and Preparation Time

According to Eq. 11.1, the instantaneous luminosity is constant as long as the number of particles per bunch remains constant over time. In reality, the beam current in storage rings decays with time: the beam lifetime, or τ (see e.g. Eq. 8.65), can be at the level of a fraction of hour to several hours. If the run time interval t_r starts only after a beam time preparation long t_p, what is the most convenient combination of time intervals which maximizes the integrated luminosity?

Assuming an exponential decay $L(t) = L_0 e^{-t/\tau}$, the average integrated luminosity available for data collection is [1]:

$$< \mathbb{L} >= \frac{\int_0^{t_r} L(t)dt}{t_r + t_p} = L_0 \cdot \tau \frac{1 - e^{-t_r/\tau}}{t_r + t_p} \tag{11.4}$$

The quantity $\langle \mathbb{L} \rangle$ is therefore maximized as function of the run time by:

$$\frac{d<\mathbb{L}>}{dt_r} \equiv 0 \quad \Rightarrow \quad \frac{L_0 \tau}{(t_r + t_p)^2}\left[\frac{1}{\tau}(t_r + t_p)e^{-\frac{t_r}{\tau}} - (1 - e^{-\frac{t_r}{\tau}})\right] = 0;$$

$$t_r = \tau \ln\left(1 + \frac{t_r + t_p}{\tau}\right) \equiv f(t_r) \tag{11.5}$$

Figure 11.1 illustrates the solution of Eq. 11.5 for three values of the ratio t_p/τ. The maximum of $< \mathbb{L} >$ is obtained for $t_p < \tau$. Indeed, this goes in the direction of the ideal case of very short preparation time and extremely long lifetime.

Fig. 11.1 Representation of Eq. 11.5 for $t_p = 8, 16$ h, $\tau = 8, 16$ h. The maximum value of $< \mathbb{L} > /L_0$ corresponding to the 3 ratios 8/8 (black), 8/16 (blue) and 16/8 (red) is, respectively, 0.32, 0.42 and 0.22. The optimum $(t_p/\tau = 0.5)$ corresponds to $t_r \approx 14$ h

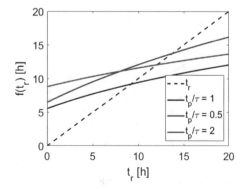

11.2   Crossing Angle

Circular colliders usually adopt a non-zero crossing angle at the IP to avoid unwanted collisions between bunches of the two counter-propagating trains, which could dilute the charge density over time, generate spurious events, etc. The immediate geometric effect, however, is that of reducing the luminosity by virtue of an effective larger interaction area [1,2].

We discriminate two contributions to the effective area. The first one is shown in Fig. 11.2-left plot: the projection of the longitudinal size of beam-2 ($\sigma_{z,2}$) onto the axis orthogonal to the motion of beam-1 (r_1) adds in quadrature to the natural size of beam-1 ($\sigma_{x,1}$). The second one is shown in Fig. 11.2-right plot: the projection of the horizontal size of beam-2 ($\sigma_{x,2}$) onto the longitudinal axis of the same beam (r_2) is seen by beam-1 as a longer interaction interval, i.e., it adds in quadrature to the natural length of beam-2 ($\sigma_{z,2}$).

In short, the two terms modify the instantaneous luminosity in Eq. 11.2 with an effective enlargement of the horizontal beam size and bunch lengthening. The latter effect is usually negligible with respect to the former one because strong focusing generates "pencil beams" at the IP, i.e., $\sigma_x \ll \sigma_z$:

$$\sigma_x \rightarrow \sqrt{\sigma_x^2 + \sigma_z^2 \tan^2 \tfrac{\phi}{2}} \cdot \frac{\sqrt{\sigma_z^2 + \sigma_x^2 \tan^2 \tfrac{\phi}{2}}}{\sigma_z} =$$

$$= \sigma_x \sqrt{1 + \left(\tfrac{\sigma_z}{\sigma_x}\right)^2 \tan^2 \tfrac{\phi}{2}} \sqrt{1 + \left(\tfrac{\sigma_x}{\sigma_z}\right)^2 \tan^2 \tfrac{\phi}{2}} \equiv \tfrac{\sigma_x}{S}; \qquad (11.6)$$

$$\Rightarrow L = \frac{n_1 n_2}{4\pi \sigma_x \sigma_y} n_b f_{ip} S, \qquad S(\sigma_x \ll \sigma_z) \approx \frac{1}{\sqrt{1 + \left(\tfrac{\sigma_z}{\sigma_x}\right)^2 \tan^2 \tfrac{\phi}{2}}}$$

By definition, $S \leq 1 \; \forall \phi \geq 0$. For example, for beams stored to produce 7 TeV invariant mass in LHC, $\phi \approx 0.3$ m rad, $\sigma_x \approx 20 \mu$m and $\sigma_z \approx 8$ cm, giving $S \approx 0.85$.

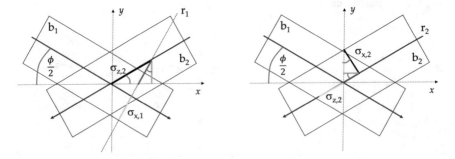

Fig. 11.2 Beam-1 (b_1) and beam-2 (b_2) interacting at a total crossing angle ϕ (angles and beams' transverse and longitudinal sizes are sketched for illustrative purpose only). The longitudinal (left) and transverse size (right) of beam-2 is seen by beam-1 as an additional contribution to the transverse interaction area ($\sigma_{z,2}$ and $\sigma_{x,2}$ in bold black is projected onto the red segment)

11.3 Hourglass Effect

So far, the luminosity was defined assuming that the betatron functions at the IP are constant along the interaction region. The length of the effective interaction region is the bunch duration, which can span from tens' to few hundreds' of millimeters. In reality, since the IP is a drift section internal to a particle detector and, as said, the betatron functions have minima at the IP, the section is a low-β insertion, and $\beta(s)$ varies according to Eq. 4.138 [1,2]. It is easy to see that, for example, $\beta^* \approx 3$ cm corresponds to a 2-fold larger β at the edges of an interaction region long 3 cm. Owing to Eq. 11.3, such growth of $\beta(s)$ is expected to reduce the luminosity. Indeed, we show below that the effect becomes important for small β^* in the presence of long bunches, i.e., in proportion to the ratio σ_s/β^*. In the literature, the enlargement of the beam size in proximity of a waist is called *hourglass effect*, by virtue of the shape of the beam envelope around the waist (see Fig. 4.15).

For simplicity, symmetric beams at the interaction region are considered. The variation of the beam size along the interaction region is:

$$\sigma_u^* = \sqrt{\epsilon_u \beta_u^*} \rightarrow \sigma_u = \sqrt{\epsilon_u \beta_u^* \left(1 + \frac{s^2}{\beta_u^{*2}}\right)} = \sigma_u^* \sqrt{1 + \left(\frac{s}{\beta_u^*}\right)^2} \tag{11.7}$$

We re-define for brevity $n_d = 2n_1 n_2 n_b f_{ip}$, and we introduce $w_u(s) = s/\beta_u^*$. Then, the luminosity is calculated by integrating first in dx, dy, then in ds_0 and finally in ds:

$$L = \frac{n_d}{(\sqrt{2\pi})^6 \sigma_x^2 \sigma_y^2 \sigma_s^2} \int\int\int\int e^{-\frac{x^2}{\sigma_x^2}} e^{-\frac{y^2}{\sigma_y^2}} e^{-\frac{s^2}{\sigma_s^2}} e^{-\frac{s_0^2}{\sigma_s^2}} dx\,dy\,ds\,ds_0 =$$

$$= \frac{n_d}{(\sqrt{2\pi})^6 \sigma_x^{*2} \sigma_y^{*2} \sigma_s^2} \int\int\int\int \frac{e^{-\frac{x^2}{\sigma_x^{*2}[1+w_x(s)^2]}} e^{-\frac{y^2}{\sigma_y^{*2}[1+w_y(s)^2]}}}{[1+w_x(s)^2][1+w_y(s)^2]} e^{-\frac{s^2}{\sigma_s^2}} e^{-\frac{s_0^2}{\sigma_s^2}} dx\,dy\,ds\,ds_0 =$$

$$= \frac{n_d \pi^{3/2} \sigma_x^* \sigma_y^* \sigma_s}{(\sqrt{2\pi})^6 \sigma_x^{*2} \sigma_y^{*2} \sigma_s^2} \int \frac{e^{-\frac{s^2}{\sigma_s^2}}}{\sqrt{1+w_x(s)^2}\sqrt{1+w_y(s)^2}} ds = \dots \left[\frac{s}{\sigma_s} \rightarrow \zeta\right] \dots$$

$$= \frac{n_d \pi^{3/2}}{(\sqrt{2\pi})^6 \sigma_x^* \sigma_y^* \sigma_s} \int \frac{e^{-\zeta^2}}{\sqrt{1+\left(\frac{\zeta}{\beta_x^*/\sigma_s}\right)^2}\sqrt{1+\left(\frac{\zeta}{\beta_y^*/\sigma_s}\right)^2}} \sigma_s d\zeta = \dots \left[r_u := \frac{\beta_u^*}{\sigma_s}\right] \dots$$

$$= \frac{n_1 n_2 n_b f_{ip}}{4\pi \sigma_x^* \sigma_y^*} \int \frac{1}{\sqrt{\pi}} \frac{e^{-\zeta^2}}{\sqrt{1+\left(\frac{\zeta}{r_x}\right)^2}\sqrt{1+\left(\frac{\zeta}{r_y}\right)^2}} d\zeta \equiv L^* \cdot \mathbb{H}(r_x, r_y)$$

$$\tag{11.8}$$

For optics of the interaction region symmetric in the two transverse planes ($\beta_x(s) = \beta_y(s)$), the hourglass effect reduces the nominal luminosity L^* by the quantity:

$$\mathbb{H}(r) = \int_{-\infty}^{+\infty} \frac{1}{\sqrt{\pi}} \frac{e^{-\zeta^2}}{1+\left(\frac{\zeta}{r}\right)^2} d\zeta = \sqrt{\pi} r e^{r^2} [erf(r)] < 1 \quad \forall r \tag{11.9}$$

$\mathbb{H}$ is a monotonic function of r with asymptotic value 1 for $r = \frac{\beta^*}{\sigma_s} \to \infty$. For example, $H(0.5) \approx 0.53$, $H(1) \approx 0.75$, and $H(2) \approx 0.9$.

11.3.1   Discussion: Luminosity of a Compton Source

The definition of L in Eq. 11.2 can be identically applied to the interaction of a charged particle beam and a photon beam, such as in an ICS light source. In this case, the number of events is the number of scattered photons, and the cross section in the no recoil approximation is the Thomson cross section: $N_{ph} = L\sigma_T$. How does the luminosity depend from the laser parameters and the electron beam energy, for any given energy of the scattered photons?

For simplicity, we assume both beams to be round and perfectly matched, with identical beam sizes at the IP. The laser wavelength is λ_L, and the electron beam is intended to be at the diffraction limit. If Q_b and U_L are the bunch charge and the laser pulse energy, respectively, the luminosity for a single interaction is:

$$L_{ICS} = \frac{N_e N_L f_{ip}}{2\pi \sqrt{\sigma_{x,e}^2 + \sigma_{x,L}^2} \sqrt{\sigma_{y,e}^2 + \sigma_{y,L}^2}} = \frac{1}{4\pi\sigma_L^2} \frac{Q_b}{e} \frac{U_L}{E_L} = \frac{1}{4\pi} \frac{Q_b}{e} \frac{U_L}{hc} \lambda_L \left(\frac{4\pi}{\beta_L \lambda_L}\right) =$$

$$\tag{11.10}$$

$$= \frac{1}{ehc} \frac{Q_b U_L}{\beta_L} \approx \frac{Q_b U_L}{\sigma_L^2} \frac{\gamma^2 \lambda_s}{\pi ehc}$$

The very last expression is derived by imposing the resonance condition $\lambda_s \approx \lambda_L/(4\gamma^2)$ for the ICS radiation collected on-axis and for head-on collision (see Eq. 10.57).

As expected, the number of ICS photons is proportional to the total bunch charge and the laser total pulse energy. Owing to the local interaction, L is maximized by small laser and electron beam transverse sizes at the IP. However, very tight waists would enhance the hourglass effect, thus degrading the effective luminosity. A trade-off in the control of the transverse sizes has to be reached, eventually.

The luminosity of the ICS light source is proportional to the electron beam energy. This means that, for any target λ_s, λ_L could be made longer (in correspondence of which a larger laser power can be available) at higher electron beam energies. This, however, is at the expense of larger RF power as required by a longer accelerator or higher accelerating gradients. The luminosity is larger at higher energies of the scattered photons, which implies again higher electron beam energies.

11.4 Beam-Beam Tune Shift

The mutual penetration of two colliding beams at the IP makes each beam subject to the e.m. field generated by the other beam [1,3]. The effect is equivalent to focusing in both transverse planes. Assuming infinitely long ("coasting beam") approximation, transversely "round" Gaussian beams of identical sizes at the IP ($\sigma_{x,1} = \sigma_{x,2} = \sigma_{y,1} = \sigma_{y,2}$), the angular divergence ("kick") induced by the beam-beam interaction is:

$$\Delta r' = \frac{\Delta p_r}{p_z} = \frac{1}{p_z} \int_{-\infty}^{+\infty} \rho(r,s) E_r(r) ds = \frac{2 r_0 n_1}{\gamma} \frac{1}{r} \left(1 - e^{-\frac{r^2}{2\sigma_r^2}}\right) \approx \left(\frac{r_0 n_1}{\gamma} \frac{r}{\sigma_r^2}\right)_{r \ll \sigma_r} \tag{11.11}$$

The approximation is for a kick in proximity of beam's axis, where the largest portion of the charge distribution lies. Specialized to the case of flat beams, i.e., $\sigma_x \gg \sigma_y$, the kick in each transverse plane becomes:

$$\Delta u' \approx \frac{2 r_0 n_1}{\gamma} \frac{u}{\sigma_u(\sigma_x + \sigma_y)} \equiv \delta_{bb} u, \quad u = x, y \tag{11.12}$$

The linearized *beam-beam effect* in Eq. 11.12 is in fact the inverse of a focal length, $\delta_{bb} = 1/f_{bb}$. Such parasitic focusing is typically weaker in circular colliders with respect to linear colliders, by virtue of a larger interaction area. The perturbation has to be small enough not to disrupt the beams' quality over many consecutive hours of operation.

If the effect is strong enough, the transverse kick leads in turn to a variation of the particles' lateral position already internally to the interaction region:

$$\begin{cases} \Delta r' = -\frac{r}{f_{bb}} \\ \Delta r = \Delta r' \cdot \sigma_z \end{cases} \Rightarrow \Delta r = -\frac{\sigma_z}{f_{bb}} r \tag{11.13}$$

$$\Rightarrow D_{bb,u} := \frac{\Delta u}{u} = -\frac{\sigma_z}{f_{bb,u}} = \frac{2 r_0 n_1}{\gamma} \frac{\sigma_z}{\sigma_u(\sigma_x + \sigma_y)} \tag{11.14}$$

D_{bb} is called *disruption parameter*. When the kick is relatively "gentle" or $D_{bb} < 1$, the effect can be evaluated as a perturbation to the single-turn beam matrix. For flat beams, the larger contribution is in the vertical plane (see Eq. 11.12):

$$\tilde{M}_t = B \cdot M_t \cdot B = \begin{pmatrix} 1 & 0 \\ -\frac{\delta_{bb}}{2} & 1 \end{pmatrix} \begin{pmatrix} \cos \Delta\mu_y & \beta_y^* \sin \Delta\mu_y \\ -\frac{\sin \Delta\mu_y}{\beta_y^*} & \cos \Delta\mu_y \end{pmatrix} \begin{pmatrix} 1 & 0 \\ -\frac{\delta_{bb}}{2} & 1 \end{pmatrix} =$$

$$\approx \begin{pmatrix} \cos \Delta\mu_y - \frac{\delta_{bb}}{2} \beta_y^* \sin \Delta\mu_y & \beta_y^* \sin \Delta\mu_y \\ -\frac{1}{\beta_y^*} \left(\sin \Delta\mu_y + \frac{\delta_{bb}}{2} \cos \Delta\mu_y\right) & \cos \Delta\mu_y \end{pmatrix} = \tag{11.15}$$

$$\approx \begin{pmatrix} \cos 2\pi(Q_y + \xi_y) & \beta_y^* \sin 2\pi(Q_y + \xi_y) \\ -\frac{1}{\beta_y^*} \sin 2\pi(Q_y + \xi_y) & \cos 2\pi(Q_y + \xi_y) \end{pmatrix},$$

where we introduced the *beam-beam tune shift*:

$$\xi_y = \frac{\delta_{bb}\beta_y^*}{4\pi} = \frac{r_0 n_1 \beta_y^*}{2\pi \gamma \sigma_y (\sigma_x + \sigma_y)}, \qquad (11.16)$$

we assumed $\xi_y \ll 1$, and neglected higher orders in δ_{bb}. The tune-shift is usually at the level of ~ 0.01 for collisions $e^- e^+$, and < 0.05 for pp. An analogous expression is obtained for the horizontal plane, but smaller in proportion to the ratio of the transverse beam sizes at the IP.

The beam-beam tune shift should be small enough to guarantee stable particles' motion. Firstly, dangerous resonances should be avoided. Secondly, the overall stability of the periodic motion should be ensured by satisfying the condition $|Tr(\tilde{M}_t)| < 2$. In the approximation of small beam-beam tune shift, the latter condition gives (in each transverse plane):

$$2|\cos 2\pi(Q+\xi)| = 2|\cos \Delta\mu \cos(2\pi\xi) - \sin \Delta\mu \sin(2\pi\xi)| \approx$$

$$\approx 2|\cos \Delta\mu - 2\pi\xi \sin \Delta\mu| < 2 \qquad (11.17)$$

$$\Rightarrow \xi < \frac{|\cot \Delta\mu|}{2\pi}$$

At the same time, a small beam-beam tune shift limits the instantaneous luminosity. If ξ_{th} is the maximum beam-beam tune shift allowed in the vertical plane by beam stability, such that $\xi_y \leq \xi_{th}$, then the maximum value of n_1 is found by plugging Eq. 11.16 into Eq. 11.2, to get the tune-shift-limited luminosity:

$$\begin{cases} n_1 < \xi_{th} \frac{2\pi \gamma \sigma_y (\sigma_x + \sigma_y)}{r_0 \beta_y^*}, \\ L = \frac{n_1 n_2 n_b f_{ip}}{4\pi \sigma_x^* \sigma_y^*} < \frac{\gamma \xi_{th}}{2 r_0 \beta_y^*} \left(1 + \frac{\sigma_y^*}{\sigma_x^*}\right) n_2 n_b f_{ip} \end{cases} \qquad (11.18)$$

Unlike in storage rings, usually $D_{bb} \gg 1$ in linear colliders, where the bunches are used once and therefore a stronger beam-beam effect can be tolerated. In practice, the deformation of the bunch transverse size along its duration profits of the parasitic e.m. focusing, which is stronger in the vertical plane for flat beams. This brings to even lower beam sizes in correspondence of the IP, and eventually to an enhancement of the luminosity (*pinch effect*) which counteracts the hourglass effect.

11.5 Beam-Beam Lifetime

Colliding beams interact not only through the expected channels to produce new events, but also via Coulomb force [3]. This generates an energy change (beam-beam Bremsstrahlung or radiative Bhabha scattering in $e^- e^-$ and $e^- e^+$ colliders) which can exceed the RF or the momentum acceptance. In such case, particles get lost and the beam current reduces with time. If σ_{loss} is the cross section describing the interaction and we define the event rate as the relative loss of beam particles per unit of time, then the *beam-beam lifetime* can be expressed as function of the single bunch luminosity (see Eq. 11.3, $n_b = 1$):

$$\frac{1}{\tau_{bb}} = \frac{1}{n_1}\frac{dn_1}{dt} = \frac{L\sigma_{loss}}{n_1} \approx \frac{\sigma_{loss}}{n_1}\frac{n_1 n_2}{4\pi\sigma_x^*\sigma_y^*}f_{ip} = \frac{n_2 f_{ip}}{4\pi\sigma_x^*\sigma_y^*}\sigma_{loss} \qquad (11.19)$$

Since the r.h.s. of Eq. 11.19 is approximately constant with time, Eq. 11.19 leads to an exponential decay with time of the bunch population, $n_1 = n_0 e^{-t/\tau_{bb}}$. In general, $\sigma_{loss} \propto 4r_0^2\alpha$ (dominated by Coulomb interaction), and it typically results $\sigma_{loss} \lesssim 0.2$ barn both at electron and proton colliders over a wide range of beam energies.

References

1. W. Herr, B.J. Holzer, B. Muratori, Concept of luminosity, in *Elementary Particles–Accelerators and Colliders*, vol. 21C, ed. by S. Myers, H. Schopper (Published by Springer Materials, 2013). ISBN: 978-3-642-23052-3
2. K. Potter, Luminosity measurements and calculations, in *Proceedings of CERN Accelerator School: 5th General Accelerators Physical Course*, Geneva, Switzerland. CERN 94-01 vol. I, ed. by S. Turner (1994), pp. 117–130
3. H. Mais, C. Mari, Introduction to beam-beam effects, in *Proceedings of CERN Accelerator School: 5th General Accelerators Physical Course*, Geneva, Switzerland. CERN 94-01, vol. I, ed. by S. Turner (1994), pp. 499–524

Additional Bibliography

Manuals and comprehensive textbooks on particle accelerators:

- S.Y. Lee, *Accelerator Physics*, 4th edn., published by Word Scientific, ISBN-10: 9813274786 (2018).
- H. Wiedemann, *Particle Accelerator Physics*, 4th edn., published by Springer, ISBN: 978-3-319-30759-6 (2015).
- E. Wilson, *An Introduction to Particle Accelerators*, published by Oxford University Press, ISBN-13: 9780198508298 (2001).
- S. Humphries Jr., *Principles of Charged Particle Acceleration*, published by Wiley, ISBN-10: 0471878782 (1986).
- T.P. Wrangler, *RF Linear Accelerators*, 2nd edn., published by Wiley-VCH, ISBN-10: 3527406808 (2008).
- *Handbook of Accelerator Physics and Engineering*, 3rd printing, ed. by A.W. Chao, M. Tigner, published by World Scientific, Singapore, ISBN: 9810235005 (2006).

Textbooks of beam physics with an accent on Hamiltonian dynamics:

- A. Wolski, *Beam Dynamics in High Energy Particle Accelerators*, published by Imperial College Press, London, UK, ISBN: 978-1-78326-277-9 (2014), pp. 3–80.
- J. Rosenzweig, *Fundamentals of Beam Physics*, published by Oxford University Press, ISBN-13: 9780198525547 (2003).
- G. Stupakov, G. Penn, *Classical Mechanics and Electromagnetism in Accelerator Physics*, published by Springer International Publishing AG, 6330 Cham, Switzerland, ISBN: 9783319901879 (2018).
- L. Michelotti, Phase space concepts. AIP Conf. Proc. **184**, 891 (1989).

© The Editor(s) (if applicable) and The Author(s), under exclusive license to Springer   261
Nature Switzerland AG 2022
S. Di Mitri, *Fundamentals of Particle Accelerator Physics*, Graduate Texts in Physics,
https://doi.org/10.1007/978-3-031-07662-6

Beam brigthness, emittance, and collective effects:

- A.W. Chao, *Physics of Collective Beam Instabilities in High Energy Accelerators*, published by Wiley, ISBN: 9780471551843 (1993).
- C. Lejeune, J. Aubert, Emittance and brightness: definitions and measurements. Adv. Electron. Electron Phys., Suppl. A **13**, 159 (1980).
- S. Di Mitri, M. Cornacchia, Electron beam brightness in linac drivers for free-electron lasers. Phys. Reports **539**, 1–48 (2014).

Insertion devices, undulator radiation and free-electron lasers:

- *Undulators, Wigglers and Their Applications*, ed. by H. Omaki, P. Ellaume, published CRC Press, ISBN: 9780415280402 (2003).
- J.A. Clarke, *The Science and Technology of Undulators and Wigglers*, published by Oxford University Press, ISBN-13: 9780198508557 (2004).
- G. Dattoli, A. Renieri, A. Torre, *Lectures on the Free Electron Laser Theory and Related Topics*, published by World Scientific, ISBN: 978-981-02-0565-2 (1993).
- P. Schmuser, M. Dohlus, J. Rossbach, C. Behrens , *Free-Electron Lasers in the Ultraviolet and X-Ray Regime*, 2nd edn, published by Springer, ISBN: 978-3-319-04080-6 (2014).

For a review of particle colliders, nonlinear dynamics, and beam-beam interaction:

- V. Shiltsev and F. Zimmermann, Modern and future colliders, Review of Modern Physics, Vol. 93, 015006, (January–March 2021).
- *Theoretical aspects of the behaviour of beams in accelerators and storage rings*, in *Proceedings of the Fifth Course of the International School of part. Acceleration of the "Ettore Majorana" Centre for Scientific Culture*, CERN 77–12, ed. by A. Zichichi, KK. Johnsen, M. H. Blewett. Geneva, Switzerland (1977).
- *Nonlinear Dynamics Aspects of Particle Accelerators*, in Lecture Notes in Physics Vol. 247, ed. by J.M. Jowett, M. Month, S. Turner. Springer, New York, USA (1985).

For an introduction to non-conventional particle accelerators:

- R. Bartolini, *Accelerator Physics*, published by CRC Press LLC, ISBN: 9781482240924 (2017).

Some breakthrough articles (in chronological order):

- J. Liouville, Sur la Theorie de la Variation des constantes arbitraires. Journal de Math. Pures et Appl. **3**, 342 (1838).
- A. Liénard, Champ électrique et Magnétique. L'éclairage électrique **16**, 27–29, 5 (1898).

- A.A. Michelson, E.W. Morley, On the relative motion of the earth and the luminiferous ether. Am. J. Sci. **34**(203), 333–345 (1887).
- R. Wideröe, Uber ein neues Prinzip zur Herstellung hoher Spannungen. Arch. Electrotech. **21**, 387–406 (1928).
- L.W. Alvarez, The Design of a Proton Linear Accelerator. Phys. Rev. **70**, 799–406 (1946).
- J. Schwinger, On the Classical Radiation of Accelerated Electrons, Phys. Rev. **75**(12), 1912–1925 (1949).
- R.Q. Twiss, N.H. Frank, Orbit Stability in a Proton Synchrotron. Rev. Sci. Instrum. **20**(1), 1–17 (1949).
- J.R. Pierce, Travelling-wave tubes. The Bell Syst. Tech. J. **29**(2), 189–250 (1950).
- E.D. Courant, M.S. Livingston, H.S. Snyder, The strong-focusing synchroton–a new high energy accelerator, Phys. Rev. **88**, 1190–1196 (1952).
- W.K.H. Panofsky, W.A. Wenzel, Some considerations concerning the transverse deflection of charged particles in radio-frequency fields. Rev. Sci. Inst. **27**, 967 (1956).
- K.W. Robinson, Radiation Effects in Circular Electron Accelerators. Phys. Rev. **111**(2), 373–380 (1958).
- L.J. Laslett, On intensity limitations imposed by transverse space charge effects in circular particle accelerators, in *Proceedings of the Summer Study on Storage Rings*, BNL Report 7534 (1963), pp. 324–367.
- C. Bernardini, G.F. Corazza, G. Di Giugno, G. Ghigo, J. Haissinski, P. Marin, R. Querzoli, B. Touschek, Lifetime and beam size in a Storage Ring. Phys. Rev. Lett. **10**, 407–409 (1963).
- S. van der Meer, Calibration of the effective beam height, in the ISR, CERN Report No. ISR-PO-68-31 (1968).
- J.M. Madey, Stimulated emission of bremsstrahlung in a periodic magnetic field. J. Appl. Phys. **42**, 1906 (1971).
- R. Chasman, G.K. Green, E.M. Rowe, Preliminary Design of a dedicated synchrotron radiation facility. IEEE Trans. Nucl. Sci. NS-22, 1765–1767 (1975).
- K. Brown, A First- and Second-Order Matrix Theory for the Design of Beam Transport Systems and Charged Particle Spectrometers, SLAC-75-rev-4 (1982), pp. 72–134.
- G. Vignola, Preliminary design of a dedicated 6 GeV synchrotron radiation storage ring. Nucl. Instrum. Meth. Phys. Res. A **236**, 414–418 (1985). A. Jackson, A comparison of the Chasman-Green and triple-bend achromat lattices. Part. Accel. **22**, 111–128 (1987).
- K. Halbach, Specialty magnets. AIP Conf. Proc. **153**, 1277 (1987).
- R. Bonifacio, F. Casagrande, G. Cerchioni, L. de Salvo Souza, P. Pierini, N. Piovella, Physics of the high-gain FEL and superradiance, in La Rivista del Nuovo Cimento (1978–1999) Vol. 13 (1990) pp. 1–69.

Index

S. Di Mitri, *Fundamentals of Particle Accelerator Physics*, Graduate Texts in Physics,
https://doi.org/10.1007/978-3-031-07662-6

Printed in the United States
by Baker & Taylor Publisher Services